CE-5

Close Encounters of the Fifth Kind

242 case files
exposing alien contact

Richard F. Haines, Ph.D.

Sourcebooks, Inc.
Naperville, IL

Published by Sourcebooks
P.O. Box 372, Naperville, Illinois 60566
(630) 961-3900
FAX: 630-961-2168

 Sourcebooks, Inc.

Cataloging-in-Publication Data
Haines, Richard F.
      CE-5: close encounters of the fifth kind: 242 case files exposing alien contact/Richard F. Haines.
      p.   cm.
   Includes bibliographical references and ind
   ISBN 1-57071-427-4   (alk. paper)
   1. Human-alien encounters. I. Title.
   BF2050.H35   1998
   001.942—dc21   98-28651
                                                               CIP

3 1571 00180 2522

Printed and bound in the United States of America
10   9   8   7   6   5   4   3   2   1

Contents

Part I

Part II

Part III

Figures

Tables

List of Abbreviations

APRO ....Aerial Phenomena Research Organization
CE-5 ....Close Encounter of the Fifth Kind
CIA....................Central Intelligence Agency
CTS ....coherent-thought sequencing
CUFOS ....Center for UFO Studies, J. Allen Hynek
CSETI ....Center for the Study of Extraterrestrial Intelligence
E.T. ....extraterrestrial
km ....kilometer
kt. ....knot
mph ....miles per hour
MUFON............Mutual UFO Network
NASA ....National Aeronautics and Space Administration
NICAP ....National Investigations Committee on Aerial Phenomena
nm. ....nautical mile (6,080 feet)
NSA...................National Security Agency
OVNI ....French for UFO (Objets Volants non Identifies)
PIO ....public information officer
smi......................statute mile (5,280 feet)
UFO ....unidentified flying object

Preface

The subject of extraterrestrial (E.T.) life has fascinated man for centuries and has generated scores of theories and thousands of books. We have pondered what might be true about E.T., always considered through the invisible biases of our own planetary cultures, languages, beliefs, hopes, and numerous unconscious fears. "They must look like us, act like us, and think like us," we may reason, not thinking very logically about the creativity of God, the vastness of the universe, or the awesome span of time of the existence of matter. Why can't E.T. exist? Why can't "they" be different from "us?"

Over the years, I have become increasingly interested in what evidence exists to support the contention that anomalous aerial phenomena are not only intelligently controlled but intimately related in some way to E.T. In researching an earlier book, *Project Delta*, I found myself carried quite a distance down this particular evidential road. The book presents over four hundred UFO sightings from among the more than one hundred thousand others now available. Each case involves two or more objects seen at the same time and place. I concentrated on the appearance and flight behavior of the objects themselves. What I found was compelling evidence to claim that most of these aerial objects far exceeded the terrestrial technology of the era in which they were seen. I was forced to conclude that there is a great likelihood that Earth is being visited by highly advanced aerospace vehicles under highly "intelligent" control indeed. However, there are few humanoids mentioned in these reports. This is as much due to my deliberate selection of cases (i.e., most cases involve objects seen at a distance from the witness) as from my own unconscious bias against the possibility that E.T.s do exist. But in my research for the present book, I had to confront the logical possibility that some advanced flying vehicles must also have pilots or at least be intelligently controlled. In this book, I include selected cases involving both the vehicles and their alleged occupants, most of which were encountered at close range.

It is one thing for a witness to claim that strange-looking objects flew across the sky. But it is quite another for him to claim that he signalled to them and they signalled back, or that he fired a rifle at one and heard the bullet ricochet off its surface. These are but two of the fascinating subjects presented here.

There may well be reason to fear visitors from outer space if the writings of Edwards (1966), Pratt (1996), Wilkins (1954), and others are taken seriously. But here I focus not as much on the alien's behavior as upon what may have provoked it. This book is about human behavior as much as it is about alleged UFO and humanoid responses. Perhaps we can learn something about ourselves by looking at these close-encounter narratives in this way.

This book raises truly profound questions if the supporting data is found to be factual and true. Is mankind alone in the universe? Who is man? Who is the aggressor? What best characterizes the "visitor's" behavior? We must be passionate about seeking the truth, allowing no dogma or entrenched view, scientifically based or otherwise, to inhibit our search. In today's vacuum of values and plethora of beliefs we each must decide for ourselves whether we are alone in the vast universe or are only one more species among millions. While governments, universities, and churches alike have largely opted out of this critically important debate, the examination is left up to us.

Always seek the truth no matter where it takes you. This exhortation is all the more valuable when you investigate anomalous phenomena. You will probably be led down crooked, sometimes difficult and dangerous paths that only a few others have explored. You may find yourself quietly avoided by some former friends. You may want to quit out of frustration. Yet, you will probably go on for some unexplained inner reason as I have. Let the phenomenon speak for itself. Don't listen to anyone else but your God. Keep an open mind concerning all possible and impossible explanations. And then...enjoy your search.

Richard F. Haines
Los Altos, California
September 1998

Acknowledgments

When one has the kind of friends I have—who will give so lavishly of their time and energy—it is not only an honor to acknowledge their help but the right thing to do. The following people provided detailed information from their own research files, information which has contributed both to the breadth and depth of this book. Many also shared their personal support for this effort during times when I was on the verge of dropping the whole project. My particular thanks go to: Jan Aldrich, Loren Gross, Larry Hatch, Jim McCampbell, and Bradley Sauer.

In addition, the following individuals provided valuable assistance by loaning me CE-5 reports and illustrations. Many also commented on these ideas, and provided me with important new insights. To these friends and colleagues I say thank you sincerely, Bill Banks, Fred Beckman, Eddie Bullard, Gordon Creighton, Ray Fowler, Timothy Good, Dale Goudie, Steven Greer, John Greenleaf, Jon Griffith, Dick Hall, Derek Lauder, Georges Leftheriotis, Larry Lemke, Illobrand von Ludwiger, Bob Pratt, Fran Ridge, Mark Rodeghier, Vladimir Rubtsov, John Spencer, Kurt Story, Jeff Talbert, Bob Teets, Brian Smith, Willy Smith, Reuben Uriarte, Jacques Vallee, John Whittenbury, and Don Worley.

Both Sandy Gordon and Dave Faust provided their usually high quality of editorial and artistic services. My sincere thanks to both of them.

Finally, I am most grateful all of the investigators and authors whom I have referenced here. Their past research has built a firm foundation upon which *CE-5: Close Encounters of the Fifth Kind* has been written.

"To attain perfection, one must first
of all be able not to understand many
things. For if we understand things
too quickly, we may perhaps fail to
understand them well enough."

—Dostoevsky, *The Idiot*

Chapter 1

Introduction

Almost everyone today has read about (and seen in movies and on television) fantastic accounts and graphic displays of strange flying discs that hover silently, rise up vertically at fantastic speed, and perform breathtaking maneuvers in the air and even below water. We are captivated by their dazzling appearances and antics, and wonder whether they could be real. To many people, these images came only from creative imaginations of UFO fanatics in Hollywood or Sci-Fi computer-graphics, while to a growing number of others they represent actual three-dimensional objects.

Usually these objects seem to mind their own business. According to many thousands of eye witness reports, press clippings, movies, and occasional radio and television accounts, these aerial objects act as if they are oblivious of their witnesses.

Yet, once in a while we read about someone who waves at the craft or shines a light at it at night. Seldom do we find out what happens in these kinds of circumstances. This is the main subject of *CE-5: Close Encounters of the Fifth Kind*.

Occasionally someone behaves in a way that seems to be exactly repeated by the UFO. Two short flashes of a flashlight, for example, are followed by two short light flashes from the airborne object. A pilot wiggles the wings of his airplane slightly, and an oval metallic disc nearby wavers soon afterwards. A driver is exactly paced by a nearby unidentified aerial object mile after mile of curving roads. Do such mimicking responses by the unusual phenomenon represent some type of deliberate communication, a pure coincidence, or something else we know nothing about? One purpose of this book is to explore these possibilities.

The literature contains many scores of accounts in which the UFO or alien being elicits some kind of a response from a human. An interesting literature review on this subject by B.C. Sparks (1971) uncovered ninety-seven cases spanning the period 1838 to 1970. He subdivided these cases into five categories: rays, vapors, emanations (non-directional radiation, thermal, chemical), incidents and accidents caused by the UFO and/or E.T., and abduction without return. The only human behavior listed that might be interpreted to the visitors as provocative or threatening is the individual's approach. He found no correlation between year (between 1944 and 1969) and number of cases for any of the five categories cited above, i.e., they occurred almost at random. His plot of the number of cases by geographic location and year also showed no clear-cut patterns. Three such cases are summarized below.

Case A. 54-09-26 [year-month-day] This terrifying incident took place in Chabeuil, France. Mr. Lebouef was approached by a being about three feet tall. He saw a UFO nearby. During the following two days he suffered a severe fever (40° C) and nervous collapse (Michel 1958, 82-84).

Case B. 70-09-03 A woman living in Belo Horizonte, Brazil was chased by a UFO which got within about sixty-five feet of her. She felt great heat and later experienced burns on her back, severe headaches, and impairment of vision two hours later (*APRO* 1970).

Case C. 75-10-27 This event allegedly took place near the Atlantic ocean port of Fortaleza, Ceara, region of Brazil about 415 miles northwest of Recife. The inhabitants of the village of Sao Gondolo de Amarante hid in their houses each evening because of the regular appearance of a disc-shaped aerial object which "attacked folk with a ray." It emitted a powerful blue light, paralyzing those who were unfortunate enough to be caught in it. Allegedly, a man almost died from the severity of the burns he suffered. Lawyer Joao Luciano Gualberto, who lives in the district and gave the news, intimated that the saucer made its appearance just after 6:00 P.M. and stopped at frequent intervals above the same place. Several other witnesses confirmed this.

"A little girl who was bathing in the nearby river felt a sort of burning and saw what the lawyer called an 'oscillating blue light.' Frightened, she ran back home where she arrived with extremely bloodshot eyes and a high body temperature." [Bowen 1976, iii; *La Razon* (Argentina) 1975; Also cf. a very similar case which took place in October 1974 at a lake in northern Ceara (Pratt 1996, 149-50).]

A major problem in dealing with accounts like these is that we do not have all the information we need to assess their reliability nor the motivations of the humans involved. Did the UFO really cause the effect reported by the human? In most of the cases presented here, this is precisely what happened. The main focus of this book is from the opposite vantage, i.e., *what the UFO did in apparent response to various overt human behaviors.*

Communication-Information-Uncertainty

Taken in its simplest sense, communication is simply the transmission of information between two or more people, animals, or living creatures. It may be deliberate, but I suspect that most is not. Thus, if I do something which permits you to become more informed about something then communication has occurred.

We must define what information is if we are to understand and apply this definition of communication properly. I use the general definition suggested by Shannon and Weaver (1949) which, put simply, states that information has been transmitted if the receiver's uncertainty has been reduced as a result.

So what is uncertainty? If I think that some action I take will have two or more possible outcomes, then my confidence in expecting each result is less than if I know for sure there will be only one outcome. In short, I am uncertain which outcome will occur. Thus, if I'm outdoors at night and see a strange luminous object hovering nearby and I happen to aim my flashlight directly at it, I face a dilemma. Should I switch my flashlight on or not? I am uncertain. Perhaps my light signal will not produce any effect, but then what if...? It is this array of "what ifs" that leads to my state of uncertainty. In this illustration, my uncertainty may be based upon an unconscious fear of danger or physical pain originating from the object or, perhaps, upon a conscious knowledge that other people have been burned or electrically shocked by such unidentified objects. And so the act of communicating is not as simple as one might assume because it rests upon consequences to one or both parties.

3

Using the above definition, information will have been transmitted if, because of my communicative behavior, my uncertainty has been reduced. In situations where there is uncertainty, such as a UFO close encounter, the response of a UFO to some preceding human behavior may or may not reduce one's uncertainty. It will depend upon what the person is uncertain about, e.g., if one is uncertain whether the object is capable of responding at all then any response by the UFO will settle the matter. But if one is uncertain whether he will be injured by the object's response and all he sees is a light signal corresponding to his own, he will feel relieved temporarily, but he will be just as uncertain as before. Little useful information will have been transmitted.

At this time, we have almost no reason to believe that UFOs (or at least their controllers) need or want to communicate with the human race! Thus, we have no a priori reason to believe that the response of a UFO to a prior human behavior will reduce any but the most simple human uncertainty about "them." It follows that attempts to communicate with them may be fruitless from mankind's limited point of view. On the other hand, UFO attempts to communicate with us, if such can be shown to occur reliably, either may be deliberate or artifactual, i.e., deliberate attempts to help reduce our uncertainty in some way or only unintentional artifacts of their physical presence. The likelihood of both occurring is small if they are as advanced as they seem.

Coincidence-Causality

The misunderstood subject of coincidence often is associated with mysticism and muddy thinking. The dictionary defines coincidence as "1. The state or fact of coinciding 2. A sequence of events that although accidental seems to have been planned or arranged." (The American Heritage Dictionary, Pg. 289, 1982). Here, the term "accidental" is defined as "occurring unexpectedly and unintentionally; by chance." The word "planned" means "to formulate a scheme or program for the accomplishment or attainment (of something)." But the word "coincidence" hinges on the word "seems." It is the subjective assessment of the person present that imputes causality to a sequence of events. And it is precisely for this reason that one person's claim of coincidence can be challenged by others at a later time.

The subject of coincidence also is linked closely to the issue of causality; the two are usually thought of as lying at opposite ends of the same continuum.

Causality implies a direct or indirect relationship between two events. For example, if I strike a cue ball sitting on a pool table with the tip of my cue stick, kinetic energy will be imparted to the cue ball, causing it to roll across the top of the table. Before I contacted the ball, it did not move; now it does. Barring a simultaneous earthquake or some sharp physical contact with the edge of the pool table, the ball's motion is not coincidental but causal in nature. Kinetic energy was added to its own potential energy; the result was cue-ball motion.

Now consider the following situation. I am standing some distance from the pool table and concentrating on the cue ball, indeed, commanding it to start rolling. Suddenly it begins to roll across the table. Most people would conclude that the motion was purely coincidental with the mental activity and certainly not caused by it. A handful of people today would think conversely. The word "coincidence" has become little more than a convenience to permit one to escape from having to understand an invisible cause and effect event. In fact, controversy continues to this day about whether man's cognitive processes can produce physical effects at a distance (psychokinesis). In this situation, most people quickly look for another source of energy which caused the ball to move. Here lies the real dilemma for ufologists faced with many good examples of invisible links between human behavior and subsequent UFO responses, as we shall soon see. Are all of these events coincidental? Their large and ever-growing number and surrounding circumstances stretches this hackneyed explanation to the breaking point. Consider the following event that took place near Rebouillon, France.

5

Case D. 66-07-19 Three men were on their way to the village of La Granegone at about 3:45 A.M. when they noticed a light hovering some seventy meters above the ground over the Rebouillon Valley, opposite the small village of Lentier. Several searchlight beams shown downward from it. The men pulled their car over to the shoulder of the road and spotted a totally silent, grey, oval, metallic object. *The searchlight beams suddenly went out,* making it possible to see more details on the object. It had a number of "portholes" along its circumference and something like a bent stovepipe at one end. They watched it for about twenty minutes during which time the object seemed to be attempting to stabilize itself; it flew in half-circles and came to within one hundred fifty to two hundred meters of the astonished men. The driver started his engine and started travelling down the road again. As he did so, *the lights on the object also came on.* If the first event was coincidental what is the liklihood that the second one was also?

Rules of evidence for the causal encounter

The physics of cue-stick-to-cue-ball contacts are well understood. Few physicists would debate whether or not the motions imparted from one to the other are causal. They are! But they would debate long and loudly about whether alleged light flashes seen coming from a UFO were the direct result of some prior human behavior, even assuming that the physicist accepted presence of a UFO in the first place. And so, what kinds of evidential rules should we use to support the assertion that a UFO's response is caused by human behavior? If we are to establish that a particular human behavior caused a subsequent corresponding response by a UFO with high certainty, we must employ a logical, consistent, and extremely conservative set of rules.

I propose the following four simple rules of evidence for an encounter to be considered causal:

1. The human must act initially and deliberately, by free will, rather than out of compulsion or somehow provoked by the UFO or alien being. Cases involving fear-based human behavior are included deliberately for what they can teach us.

2. The human's behavior must precede the UFO's response in time. Indeed, there are hundreds if not thousands of documented sightings where humans reacted to something the UFO did. A few examples are included here, but they really don't teach us very much.

3. The UFO's response must follow the human's behavior within a certain critical span of time which is referred to here as response delay (RD). For practical purposes, RD should be from about one second to several minutes at most.

4. The termination of the human's behavior (to achieve some deliberate communication with the UFO) should soon be followed by the termination of the UFO's response, i.e., the response must not continue on very long after the human has stopped acting.

Chapter 2 provides several other rating criteria which are used to compare one case with another and to conduct statistical tests.

Before we begin a review of actual sighting reports of alleged causal encounters, let us review the field of ufology (the study of UFO phenomena) to see what might be learned and where *CE-5: Close Encounters of the Fifth Kind* fits into the history of this field's work.

A brief overview of the development of ufology in America

The serious study of unidentified aerial phenomena in America has progressed through several phases which can be characterized as developmental growth periods. The metaphor of the human being is used.

Preconception Period (from pre-recorded history to early 1900s). Anyone familiar with the writings of Charles Fort, fascinating articles compiled by William R. Corliss (e.g., Strange Phenomena: A Sourcebook of Unusual Natural Phenomena, 1974), folklorists like Thomas (Eddie) Bullard, Joseph Campbell, and others realize that strange luminous phenomena have appeared in the sky since the emergence of human memory. Some of these wonderful reports have survived in written form, while others were kept alive only as folkloric verbal tales. Other pictures were chiseled in stone walls and painted on canvas. And so, our age has an historic legacy concerning UFOs that stretches back hundreds of centuries. Does the consistency of these stories indicate little more than a working out of the human desire that these details be true, or that humans are reasonably reliable observers and reporters? I have argued for the latter in my book *Observing UFO* (1980).

While the industrial revolution brought with it a new way of looking at ourselves as mass producers and mass consumers—rather than small groups of individual workers and artisans—it also brought with it many new human expectations. Civilization quickly came to expect to be able to travel; communicate over long distances; mass-produce goods; record and store information; generate, store, transmit, and modify energy; live longer and more healthy lives; understand our own nervous system; and numerous other expectations.

Our expectations also encouraged us to leave the surface of Earth. Ours is the first generation to set foot on the Moon and at the same time to extend our senses far out into the solar system. As a contributor to a few of these aerospace accomplishments, I have come to appreciate the truly great impact these events have had on our culture. I see the continuing and escalating insertion of UFO phenomena into our times as an integral part of our own outward migration to the stars.

7

Infancy Period (approximately 1910 to mid-1940s). Born of man's first successful powered flight at Kitty Hawk on December 17, 1903, and raised by commercial and military parents, our fascination with flying and things that fly has continued to mature over the years. While there were no ufologists yet, there were many astronomers, meteorologists, physicists, engineers, and others who studied Earth's atmosphere to discover its valuable secrets, secrets which were put to use in fighting two world wars. It was during both the Pacific and European war theatres that pilots and ground personnel began reporting strange luminous balls which approached, paced, and cavorted around their airplanes. Indeed, these phenomena continue to be seen. The infancy period of ufology resembles the neonate who doesn't yet know its own identity nor what it will mature into.

As Gross (1976) has so clearly documented, there was a small, almost cultic following of the American writer Charles Fort beginning during this period. He collected, categorized, and publicized many bizarre events around the world, some of which are identical to modern-day UFO. But it wasn't Fort's book, *Book of the Damned,* published in 1919 as much as Jacques Vallee's *Passport to Magonia,* published in 1969, that clinched the true historical reality of UFO.

In a thought provoking article, Clark argues that there is no real support for the contention by some that by the summer of 1947, Americans had become so frightened by the threat of thermo-nuclear war that they began "watching the skies in an anxious search for salvation from elsewhere" (1996b, 3). He surveys press reports from the summer of 1947 to show that very few Americans linked the flying discs sighted mainly in the Pacific Northwest with extraterrestrial spacecraft "much less benevolently intentioned ones....The ETH [extraterrestrial hypothesis] became a popular interpretation only when the perceived failure of other explanations to account for the phenomenon's persistence, its appearance, and its behavioral characteristics became apparent to all but the most committedly hostile Americans." Like an infant, American ufology's thirst for new experiences and ideas led it to begin to explore the fringes of its tiny world, a world still strongly controlled by traditional scientific concepts.

Childhood Period (approximately late 1940s to 1950s). This period of American ufology was truly one of innocence and development of basic understandings. Most Americans were as open-minded about UFO phenomena as are young children who see a butterfly for the first time. Like infants who have difficulty in making a decision, we were confronted by a dazzling array of awe-inspiring luminous phenomena which challenged us either to accept or reject the evidence. This very ambivalence remains to this day.

This period also saw the rise of several serious, private study groups, such as the Aerial Phenomena Research Organization (APRO) founded in January 1952, and many other so-called contactee groups. By the mid 1950s, there were over 150 contactee-oriented clubs in America. Jacobs points out that both Donald Keyhoe and the founders of APRO, Coral and Jim Lorenzen, spent much of their time trying to "correct the damage to the legitimacy of UFO research" (1975, 124). Somehow the contactee claims touched a nerve in America which made contactee groups far more popular than the more scientifically oriented study groups. The first annual "Giant Rock Convention," held near Yucca Valley, California, in 1954, brought more than five thousand curious souls together. Perhaps it was the bold, if often absurd, descriptions of direct physical contact with extraterrestrials that captured imaginations, seemingly turning science fiction into science fact.

It was also during this period that the Federal Bureau of Investigation, Central Intelligence Agency (CIA), and several branches of the Department of Defense also became involved (Fawcett and Greenwood 1984). Indeed, our nation's policy on UFOs changed significantly in 1953; it was probably due to the huge UFO wave in the summer of 1952 over Washington, D.C., and continuing UFO reports from Korea that the CIA became interested in UFOs. Jacobs (1975) and Corso (1997) provide illuminating commentary on this. Numerous declassified memos have been released through Freedom of Information Act requests over the years.

Ongoing cold war fears following World War II led some to suggest that the UFO phenomena was merely some kind of propagandist activity conducted by the Soviet Union. In effect, the public (i.e., the young child) was told by government spokesmen (i.e., the parent figure) to grow up! A long list of reasons was conveniently provided to the American public as to why belief in extraterrestrials was far-fetched. UFOs, the public was told, must be a natural phenomenon, if not Russian. And so a part of America grew up believing that UFOs were not to be taken seriously, while another part were seeing them in the skies for themselves! One result of this paradox was a growing distrust of what the government told the general public.

Many ufologists were like little children, awed by something totally new to them, and largely unaware that the phenomenon really had existed for a very long time. Ufologists had no consistent terminology, textbooks, field research methods, financial support, nor governmental support. Nevertheless, like innocent children, ufologists had much enthusiasm and joy of involvement.

As Jacobs clarifies, the main emphasis during the 1940s and 50s clearly was focused on the aerial phenomena itself in an attempt to find any physical explanation for it (1975). The mention of UFO crewmembers, however, was studiously avoided by all but a small group of devotees. The prominant and well-publicized existence of contactees, who claimed to have met and travelled with extraterrestrial beings aboard their ships, was largely ignored by serious UFO investigators and the public.

The childhood innocence period enjoyed books, magazines, and thrilling radio programs about extraterrestrial life. Many tabloid books had colorful covers showing people on the ground waving or shooting at rocket-shaped spaceships with cadillac-like tail fins. Thousands looked forward to their weekly radio adventure programs which described vivid encounters with bug-eyed monsters that carried humans off into the woods, or worse, into outer space.

Consider the drawings and photographs of alien beings depicted on the dust jackets of books of this early period. There were one-eyed, bald-headed creatures with many tubular arms sticking out of a thin torso with several short legs and no feet. There were hairy creatures with firey red eyes which blazed forth at anyone brave enough to stare into them. And there were metallic robots looking remarkably like humans while performing wondrous feats. Perhaps unconsciously, some people assumed that these kinds of creatures actually existed and behaved the way they were said to have behaved in these fictionalized stories. Was there a remnant of belief in E.T. from these early years which no amount of carefully implemented social conditioning could dislodge? Little by little, the child was beginning to grow up.

This period was also a very creative one as compared with today. Many hypotheses concerning UFOs were proposed by official government commissions, the works of science fiction and motion picture writers, and some ufologists as well. It seemed to be a time of profound social change when post-war feelings of freedom influenced the perceptions of unusual flying objects. Take, for example, the Americanism "little green men from Mars." This phrase caught on and was used so often that some witnesses saw them exactly that way—short, green Martians. Indeed, Mars remained the home planet of choice of these "visitors" for decades in the mind of the American public. Only recently have other planets been discovered that are within reasonable travel distance from Earth. How long will it take for people to coin a new phrase such as "tall greys from Episilon Bootie?"

But a significant event was to occur which turned ufological history.

As Jacobs and others have noted, the American military establishment and certain elements within the CIA were deeply involved in analyses of UFO reports during this same period. According to Zechel (1979), Admiral Roscoe Hillenkoetter, first director of the CIA, told Donald Keyhoe that the CIA had been interested in UFO activity "from the very beginning of reports (i.e., June 1947) and kept a watchful eye on the subject despite the lack of directives to do so." Zechel (1978) recalls a memo dated July 29, 1952, from the CIA to Ralph L. Clark, then Deputy Director of Intelligence. This memo presumably arose as a direct result of the extensive UFO activity over Washington, D.C. in July 1952. It states:

> In the past several weeks, a number of radar and visual sightings of unidentified aerial objects have been reported. Although this office maintained a continuing review of such reputed sightings during the past three years, a special study group has been formed to review this subject to date. O/CI will participate in this study subject to date. O/CI will participate in this study with O/SI and a report should be ready August 15, 1952.

The group of emminent scientists who were assembled for this task (along with their largely unsung support teams) came to be known as the Robertson Panel, named after Dr. H.P. Robertson of the California Institute of Technology. Other members of this prestigeous group of scientists included Luis W. Alvarez (University of California), S. A. Goodsmit (Brookhaven National Laboratory), and Thornton Page (Johns Hopkins University). The group was convened by the CIA (Jacobs 1975) to evaluate the possible threat to national security posed by UFOs and also to make recommendations about what to do about them. The panel issued their summary of findings to the U. S. Air Force, and Major L.J. Tacker issued the following public news release on April 9, 1958—although it was prepared on January 17, 1953, five years before.

> As a result of its considerations, the Panel concludes: That the evidence presented on unidentified flying objects shows no indi-

11

cation that these phenomena constitute a direct physical threat to national security.

We firmly believe that there is no residum of cases which indicates phenomena which are capable of hostile acts, and that there is no evidence that the phenomena indicate a need for the revision of scientific concepts.

In the light of this conclusion, the Panel recommends: That the national security agencies take immediate steps to strip the unidentified flying objects of the special status they have been given and the aura of mystery they have unfortunately acquired.

We suggest that this aim may be achieved by an integrated programme designed to reassure the public of the total lack of evidence of inimical forces behind the phenomena.

This press release raised many more questions than it answered. The absence of a denial of UFO existence, in fact, implied that this noted panel of experts believed E.T. possible.

But it was the American public's attitude toward UFOs, and not their existence, that was of most concern to the government. As mentioned below, more recent evidence obtained through the Freedom of Information Act, has indicated an almost opposite conclusion from that reached by the Robertson Panel. Many people believe that a carefully crafted, psychologically-based, public re-education program was initiated in the early 1950s which continues to this day. Was public opinion about UFOs being systematically molded to hide actual evidence?

One might also question whether the Robertson Panel prepared one summary for public consumption, and a different version for those within the government with a special need to know *the truth*. Indeed, the five years which elapsed between the press release writing and and its public release would be sufficient time to do so. As this book shows, humans have shown aggressive behavior toward UFOs over many years. But equally true is the documentation that alien beings show extreme self-control following these hostile provocations.

It appears as if the findings of the Robertson Panel brought some people in the government into ufology's Adolescent Period—although somewhat earlier than the American citizenry. This was very likely due to the hundreds of high-quality reports made by pilots, ground troops, air combat controllers, naval personnel, and others fighting in Korea from 1950 to 1953 (cf. Haines 1990), and also from the many hundreds of high-quality sighting reports arriving at the Pentagon from elsewhere around the world.

Today, some people believe that in mid-1947, before the Korean War began, a small group of scientists, military, and politicians obtained physical proof that extraterrestrial beings had visited Earth and possessed highly advanced, i.e., energy management, technology (Friedman 1996). In this regard Greer states, "After all, at the time, we were working on the development of the hydrogen bomb—and our arch enemy the Soviets were hot on our tail. What could be more destabilizing to an already fragile world order than the introduction of inter-stellar propulsion technology to a world of vacuum tubes and internal combusion engines? To say we were facing a quantum leap in technological capability is an understatement. And we wanted it safely for ourselves" (1996, 3).

To this possibility I suggest that the reason why American leaders did not disclose what they knew about these "visitors" was not because it would destabilize any relations we had with the Soviet Union but that government leaders were afraid such information might upset almost every segment of American society and thereby weaken the fabric of our nation to the point of turmoil and possible anarchy. Americans would realize that their armed forces could not really protect them against such technologically advanced foes. Many citizens would do whatever was necessary to protect themselves. A nation without government protection is a nation out of government control.

If America did capture a crashed flying saucer in the mid- to late 1940s, as some claim (cf. Randle and Schmidt 1991, 1994; Friedman and Berliner 1992), it is likely that our government would have subjected it to intense scrutiny. Yet, because of our relatively infantile understandings in key areas of science, we probably were not able to understand how it flew or most of its other technical details. Under these circumstances the best course of action would be to keep whatever was found top secret while continuing to study it over the years as our science and technology advanced. In this way, our enemies would not gain an advantage, our civilian society would not panic or become disorganized, and our technologically based economy could continue in the same direction it had for decades. Interested readers also should read Sider and Scott (1992) for their assessment of this particular era.

According to an article in *The UFO Investigator* (July 1957 1:1), a special order was issued by the U.S. Army on January 31, 1957, that effectively prohib-

13

ited anyone in the military from releasing UFO-related information to the general public. Why did the Pentagon feel this was necessary?

Adolescent Period (1960s to 1970s). This decade was filled with high-quality UFO sightings occurring around the world; a check of several large computerized databases of sightings will quickly confirm this. These sightings helped several UFO organizations to grow, including the National Investigations Committee on Aerial Phenomena (NICAP) and APRO, the two largest (cf. Clark 1996a; Jacobs 1975). So-called "flap" periods—significant increases in the number of reports as compared with preceeding or following periods—also occurred during the 1960s.

While a small group of prominant American scientists and Department of Defense spokesmen denied UFO existence, many Americans knew differently. The credibility gap between citizens and their leaders continued to widen, partly because of the absurd stand adopted toward UFOs by these leaders. Interestingly, almost no prominant political figures took a public stand on the subject; this also is typical of adolescent behavior.

It was also during this period that foreign military and political leaders were encouraged to adopt the same public stance on UFOs as was adopted in the United States. Before the 1960s many foreign generals and admirals openly admitted that their troops had seen and photographed UFOs. After the mid-1960s they denied such events and unabashedly referred inquiries to American embassies (cf., Pitt 1956).

High-level military and intelligence personnel continued their deliberate, carefully planned program to debunk the subject of UFOs during this same period. Some of the psychological warfare tactics learned during World War II and the Korean War were employed to try to convince Americans that the so-called truth about UFOs had been fabricated. A careful, retrospective reading of the Air Force's Project Blue Book files and more recent Freedom of Information Act releases make this abundantly clear. But at the time, average citizens soon came to understand that they faced ridicule if they reported a UFO sighting. Only very brave souls who felt a sense of patriotic pride came forward with their sighting experiences. And so it was that most UFO reports made public during this period were officially denied or disproven by such learned spokesmen as Drs. Donald Menzel, Edward Condon, and Carl Sagan. Most Americans simply accepted the pronouncements of these experts without question. Only a handful of ufologists challenged them.

Yet, this program to silence America also worked against our country's obvious need to remain vigilant and to report any unusual aerial vehicle from behind the Iron Curtain which might be spying on us. Indeed, America and the Soviet Union were still locked in the cold war. The Department of Defense issued a new reporting requirement known as AF-200-2. According to this regulation, all military personnel (and also civilians) must report any unidentified flying object to them.

Also typical of this period of ufology was a growing acceptance for the logical view that UFOs had to be piloted. Humanoids, probably possessing great intelligence, most likely were associated with flying objects. This logic led some investigators to drop their UFO studies. A few began to specialize in contactee cases while others concentrated only on the so-called "nuts and bolts" sightings, i.e., where there was evidence of some kind which could be examined in laboratories. The fine work of Phillips (1975, 1980), McDonald (1967, 1970), and Rutledge (1981) are some prominant examples.

The literature of this period included new experimental ideas often associated with the novelty and adventure-seeking period of adolescence. Published works could be found spanning a wider range of subjects than before (including science fiction themes) such as world-in-crisis, ecology, demonology, past lives, life after death, communication with dolphins, remote viewing, neurobiology, and various psycho-sociological themes. This was also the period during which, with some fanfare, the U.S. Air Force abandoned its own official UFO study, Project Blue Book, in December 1969, leaving the task to the private sector. Nevertheless, the rather terse explanation that was given for terminating this project *never denied the existence of UFOs!* At least historians could not convict the Air Force of lying.

Adult Period (1970s–1980s). Ufology seemed to mature for a short period in America in terms of the way it approached field research and interpreted the collected data. Mimicing adult behavior, there was open communication among its devotees, interesting meetings held in which high-quality sighting data was shared, and there was a plethora of well-written books. There was relatively little personality assassination at this time.

It was also during this period that ufologists became bolder and more adult toward their parental government. They asked for the release of thousands of previously withheld documents on UFO activity. P. Gersten wrote the following in an article titled, *"What the U.S. Government Knows about Unidentified Flying Objects."*

At last! New evidence for the existence of unconventional aerial objects...can be found in three thousand pages of previously classified documents on UFOs released...over the past few years by the Departments of State/Army/Navy/Air Force/Defense, the Federal Bureau of Investigation, the National Security Agency, the Defense Intelligence Agency, and the Central Intelligence Agency.... This overwhelming evidence indicates that Unidentified Flying Objects do exist, and that some of them are unconventional craft that (1) *pose a threat to national security* and (2) perform beyond the range of present-day technological development (1981).

Comparing the earlier release of findings from the Robertson Panel with the more recent evidence uncovered by Gersten and others, we have to ask ourselves: where does the truth really lie?

Skeptics continued to demand physical proof of UFO existence, a demand that far exceeds the scientific rigor required by mainstream science. As suggested by Zeiler, UFO evidence to date has satisfied the "evidential threshold of normal scientific protocols; [but]...the evidence has been rejected by the dogmatic, specious demands for physical proof" (1996).

Nevertheless, throughout this same period dedicated individuals worked largely in isolation on different pet projects, largely ignoring the skeptics. Particularly important for our present review of ufology was the development by Dr. J. Allen Hynek of a new terminology for classifying UFO sightings. He suggested the following taxonomy, although it is not without its weaknesses. Its subsequent use has shown that there is indeed a large body of evidence which fits well within it.

> *Daylight Disk (DD):* An apparently solid, three-dimensional object seen during daylight hours.
>
> *Nocturnal Light (NL):* An anomalous self-luminous source of light seen after dark.
>
> *Radar-Visual (RV):* A sighting in which both radar contact and visual contact are made.
>
> *Close Encounter of the First Kind (CE-1):* A report of a structured object or craft seen at relatively near distance, approximately four hundred feet or less.

Close Encounter of the Second Kind (CE-2): A report where some measureable evidence is found at the site of a UFO after it has departed.

Close Encounter of the Third Kind (CE-3): A report including both a structured craft seen at near distance and also creatures of some kind somehow associated with the craft.

Another category of sighting was introduced after Dr. Hynek passed away. It was called the:

Close Encounter of the Fourth Kind (CE-4): A report in which there is evidence for some form of mental and/or physical transport of the human being from one location to another. Akin to the psychological fuge state, there is seldom any conscious recollection of the travel or surrounding details during a CE-4. Such encounters have been reported at least a thousand times by people in most nations of the earth. Their stories are remarkably consistent (Bullard 1987, 1991; Pritchard et. al. 1994).

The very establishment and use of the CE-4 category was yet another sign that ufology had begun to mature and was able to deal less emotionally with this extremely bizarre body of evidence. But times and people change rapidly; we seemed to have already slipped into another phase.

Pre-Senile Period (1980s to 1990s). This current stage of development of ufology is characterized both by irrational claims about what is going on, as well as by some outstanding work. The irrational claims are made by individuals who are largely unqualified to make such statements about UFO phenomena. Most of them have done no research of their own but only seem to want notoriety and/or money. Some may even be participating in a deliberate disinformation campaign.

Like people suffering the early stages of senility, some UFO researchers have selectively forgotten the history of this field and have repeated earlier errors; one being the belief that self-funded research is adequate to discover the core identity of the phenomenon. Dr. J. Allen Hynek asked a most timely and relevant question years ago. Could our nation have placed men on the Moon through the expenditure of out-of-pocket change alone?

Another error of this period is to continue working individually, rather than in concert, on the challenges posed by UFO phenomena. There is much more creativity, productivity, and mutual challenge generated within well-directed teams, as the famous Lockheed "Skunk Works" has clearly demonstrated over the years (cf., Rich and Janos 1994). Other trends include jumping to premature conclusions and misinterpreting events to suit one's own ends. Nevertheless, highly qualified people have made some important contributions. The outstanding work of investigators like Bullard (1991), Clark (1996), Devereaux (1982), Druffel (1988a, 1988b), Hall (1964), Rodeghier (1981), Thompson (1991), Vallee (1990, 1991), and Webb (1994) are a few cases in point.

An international event of note was an attempt to create within the Energy, Research and Technology Committee of the European Parliament a "European Observation Station for Sightings of Unidentified Flying Objects." As Doughty (1995) sarcastically said in his newspaper article, "It is the achievement of a dream for determined UFO watchers. But more skeptical earthlings are bug-eyed over a bizarre scheme from Brussels for spending even more money." Almost predictably, the proposal failed.

During the early 1990s two quite different definitions were proposed for the next available close encounter category:

Close Encounter of the Fifth Kind (CE-5).

The definition of this particular category is vitally important to its proper study. Vallee (1990) suggested that it refer to physical injury cases arguing that many do not occur at close range. Greer (Pg. 10, 1992a) defined a CE-5 event as "any human initiated interactive encounter with a UFO or their occupants." Also see his foreword to Part II in this book.

However, even Greer's definition seems too narrow to allow for the inclusion of such a broad range of reported experiences which may well originate from a common dynamic process. Therefore, the following definition of a CE-5 is used here:

> A Close Encounter of the Fifth Kind refers to any reported experience of deliberate human behavior that was soon followed by an obvious response from a UFO and/or humanoid and which included other effects suggesting it was not coincidental...in short, *human-initiated contact.*

Again, I suggest that the four simple rules of evidence listed earlier need to apply before a case is accepted as a genuine CE-5. For example, the human must participate by free will in a bilateral, seemingly mutual interaction with the UFO and/or creatures associated with it. This requirement clearly separates this category from a CE-4 where one may not behave volitionally at all. The CE-5 event represents a clear advance in ufology since it adopts a proactive approach toward the phenomenon—attempting to evoke a measureable response from the UFO and/or its occupants. Adoption of a CE-5 approach seems to be a logical extension of what ufologists have attempted to achieve in earlier years—from passive monitoring (with the hope of being at the right time and place to see a UFO) to proactive human involvement; beckoning a UFO into a friendly encounter, if possible.

However, the experiences are not as clear-cut as the above rules would demand. Actually, there is a broad spectrum of human behavior which is followed by a reasonably broad array of UFO responses. And so it becomes difficult to tell whether or not a particular incident qualifies as a "genuine" CE-5. This fact is illustrated by the following six cases:

Case E. 52-07-23. This U.S. Air Force interception took place off the coast of Hokkaido, Japan, at 8:15 P.M. Captain Norman Lamb was piloting an F-94 jet interceptor on routine patrol duty having taken off from Misawa Air Force Base. He was at about eighteen thousand feet altitude on a heading of 015 degrees when he saw "a strange phenomenon" approaching him rapidly from his five o'clock position and fifteen degrees high. It emitted dazzling, greenish-blue rays of light. It had no red navigation lights which airplanes are required to have. He became alarmed. His curiosity grew as the object reached his level and then took up position about five hundred feet off his right wing for an undisclosed period of time. He decided to bank toward it. When he did the glowing mass of light braked and made an erratic move toward the earth, passing out of his range of vision. Did the UFO react because of a change in the airplane's flight path toward it or was it coincidental? Of course there is no way to be sure either way (Sidenbery 1952).

Case F. 52-07-26. Typical of many others, this incident, which is included in the U. S. Air Force Project Blue Book files, involved an airplane approaching Aberdeen, Maryland, at 10:46 P.M. on one of the busiest days of UFO activity in American history. The pilot, a Civil Aeronautics Administration Flight Inspector, saw five orange-white points of light in the general vicinity of Washington, D.C. There had been numerous visual and radar contacts with UFO in the area for days before this, so air traffic controllers, pilots, ground personnel, the press, and the public were on the lookout for anything unusual in the sky. As Gross (1986,

25-26) points out, "at 10:52 P.M., all suspicious targets disappeared from Washington National radar." What is interesting is the fact that the Air Force was in the process of deciding whether to launch interceptor jets at this same time. They did so at 11:00 P.M. when two F-94 jet airplanes left Newcastle Air Force Base to the north of Washington, D.C. (Project Blue Book; 26 July 1952). This file contains a minute-by-minute Air Route Traffic Controller record of events similar to a tape transcript.

Officers at the Pentagon's Combat Command Post called Washington National Airport at 11:12 P.M. to verify if radar returns were still present. Their reply isn't known, but radar controllers continued to vector the jets' flight. At 11:13 P.M. Air Route Traffic Control personnel received a radio call from an aircraft referred to as "N606," asking about information on any air traffic near Andrews Air Force Base, about ten miles southeast of Washington National Airport. The pilot had seen "an odd light" in the sky. However, neither the ground radar team nor the approaching military interceptors could locate it. Radar contacts with the UFO began again at 11:30 P.M. at Washington National Airport so that air traffic control personnel could assist the jets.

Lt. William Patterson, one of the F-94 pilots sighted "four bright glows in the dark" that were "really moving." He was badly frightened *when the group of glowing objects surrounded his plane.* With the CAA radar operators monitoring his airplane and the nearby objects, Patterson "asked them in a scared voice what he should do. There was stunned silence; no one answered. After a tense moment, *the UFOs pulled away and left the scene.*" (Hall 1964, 159). Our interest here is with the rather typical cat-and-mouse game played out many different times over the years between UFOs and military interceptor airplanes. Such games must have given officials in charge heartburn. Was this a CE-5 case or not? It is likely that the UFOs were fully able to detect the presence of the approaching military airplanes. But we don't know what interpretation they placed on the attempted intercept by the Air Force airplanes (Gross 1986, 25-26).

Case G. 56-07-30 An eye witness on the ground at Utica, New York, watched for about fifteen minutes as a round, black object with an estimated twenty-five-yard diameter maneuvered up and down through clouds and appeared to hover at one point. According to an Air Force report, it "ascended into clouds at extremely fast rate of speed. *Object seemed to ascend into clouds whenever any aircraft came within four miles of it.*" (U.S.A.F. Project Blue Book file, 30 July 56; italics mine.) Were the reported flight dynamics of this object somehow linked to the approach of an airplane—as has been reported many times before and after this event? If so, it suggests a high degree of intelligence and aeronatucal capability.

It is also interesting to note the Air Force's reference to the phrase "any aircraft" which suggests that more than one was involved while the sighting lasted only about fifteen minutes. Is it possible that in its own investigation of this event, the Air Force learned of other airplanes that had been in the area at the time? Another interesting detail is their reference to a distance of "four miles." Why would four be cited rather than "several," "five," or other more standard measure? This suggests that the Air Force may have found some evidence for a range of exactly four miles to the unidentified object.

Case H. 57-06-17 A strong stationary radar return was received at the Ohakea Air Base in New Zealand and a Vampire jet interceptor in the area was vectored by a radar controller to try to locate and identify the source of the unknown return. The pilot sighted a "bright blob of light at the reported location of the radar target but the light zoomed away as the Vampire jet approached" (Civilian Saucer Investigation 1957, 15). The official response of the New Zealand Air Force was that the return was "due to a radar reflector descending from a burst balloon becoming temporarily retarded in its descent upon reaching a temperature inversion layer." This explanation seems almost more far fetched than an extraterrestrial account.

Case I. 57-11-02 Mr. Jose Alvarez was driving on Route 51 near Whitharral, Texas, about midnight when he saw a huge object straddling the roadway ahead of him. He believed it to be about two hundred feet long. As he approached it, his engine stopped and his headlights faded out with decreasing distance. *The object rose rapidly into the dark night sky and was gone* (Schopick). Was there missing time? Did Mr. Alvarez provide all of the details? Did he somehow signal to the strange object? Did the object depart because Mr. Alvarez' car was approaching? We don't know the answer to any of these questions. We do know that the sudden departure of the UFO is a very common response as noted in the following pages.

Case J. 92-07-22 This rather detailed series of events took place in southwest

England when a group of Americans and British investigators were studying crop circle phenomena and also attempting to use various procedures using what they referred to as the "CE-5 protocol." This protocol employs focused thoughts of the entire group, powerful light beam, and auditory stimuli, most often at night, to beckon a UFO to approach and to land.

Five Americans were joined by participants from other nations on Woodborough Hill near Alton Barnes, Wiltshire County, England, over several days. On July 22 the assembled team members heard a "metallic noise" at 9:30 P.M., but did not see anything unusual. The agreed upon methodology to initiate contact was to mentally concentrate upon a "shape which would be held by the group collectively, which could be painted in the sky with a high powered light, and which could be received by the crop circle makers and result in a confirmed authentic crop formation of the same shape. This, of course, would be a crop circle CE-5" (Greer 1992b, 6). This was a laudable idea if for no other reason than it was proactive regarding the unseen phenomenon and attempts to "close the response loop," i.e., to obtain evidence that a message transmitted has been a message received and acted upon.

The group agreed upon the shape shown in Figure 1, three circles joined by three lines forming an equilateral triangle. This shape happens to be the symbol adopted later by the Center for the Study of Extraterrestrial Intelligence (CSETI) for its logo.

During the night of July 22, 23, and 24 several unusual aerial lights were seen by various witnesses in the above locale. Extraordinary psychic events were also reported by two Americans present.

Figure 1

Geometric shape agreed upon
by CSETI team members on July 22, 1992

On the morning of July 25 or 26, a farmer at Roundplay, near Devizes, discovered a pattern in his field in the same shape as had been "transmitted" by the CSETI group. This location was several miles away from where the CSETI group was located. Ralph Neyes, a prominant British crop circle investigator determined that the shaped impression had probably been formed during the night of July 23 or the 24. It is not possible to date the event closer than this. Other investigators (i.e., Argost Team) indicated that the formation in the field of grain was authentic and not man-made.

However intense and convincing an experience as this might have been to those present, it is unconvincing to those who demand more scientific proof. Where are the measurements and photos of this particular formation? What proof is there that humans did not made this formation following its disclosure on July 22?

Figure 2 shows a formation found in a field of grass in Northern California in early July 1995 which may be similar to the formation described above near Woodborough Hill, England. Of course, it isn't known whether someone within the northern California CSETI group described this particular shape to anyone else who could have created it. And what evidence is available to indicate that it was not man-made? These are a few of the fundamental questions that one must answer before presenting it as proof of a CE-5.

Not until we hold all alleged evidence to equally high standards will we be in a position to claim that something truly remarkable has occurred. Personal experiences are not sufficient.

Figure 2

Photograph of geometric formation found in a field in
Santa Rosa, Calif., in July 1995 (Courtesy of Tom Page)

Signalling to UFOs

What types of human responses can be used to signal to an aerial phenomenon? The list is long and, from a methodological standpoint, can be difficult to describe. Human behaviors which may elicit UFO reaction can range from shining a million-candle-power searchlight toward it to waving, to merely thinking something (i.e., a mental projection). This range of human behavior is hard to quantify along any consistent continuum. Table 1 lists some of the many possible deliberate human means of signalling to a UFO.

Table 1

Different Kinds of Deliberate Human Projections Used to Signal a UFO

Light Beam
>Single, collimated
>Single, wide-angle
>Multiple, collimated
>Multiple, wide-angle
>Continuous or pulse-coded
>Wavelength varied

Ionizing Radiation

Radio Frequency
>Amplitude modulated (AM)
>Frequency modulated (FM)

Microwave Energy
>Radar

Other frequencies

Acoustic (air-coupled) Vibration
>Single, constant frequency
>Modulated frequency (e.g., sine wave tone)
>Other (pulse-coded)
>White noise

Physical Projectiles
>Arrows
>Bullets
>Missiles
>Vehicles
>Stones

Thought (telepathy)
>Focused/aimed
>Unfocused, vague

Many of these means have been tried with varying degrees of claimed success, as the cases presented below will attest.

What does the literature say about alien aggressiveness?

Many authors have written about alleged acts of alien aggression against humans. But few writers have considered whether the UFO or alien beings were provoked into a response.

The following works include cases of personal injury allegedly caused by UFOs and/or extraterrestrials: Binder (1970), Pratt (1996), Schuessler (1996), Wilkins (1954). This book sheds some further light on this vitally important issue as well. If our visitors were provoked by acts of human aggression and still did not react in kind, it suggests that the alleged intelligence behind the UFO is highly self-controlled.

In his review of 198 cases involving human-alien contact, J. U. Pereira (1974) found twenty-four cases (12.2%) in which aliens approached humans, and another twenty-seven (13.6%) in which the aliens departed. The other cases did not make any distinction in this regard. Of the twenty-seven cases which Pereira classed as being "hostile," twelve (44.4%) involved alien aggression shown toward humans, eight (29.6%) involved human aggression shown toward the alien(s). Three cases (11.1%) involved an actual fight of some kind, and another four (14.8%) involved some kind of "accidental violence." Within the thirty-six cases classed as "friendly behavior," the alien smiled in seven cases (19.4%), put his hand on the shoulder of the human as a friendly gesture in six cases (16.6%), extended his hand or gestured toward the human in twenty two cases (61.1%), and one case involved conversation.

What might we learn in this review?

Because this book is the first to review evidence of Close Encounters of the Fifth Kind, it is possible that we may learn whether UFOs respond in any consistent ways. If they do, it would strongly suggest an intelligence or deliberateness to their actions. We might uncover other patterns in the data which are now hidden. These patterns might tell us something about their technological capabilities. Perhaps they are not even concerned about our response to their presence.

If they are invincible, then: 1) they should not necessarily appear at any particular time of day or night (e.g., to hide in darkness); 2) they should fly anywhere they choose (e.g., directly over atomic processing plants like Hanford and Los Alamos National Laboratory or over Moscow, Fort Knox, and the Pentagon); 3) they should not need to camoflage themselves to try to hide from view; and 4) they may become extraordinarily conspicuous (e.g., returning duplicate light flashes, pacing automobiles and jet planes, hovering near naval ships).

Yet, there may be an even more important lesson to be learned. If UFOs are not merely reflections of our inner subconscious fears and hopes, they may be trying to teach us how to behave. If found to be reliable, this review can teach us a great deal about ourselves. We already know how aggressive human beings are, but we don't fully understand the reasons why.

Of course, the first challenge is to find reliable cases to study. I acknowledge that the ever-present skeptic will find much to challenge in the cases presented here. But if even a small proportion of the more than two hundred cases presented here are reliable, we face an even greater challenge: to explain how they occurred and perhaps even why. CE-5 events hold the promise of new insights into UFO presence and meaning for mankind.

References

Aerial Phenomena Research Organization. 1970. *APRO Bulletin* (September-October):1-5.

Binder, O.O. 1970. How Flying Saucers can Injure You. In *Flying Saucers Have Arrived!* 30 Documened Reports, ed. Jay David, 183-87. New York: The World Publishing Co.

Bowen, C. 1976. *Flying Saucer Review* 22, no. 2.

Bullard, T.E. 1987. UFO Abductions: The Measure of a Mystery. *Fund for UFO Research*. Mount Rainier, Md.

——. 1991. Folkloric dimensions of the UFO phenomenon. *J. of UFO Studies* 3:1-58.

Civilian Saucer Investigations. 1957. *Official Quartery Journal* 5, no. 2. New Zealand.

Clark, J. 1996a. *High Strangeness: UFOs from 1960 through 1979*. Vol. 3 of *The UFO Encyclopedia*. Detroit: Omnigraphics.

————. 1996b. The salvation myth. *International UFO Reporter* (Winter).

Corso, P.J. and W.J. Birnes. 1997. *The Day After Roswell.* New York: Pocket Books.

Edwards, F. 1966. *Flying Saucers—Serious Business.* New York: Lyle Stuart.

Devereux, P. 1982. *Earthlights: Towards an Understanding of the UFO Enigma.* Wellingborough, Northhamptonshire, England: Turnstone Press.

Doughty, S. 1995. Eurocrats' close encounter of the expensive kind. *Daily Mail.* Assoc. Newspapers Ltd.(4 September).

Druffel, A. 1988a. Abductions: Can we battle back? *MUFON UFO Journal* 247 (November).

————. 1988b. Can we battle these entities? *Flying Saucer Review* 33, no. 3 (Sept.).

Fawcett, L., and B.J. Greenwood. 1984. *The UFO Cover-Up.* New York: Prentice Hall Press.

Friedman, S.T. 1987. MJ-12: The evidence so far. *International UFO Reporter* 12, no. 5 (September-October).

Friedman, S. T., and D. Berliner. 1992. *Crash at Corona: The U.S. Military Retrieval and Cover-up of a UFO.* New York: Paragon House.

Friedman, S.T. 1989. Update on Operation Majestic - 12. In *Proceedings of 1989 MUFON International Symposium*, ed. W. H. Andrus, Jr., 81-112.

Gersten, P. 1981. What the U.S. Government Knows about Unidentified Flying Objects. *Frontiers of Science* (May-June).

Greer, S.M. 1922a. *CSETI.* Asheville, North Carolina: Privately published.

————. 1922b. *Close Encounters of the 5th Kind: Contact in Southern England* (July). Asheville, North Carolina: Privately published.

————. 1996. *Unacknowledged.* Asheville, North Carolina: Privately published, Center for the Study of Extraterrestrial Intelligence.

Gross, L.E. 1976. *Charles Fort, The Fortean Society, and Unidentified Flying Objects.* Fremont, Calif.: Published privately.

————. 1986. *UFO's: A History 1952: July 21st - July 31st.* Fremont, Calif.: Published privately.

27

Haines, R.F. 1980. *Observing UFOs*. Chicago: Nelson-Hall.

———. 1990. *Advanced Aerial Devices Reported During the Korean War*. Los Altos, Calif.: LDA Press.

Hall, R. 1964. *The UFO Evidence*. Washington, D.C.: NICAP.

Jacobs, D.M. 1975. *The UFO Controversy in America*. Bloomington, Ind.: Indiana University Press.

La Razon. 1975. 27 October.

McDonald, J.E. 1967. *UFOs: Extraterrestrial Probes? Astronautics and Aeronautics* 5 (August): 19-20.

McDonald, J.E. 1970. UFOs Over Lakenheath in 1956. *Flying Saucer Review* 16, no. 2 (March-April):9-17, 29.

Michel, A. 1958. *Flying Saucers and the Straight-line Mystery*. New York: Criterion Books.

Pereira, J. U. 1974. *Phenomenes Spatiaux*. 2nd. special edition (November): 127.

Phillips, T. 1975. *Physical Traces Associated with UFO Sightings: A Preliminary Catalog*. Northfield, Ill.: Center for UFO Studies.

Phillips, T. 1980. Landing Traces and Physical Evidence. In *Proceedings of the First International UFO Congress,* ed. C.G. Fuller, 56-65. New York: Warner Books.

Pitt. 1956. Tell us please Mr. Birch. *Flying Saucer Review* 2, no.5:10-13.

Pratt, B. 1996. *UFO Danger Zone*. Madison, Wis.: Horus House Press.

Pritchard, A., D.E. Pritchard, J.E. Mack, P. Kasey, and C. Yapp, eds. 1994. *Alien Discussions: Proceedings of the Abduction Study Conference. MIT, Cambridge, Mass, June 13-17, 1992*. Cambridge, Mass.: North Cambridge Press.

Randle, K.D., and D. R. Schmitt. 1991. *UFO Crash at Roswell*. New York: Avon Books.

Randle, K.D., and D. R. Schmitt. 1994. *The Truth about the Crash at Roswell*. New York: M. Evans & Co.

Rich, B.R., and L. Janos. 1994. *Skunk Works*. New York: Little, Brown & Co.

Rodeghier, M. 1981. *UFO Reports Involving Vehicle Interference: A Catalogue and Data Analysis*. Evanston, Ill.: Center for UFO Studies.

Rutledge, H.D. 1981. *Project Identification: The First Scientific Study of UFO Phenomena*. Englewood Cliffs, N.J.: Prentice-Hall.

Schopick, A. *UFO Electro-magnetic cases*. CUFOS.

Schuessler, J.F. 1996. *UFO-Related Human Physiological Effects*. La Porte, Texas: Published privately.

Shannon, C.E., and W. Weaver. 1949. *The Mathematical Theory of Communication*. Urbana, Ill.: University of Illinois Press.

Sidenbery, J. 1952. *Air Intelligence Report*. 35th. Fighter-Interceptor Wing, North Honshu, Japan. APO 994 (13 August) Project Blue Book files.

Sider, J., and I. Scott. 1992. Roswell and its possible consequences on American policy. *MUFON UFO Journal*, no. 296 (December): 10-11.

Sparks, B.C. 1971. *Human casualties of UFOs and Ufonauts*. Aerial Phenomenon Research Organization, Report No. 6. 8th ed. (14 April). Tucson, Ariz.

Thompson, K. 1991. *Angels and Aliens: UFOs and the Mythic Imagination*. Reading, Mass.: Addison-Wesley.

Vallee, J. 1969. *Passport to Magonia: On UFOs, folklore, and parallel worlds*. Chicago: Henry Regnery.

Vallee, J. 1990. *Confrontations: A Scientist's Search for Alien Contact*. New York: Ballantine Books.

———. 1991. *Revelations: Alien Contact and Human Deception*. New York: Ballantine Books.

Webb, W. 1994. *Encounter at Buff Ledge: A UFO Case History*. Chicago: CUFOS.

Wilkins, H. 1954. *Flying Saucers on the Attack*. New York: Citadel.

Zechel, T. 1978. The five most-asked UFO questions-1978. *International UFO Reporter* 3, no. 4 (April):4.

———. 1979. NI-CIA-AP or NICAP. *MUFON UFO Journal*, no. 133 (January-February):6.

Zeiler, B. 1996. The logical trickery of the UFO skeptic. http://www.primenet.com/~bdzeiler/trickery.htm

Always think about the substance rather than the methodology, graphic design, tables, technology, or something else.

Chapter 2
Methodology Used to Analyze Narrative Reports

In a review and analysis of this kind of subject it is very important to select cases based on explicit criteria and clearly defined, precise methodology. The objective of this chapter is to document both of these subjects.

Classes of human behaviors

Let us begin by considering various broad characteristics or features of human behavior. It is by using such behaviors that people attempt to signal UFOs. In general, they can range from:

- short to long (in duration),
- simple to complex (in detail and/or interrelationships),
- impulsive to deliberate (in premeditation),
- exactly corresponding to very different actions, and
- objectively obvious to nearly invisible.

Each of these "dimensions" can be scaled and scored to develop a useful total score to be assigned to each case. This score can be used to compare different incidents.

Duration of behaviors (DB)

How long the activities last is an important parameter in a close encounter. The approach used here yields a letter to represent the combination of both the duration of human behavior as well as the duration of the UFO's (or humanoid's) response. The rating matrix and letters are presented in Figure 3.

Figure 3

		<1 sec	1-5 sec	5-30 sec	0.5-5 min	5.1-30 min	>30 min
Duration	>30m	ee	ff	gg	hh	ii	jj
of UFO	5-30m	y	z	aa	bb	cc	dd
or Alien	.5-5m	s	t	u	v	w	x
Response	5-30s	m	n	o	p	q	r
	1-5s	g	h	i	j	k	l
	<1s	a	b	c	d	e	f

Duration of Human Behavior

Rating scores used for duration of behavior (DB)

Some narratives don't specify duration of behavior but use various words. The following general convention was used here unless the context suggested a different duration:

"Flash" was scored as < 1 second
"Instant" was scored as < 1 second
"Brief(ly)" was scored as < 1 second
"Pulse" was scored as < 1 second
"Prolonged" was scored as > 0.5 minute
"Long time" was scored as 0.5 to 5 minutes.
"Very long time" was scored as > 30 minutes
No data is scored as "kk"

One difficulty in applying this rating scale comes in deciding which particular behavior elicits the UFO response and which are secondary? Human behavior is very complex and often covert so that usually we must rely upon the testimony of the witness. Another difficulty is deciding when the behavior and subsequent response begins and ends. Often, only a rough estimate is made.

The duration cited in a report can be either relative or absolute. Relative values are considered only as approximations or rough estimates by the witness while absolute values are durations derived from clocks, wrist watches, or other time pieces. They are considered accurate to less than a minute.

Complexity of behaviors (CB)

Obviously, if someone simply flashes a spotlight once into the night sky and then notices a flash of light in return it may be astonishing but it is not nearly as important to our study of elicited UFO responses as if someone signals "What time is it now?" in Morse code, which might take a minute or more to complete, and then receives the identical message flashed back from the aerial object. The (auditory) tonal pattern employed by the human investigators to communicate with the huge UFO in the Hollywood movie *Close Encounters of the Third Kind* illustrates this idea.

Over thirty years ago, the National Investigations Committee on Aerial Phenomena (NICAP) published the now classic review of UFO evidence (Hall 1964). It included a section dealing with apparently intelligent control of UFOs. Inquisitiveness and reaction to the environment were two measures used to indicate whether UFOs were intelligently controlled. The measures included such flight behavior as approaching, pacing, circling, and signalling an airplane pilot or witness on the ground. Indeed, all of these behaviors and more are found in the following review.

The challenge in trying to rate behavioral complexity lies in developing a scale that is consistent, clearly defined, and simple to score. I propose a two-dimensional matrix where the horizontal axis represents a judgment about the complexity of the human's behavior and the vertical axis represents an estimate of the complexity of the subsequent UFO or humanoid response (see Figure 4). These examples provide some guidance in scoring complexity of human behavior.

Low ....Human sees light in sky, continues to drive vehicle along roadway, taking normal left- and right-hand turns to get home.

Medium ....Human sees light in sky and continues to drive home (as above)
but also stops several times and turns headlights on and off.

High ....Human sees light in sky and continues to drive home, stops several times, and flashes headlights on and off. Then stops, gets out
of the car, and flashes a spotlight at the UFO while attempting to
send a welcoming thought to it.

Figure 4

	high	g	h	i
Complexity of ufo or Humanoid Response(s)	med.	d	e	f
	low	a	b	c
		low	med.	high

Complexity of Human Behavior

Scoring matrix for complexity of behavior

Of course, it isn't possible to list all of the likely behaviors which humans use
to signal a UFO or an E.T. These scoring categories should be interpreted as flex-
ibly as necessary with the next lower score used when in doubt.

The following examples illustrate the general meaning of low, medium, and
high complexity of UFO responses:

Low ....UFO hovers near the ground, emits a short series of light flashes
in apparent response to the human's previous behavior.

Medium ....UFO approaches, hovers near the ground, emits longer or more
complex series of light flashes and follows the human's movement
(as in an automobile, boat, or airplane).

High ....UFO hovers near the ground, emits complex multi-colored light
flashes and a sharply defined beam of light directly at the humans
and then circles the human while changing its orientation in space.

While it is even more difficult to categorize E.T. responses than it is the actions of UFO, the following provide an idea of how E.T. responses to prior human behavior were scored:

Low ...............E.T. neither approaches nor leaves the immediate area and continues carrying out some ongoing action yet seemingly is aware of the human's behavior.

Medium ..........E.T. redirects its "attention" toward the human from some prior activity, moves or gestures toward the human, and then departs.

High ..............E.T. redirects its "attention" as above, motions or gestures, attempts to communicate with or mimics the human's behavior exactly.

Inability to estimate: Zero is inserted if it is not possible to determine the complexity of either the human, UFO, or E.T. behavior for any reason.

Premeditation of behavior (PB)

The reason a person signals a UFO in the first place is an important considera- tion in fully understanding these data. Thus, a different score is given to an incident where the person performed a specific behavior with a *visible* phenomenon already present (PB-1 scale) than when the phenomenon was not yet present, i.e., *invisible* and the person only "felt" the presence of the phenomenon intuitively (PB-2 scale). The scoring approach used here discriminates between the view that one can encourage or signal an invisible phenomenon to appear compared with the opposite point of view. Only one of these two rating scales is used per case. An "x" is inserted for the unused scale's score.

PB-1 score	Response to visible phenomenon
x	No data
a	Unplanned, reflexive human response to a visible aerial object or humanoid
b	Preplanned human response to a visible aerial object or humanoid
0	Inability to estimate

PB-2 score	Premediated response to non-visible phenomenon
x	No data
a	Unsure, but possible
b	Yes
0	Inability to estimate

UFO response delay (UFOD)

Clearly, the shorter the duration (D) between the end of the human's volitional behavior and the start of the UFO's response, the more likely it is that the response is not a coincidence. This is not to say that there are no natural (coincidental) phenomena which would yield very small values of D given the right set of circumstances.

UFOD score	Time elapsed
0	Inability to estimate
a	<3 seconds
b	3.1 - 15 seconds
c	16 - 60 seconds
d	No observed effect

The following terms received an "a" score—"abruptly," "as soon as," "at once," "immediately," "just then," "suddenly," and "the instant that."

If a human performs some behavior which seemingly elicits a corresponding behavior by the UFO, there are at least two temporal degrees of freedom involved: 1) the temporal characteristics of the acts performed by the human, and 2) the UFO's response delay (RD). What is of primary importance is whether these two are dependent or independent of one another?

Limitations of narrative accounts

As every social scientist who has conducted interviews will verify, narrative recreation of prior events has a number of limitations from the standpoint of deriving accurate understandings of what actually occurred. It is important to list the major limitations which are likely to be found here to better interpret them. They include:

Unintentional Errors

Transcription errors can occur at each retelling and rewriting. The field investigator cannot be exactly accurate in transcribing a witness's narrative account without taking extensive precautions that are costly in time, energy, and finances. Such errors are a subclass of the following selectivity error discussed below.

Inexperience-related errors occur when the field investigator does not ask the right questions of the witness and thereby distorts the narrative report.

Interpretational errors develop when the witness unintentionally projects the attributes of a familiar object onto what is being described, rather than providing an accurate description. Light refraction and diffraction effects, illusions, and mirages, can make things look very different from what their physical shape would dictate. Here, the witness is acting honestly, but naively. Another source of error lies in our unconscious biases. Davis (1991) wrote that "no judgment can ever be truly unbiased."

Intentional Errors

Selectivity errors occur when the witness selects certain details to describe and omits others. (These are also called "completeness" or "omission" errors.) Such errors may be intentional or non-intentional (e.g., caused by concussion, amnesia-related).

Fabrication errors occur when the witness adds facts on purpose. (This sometimes is done to avoid embarrassment or legal action, or to try to gain notoriety and fame.) Here, the witness is acting dishonestly from the standpoint of obtaining "the truth."

Other Errors

Publication errors occur when, for example, due to the pressure to shorten articles, editors omit material from the original report.

Of course, a major challenge facing the investigator is to determine which errors have taken place and how to correct them. To do this completely would require not only being a highly skilled field investigator who has access to the original witnesses over long periods of time, but also having adequate secretarial, technical, and financial support. After reading and thinking about each of the cases presented in this book, I have concluded that the great majority suffer only from selectivity errors.

References

Davis, J. 1991. "Combatting Mindset," *Studies in Intelligence*. (Winter):13-18.

Hall, R. 1964. *The UFO Evidence*. Washington, D.C.: NICAP.

Part I
Foreword

Is there some intelligence behind the UFO phenomenon? This is a question that scientists and investigators have wrestled with from the beginning of the modern UFO era in June 1947. It was natural to make such an inference for several reasons. Many UFOs looked like solid, flying objects, not just distant lights in the sky. What else could such objects be but intelligently guided, though by whom was an open question. The actions of the UFOs gave further support to the idea of intelligent control, as UFOs were seen to follow or buzz airplanes, sometimes flee when being observed, and other times hover in place, seemingly watching human activities.

The U.S. Air Force's UFO investigation confronted the question of intelligent control early in its work. For example, on October 1, 1948, Lt. George Gorman of the North Dakota Air National Guard engaged in an intense "dogfight" with a small light over Fargo. Investigators were perplexed by the event, and Gorman himself said, "I had the distinct impression that its maneuvers were controlled by thought or reason."

Over the years, many people have speculated about what intelligence might lie behind UFO sightings. Few, though, have closely studied the actions of UFOs to rigorously determine the characteristics of their behavior as it relates to the question of intelligence. Dr. Haines has courageously taken on this task, one that can be a slippery slope to confusion and wishful thinking if not handled properly. It is very easy to anthropomorphize the UFO phenomenon and read into it what one wishes or hopes, whether that be salvation or imminent attack.

Dr. Haines is particularly suited to the job because of his long and distinguished involvement in the field, his work as a perceptual psychologist, and, most important, his emphasis in all his work on the methodology by which we study UFOs. His book *Observing UFOs* is the key reference on the perception of unusual things in the sky, and his "Three-stage Technique" (TST), an approach to achieving unbiased hypnosis, is the key to the successful study of UFOs reactions to human behavior. I suggest the reader pay careful attention to Chapter 2, where Dr. Haines explains his methodology of studying CE-5 events.

An unanticipated benefit of this work consists in the hundreds of UFO encounters described in the text. It is too easy today, with the media's emphasis on abductions, UFO crashes, and other sensational UFO events, to ignore the vast diversity that is subsumed under the term "UFO." The UFO phenomenon is truly a challenge for science, in part because of its extensive variety. Just reading through the cases selected by Dr. Haines is an education in itself, but it is also a chance for you to get a better understanding of the dimensions of the UFO phenomenon which neither begin nor end with the latest abduction story. Put another way, the book is a reminder that, like all fields, there is a history to UFO sightings that cannot be ignored.

The central intent of Dr. Haines' study—to sort through this endless variety to see what patterns might emerge, patterns that might point to an underlying intelligence—is arguably the most important question in UFO studies. It is furthermore one of the most important scientific questions of the day, no matter how some may off handedly dismiss the UFO phenomenon and its witnesses. In this task, I judge Dr. Haines has succeeded, as you will soon read. He has isolated patterns which are intriguing and suggestive, and you can judge for yourself whether these patterns bespeak an intelligence behind the UFOs. If nothing else, they certainly argue for further study and investigation of UFOs. To not do so is to be blind to what could arguably be mankind's potential first contact with another intelligence.

I have no doubt that this work will become the standard by which we study the behavior of the UFO phenomenon. Dr. Haines has constructed a scientific framework to study UFOs and the question of intelligence that I hope others emulate and expand. I congratulate him on the determination and drive needed to carry this work to its conclusion. It reminded me of why I first began studying UFOs over twenty years ago, and also of how much there still is to know about the phenomenon. But we now know quite a bit more about the behavior of UFOs, both friendly and hostile, because of this work. And it is encouraging to note that there is more of the former than the latter. Perhaps that by itself is the best argument for an advanced intelligence at work.

Mark Rodeghier, Ph.D.
Scientific Director
Center for UFO Studies
Chicago, Illinois

"If they're there,
how come they're not here?"
— Fermi Paradox

"Since they're here,
they must come from someplace."
— Haines' logic

Chapter 3
UFO Responses to Overtly Friendly Human Behavior (91 cases)

In this chapter, I will review a large number of very interesting CE-5 events in which the human being allegedly carries out the first friendly, or at least non-hostile, behavior which is subsequently followed by some type of obvious "response" by the UFO. In the great majority of these cases, humans behave deliberately to the presence of the UFO and not to some other object or stimulus present. As Lade has suggested, "friendliness is consistent with non-interference in the development of civilization on earth" (1960).

UFO responses to humans who approach a UFO

Before reviewing the following cases, consider the instances where humans simply approached a UFO for whatever reason. Some might argue that an approach behavior could be considered either friendly or hostile by an alien. While varied UFO responses illustrate the difficulty in classifying CE-5s, they also offer us much to analyze.

From the standpoint of the physical injuries suffered, it may be immaterial whether the person approached a stationary UFO or vice versa. The interested reader should consult Pratt (1996) for many interesting examples of the latter from Brazil. Here, I present a few examples of the former.

Case A. 54-12-19 Two men spotted a round and highly unusual object sitting on the ground near Tavirona, Spain. They tried to approach it, but when they were about 50 meters from it, *it noisily rose up into the sky. (Noticiario Universal,* 20 December 1954. Olmos, V-J. and J. Vallee. *"Catalogue of 200 Type 1 Events in Spain and Portugal,"* CUFOS. (1976):4).

Case B. 54-12-29 It was 9:00 P.M. when M. Gamba witnessed a glowing red oval mass on the ground about 40 yards away from him, near Bru, France. As he walked toward it, *he felt a strange paralysis* that temporarily prevented him from getting any nearer. When his mobility returned, he ran to find other people to observe the object. According to an article in the local newspaper (*Sud-Quest,* 31 December 1954), the man went back to the site within minutes and saw *the UFO changing color from white to red. It then shot up into the sky and flew to the east.* The ground underneath the object was "found disturbed, and trees in the area had some limbs slashed." Little more is known about this incident (Lumieres dans la Nuit., no. 104).

Case C. 57-11-23 At 6:30 A.M., on a calm and clear pre-dawn morning, U.S. Air Force Lt. Joseph F. Long was driving about thirty miles west of Tonopah, Nevada. His car's engine quit suddenly, so Long pulled over and set the brake. He tried to restart his car it without success. When he got out to investigate under the hood he noticed a "steady high-pitched whining noise." It was then that he noticed four disc-shaped objects sitting on the desert sand about 300 to 400 yards away from the highway. He had never seen anything like them before and was curious. He began walking in their direction. The four objects were identical—each glowing brightly with translucent domes on top. The objects were about fifty feet in diameter, eight to twelve feet thick, and were supported by three half-round landing gear about two feet in diameter. Long later sketched the objects (see Figure 5). When Long was about fifty feet from one of the objects *the constant hum began to increase in pitch* "to a degree where it almost hurt his ears and the *objects lifted off the ground. The protruding gears were retracted immediately after take-off."* The craft leveled off at about fifty feet altitude and moved to the north slowly, crossing the highway and then following the hilly terrain to the north. (The remainder of these interesting details may be found in Microfilm Reel 31 of the official Project Blue Book Air Force files.) The official explanation for this highly detailed sighting by an officer in the Air Force was "psychological. An optical illusion or mistaken identity of a conventional craft or due to road fatigue, road hypnosis, and lighting conditions!" Once again, such specious explanations barely deserve to be cited as serious. This official explanation overlooks much of the reported evidence.

Figure 5

Eye witness sketch of one of four identical UFOs
seen in Nevada Desert on November 23, 1957
(From: USAF Project Blue Book files)

A number of sketches are provided in this book of the object that was seen. Notice the relative symmetry, horizontal orientation, width to thickness ratio, and surface details of these objects from case to case. They correspond closely with other drawings in the literature and the author's files (Haines, 1978a; 1978b; 1980). Chapter 10 presents other comments concerning these drawings.

Case D. 58-02-24 This incident involved three men driving near the village of Santo Antonio de Jesus, Brazil, at 3:05 A.M. Their car's engine began coughing and sputtering, and then stopped altogether. They got out of the car to fix the problem, but found nothing obviously wrong. Soon after deciding to camp out for the rest of the night, they sighted a self-luminous flying object between silver and blue in color approaching them. They noticed a ring around its circumference which was spinning at a high rate. The object descended toward the ground in a falling-leaf type motion and hovered only about three meters off the ground. The two passengers began to walk toward the area of illumination on the ground produced by the UFO that they estimated to be about twice the size of the UFO. At this time the craft *suddenly rose to an altitude of about 200 meters and started to perform a series of high-speed maneuvers.* It departed at 4:35 A.M., yet the car would not restart. The strange object, now silver color, returned at 6:30 A.M. and *eventually shot away out of sight in a matter of seconds.* The automobile's engine then started with ease. (Lorenzen, C., *Flying Saucers: The Startling Evidence of the Invasion from Outer Space.* New York, Signet, 1962.)

Case E. 59-10-22 During the night, a car with three people in it was near Cumberland, Maryland. Witnesses saw a metallic disc projecting a bluish-green light from its surface. They pulled over to the shoulder of the road and stopped in order to see the unusual object better. Then, the UFO descended to an altitude of about fifty feet and stopped, giving off a humming vibration. It moved somewhat farther ahead of them and stopped again. Two of the car's occupants decided to get out and walk in the direction of the UFO. As they did, the car's engine, lights, and radio all stopped at the same time. *"Within a short time, the disc shot up into the clouds* and the lights and radio came back on. The car could be restarted." (Lorenzen, C. UFOs: The Whole Story. New York: Signet, 1969.)

This incident suggests that the behavior of the humans probably was more than merely walking toward the hovering object. Perhaps they acted in some hostile or fear-filled way which they failed to mention. Nevertheless, is there a correspondence between the car's stopping and the UFO's descending and stopping? Did the UFO leave in order not to injure the humans? There is no way to know for sure and herein lies one of the problems in pursuing this type of research from a scientific point of view. The following cases illustrate some of the problems which arise from insufficient evidence.

Case F. 64-04-14 A man was driving a truck near Chico, California, when, according to a report published by the *J. Allen Hynek Center for UFO Studies* (Rodeghier 1981), the truck stalled as a domed disk hovered about 1,000 feet above. The man felt a static electricity-like sensation (his hair stood on end). The object then approached the ground and landed. The man stopped his truck and approached the object on foot. The object felt warm to the touch. *He received a small amplitude electrical-like shock and heard voices.* (MUFON UFO Journal, no. 275.)

Case G. 64-04-24 Only a small part of this classic close-approach case will be presented here since the interested reader will find it discussed in many other books. Officer Lonnie Zamorra of the Socorro, New Mexico, police department was in pursuit of a speeding automobile at 5:45 P.M., when he heard a sudden roaring sound and spotted a flame in the southwest sky. He thought that a dynamite cache had blown up, so he stopped his chase and set off in the new direction. He told Air Force, FBI, and other investigators that the sound was a roar and not a blast, changing from a higher frequency to lower before ceasing. The flame descend slowly and it soon passed out of sight behind some hills.

Driving slowly and carefully over dirt roads and paths, Zamorra finally reached a point about 150 to 200 yards from something which looked like a "car turned upside down" far from the road. He noticed "two people in white coveralls" very near the object. Zamorra says that one of them "seemed to turn and look straight at my car and seemed startled—seemed to quickly jump somewhat." Zamorra drove nearer and stopped again, radioing a message to the office that he was going to check a car "down in the arroyo." He began approaching the UFO on foot. The beings were now gone and he was about fifty feet from the unknown object. Then he heard *a "very loud" roaring sound, beginning at a lower frequency and increasing quickly. Red flames came from beneath the aluminum-white object,* whose bottom surface was about four to five feet off the ground, sitting on stilt-like legs. Thinking that it was about to explode, Zamorra turned and ran back to crouch behind his patrol car. The UFO rose slowly—about six feet in six seconds. Zamorra subsequently watched it fly away to the southwest, first with a roaring sound followed by a one-second-long "sharp toned whine from high tone to low tone" followed by complete silence. [Hall, R. "Socorro landing. Pp. 341-344, In *The Encyclopedia of UFOs.*, ed. R.D. Story, 341-44. Garden City, N.Y.: Doubleday & Co., Dolphin Books, 1980; "The Socorro Classic," *Flying Saucer Review* 10, no. 6 (November-December 1964):6-7.]

Case H. 64-04-24 Mr. Gary T. Wilcox, a dairy farmer in Tioga County, New York, was spreading manure in an open field about 9:50 A.M. when he decided to work in a nearby field. Wilcox approached the triangular plot of ground through a group of trees, noticing something strange sitting out in the field which he first took for a discarded refrigerator and then a wing fuel tank from an airplane. He walked straight toward it and soon realized it was not what he originally thought. The object

45

was shiny metal, approximately 20 feet long and 16 feet wide, and egg-shaped. He touched its surface but *felt no heat*. There were no markings on or seams in its surface, yet somehow two "human-like men suddenly appeared." They spoke in "smooth" English and wore seamless clothing. A helmet faceplate prevented him from seeing their faces. They had arms and legs yet he couldn't tell if they had feet or toes. They were allegedly interested in organic material, such as fertilizer, and questioned him as to why he was spreading manure and the different kinds of fertilizer that are used. They asked him if he would give them samples and, when he left to go get some, *the UFO rose from the ground making a noise similar to an idling automobile engine.* (Local newspaper, Binghamton, New York, 9 May 1964.)

Case I. 64-06-14 Charles Englebrecht was home at 8:55 P.M. watching television in Dale, Indiana, when he saw a light pass by outside his kitchen window. He got up to investigate. Suddenly, the electricity in his house went out. Outside, on the ground he saw a glowing, orange object about 14 inches in diameter. He went outside to investigate. As he approached it, *he felt "something pressing him back*—or his frightened state may have induced this sensation." Later he described the sensation *"much like being shocked by a small electrical charge." The thing then rose up and traveled to the west out of sight over a barn roof.* The young man called out to his parents who arrived in time to see it leave. Everyone smelled an intense, unpleasant, "sulfurous" odor near where the object had been. Police Chief Leroy Musgrave arrived later and smelled the odor at least $\frac{1}{4}$ mile from the house. A small burned area with three indentations forming a triangle was also found. Each was an inch deep and an inch in diameter. Mrs. Englebrecht said her son lost weight in the weeks after the event and "felt bad" all summer. In addition, she said that all of their cherry trees and garden plants near the affected area died soon after the sighting. Other strange electro-magnetic events happened as well. (Keyhoe, D.E., and G. Lore, Jr. *Strange Effects from UFOs: A NICAP Special Report.* Washington, D.C.: NICAP, 1969: 61-62)

Case J. 64-09-14 It was a calm, silent night when Mr. Pierre Pittou fed his dogs outside his house near the town of Valenciennes, France. Suddenly, all the dogs began to whimper and ran to their kennels. The sound of "a million bees" filled the air and Pittou saw a hovering object nearby and two beings each about four feet tall. He began to run toward them and *felt a searing pain. He fell to the ground, paralyzed.* He could see and hear but could not move. Later, a large depression was found in the ground where the UFO had been. The UFO was heard and seen by other neighbors as well

(Unnamed French newspaper, 20 July 1992). This type of incident is not uncommon in UFO literature and raises questions concerning the possible overtly aggressive behavior of humanoids encountered. Did they feel threatened by Pittou's rapid approach and take protective action?

Case K. 65-07-01 Mr. Maurice Masse, a French farmer in Valensole, watched a football-shaped object about fifteen feet long descend from the sky and land on four "crutch-like legs and a central pivot." It was only 5:00 A.M. but the sky was already quite bright as the UFO touched down 60 to 70 yards away. He saw a door slide open and a "stout, three-foot-tall humanoid" emerged wearing "something like a space suit." As the farmer approached, the figure quickly re-entered the object, which took off "with a whistling sound." Later, the police investigated the site and found four shallow depressions of a cross-shape and a hole in the middle area. (Le Journal Du Dimance, 4 July 1965; Washington Post, 7 July 1965)

Case L. 65-07-19 The witness in this incident, Mr. Denis Crowe, was a former British diplomat with the High Commission at Delhi, India, and also a commercial artist. He claimed seeing a "huge green disc resting on a beach in Vaucluse, Australia" about 5:30 P.M. As his sketch below shows (see Figure 6), its width was about twenty feet and its height about nine feet. It rested on legs and had a rim that glowed greenish-blue while its top and bottom sections were the same silver-grey. At the top was a "hollow-appearing section." Crowe approached the object to within 50 or 60 feet and then "it began to take off...with a noise like air forcibly released from a balloon.." The object was out of sight in ten seconds. Crowe had many dogs with him at the time and they all barked loudly at the object when it was stationary. But after it took off they were "strangely silent." (Keyhoe, D.E., and G. Lore, Jr. Strange Effects from UFOs: A NICAP Special Report, Washington, D.C.: NICAP, 1969:33.)

Figure 6

Sketch of UFO seen on July 19, 1965 near Vaucluse beach, N.S.W., Australia
(Adapted from: Keyhoe and Lore 1969)

Case M. 65-09-14 According to an article in the British journal *Flying Saucer Review* (Vol. 12, no. 6, 1966), a dome-shaped UFO with a blue light on it was seen by a motorcyclist near Mersea Island, Essex, England. He also heard a high-pitched humming sound coming from the object. He was surprised that his motor died as the hum changed to a buzzing sound. As he started walking toward the UFO, *he felt an electric shock and some pain* "from the light." Another motorcyclist approached and his motor also died. Nothing was reported about the aerial object's response.

Case N. 66-03-29 Charles Cozens near Hamilton, Ontario, Canada, watched in surprise as two, eight-foot-long by three-feet-thick, glowing white objects landed in a field at 9:15 P.M. Around the horizontal circumference of each object was an indentation with one-inch-diameter lights nestled inside it. The lights were about five inches apart and flashed red, green, and blue "with no two lights on simultaneously." One UFO had a four-foot-long antenna-like wire sticking downward below the lights; Cozens touched it, reporting that it was very smooth like a "polished coffee tabletop." The light and the body of the object gave off a luminescence. As he moved his hand along the wire, Cozens *saw a bright flash and heard a buzzing sound. His hand was "blown off" the UFO and he was "knocked back several feet."* A policeman later confirmed that his hand was burned and that he had "superficial cuts or scratches." One-half-inch-wide, yellowish, curved burn marks from three to four inches in length were produced on the right side of his palm. His burns improved gradually over several weeks of medical treatment. Cozens said the surface of the object was hard and at body temperature. (Keyhoe, D.E., and G. Lore, Jr. *Strange Effects from UFOs: A NICAP Special Report*. Washington, D.C.: NICAP, 1969: 33.)

Case O. 68-07 Only a small part of this complex and interesting encounter will be recounted here. It took place between the Grodner Pass and the Sella Pass traveling toward Campitello, Italy, after 1:00 A.M. Mr. W. Rizzi fell asleep in his automobile and awoke to the strong smell of something burning. He jumped out of his car, thinking it was on fire, but he found nothing unusual. Then he saw a strange light about 500 meters away shining through the swirling mist of the high mountains. As the mist parted, he saw an enormous round object on the ground emitting a strange, glowing white light. He walked toward it, "but I had no fear—I have never feared anything. I was merely terribly excited," he said.

The disc was from 70 to 80 meters in diameter, about eight to ten meters thick, and stood on three legs each about two meters tall. A rough sketch of what Rizzi saw is reproduced here (see Figure 7) from the detailed article by the witness. [*Flying Saucer Review* 26, no. 3. (September 1980):22-27.]

Figure 7

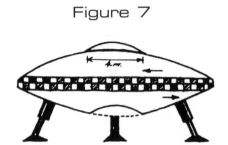

Sketch of object seen on July 1968 in the Italian Alps
(Courtesy of *Flying Saucer Review*)

The object was flooded in "fleecy" white light and Rizzi detected the intense burning smell again. He recounted, "When I had got to a distance of about three meters from the object, *I felt myself suddenly halted, blocked, with a sensation as though my body weighed 1,000 kilos. I could not move another inch and found that great effort was needed to breathe.*" Then he saw the transparent cupola on top become much brighter and two beings inside it looking down at him. A two-meter-diameter beam of light came from the center of the UFO alternating between violet and orange; from it came a humanoid who approached him. We must leave the story at this point since our interest is only in the physiological and other human experiences when approaching a UFO.

Case P. 69-02-06 This event took place in Pirassununga, Sao Paolo, Brazil, and was reported in the December 1993 issue of *Fate* magazine. More than one hundred witnesses watched Mr. Tiago Machado, age 19, approach a UFO that had landed. From across a field, the witness observed Machado encountering a number of small alien beings. As he attempted to follow them into the UFO, one of them *aimed a tube-shaped object at him. He claimed to have been hit in the legs, causing him to faint. He was unconscious for about one hour.* Later, physicians found large red marks on the backs of Machado's thighs. He was also dehydrated and unable to eat for three days, suffering insomnia and extreme retinal sensitivity to light (photophobia). (*Flying Saucer Review* 16, no. 2, 1970.)

Case Q. 75-11-05 Travis Walton, 22, Mike Rogers, 28, and five fellow forestry workers were getting ready to go home after a day of clearing underbrush and fire snags in the Apache National Forest, northeast of Phoenix, Arizona. While all seven saw the metallic disc hovering about twenty feet above the ground, only Walton got out of the truck and began to run toward "the glow." He said that he felt no particular fear until he heard or felt a powerful, cyclic throbbing sound (which suggests the presence of a very great energy). A beam of white light, about a foot across with hazy edges, flashed out of the bottom of the object, striking him in the chest. "It (felt) like an electric blow to my jaw…and everything went black," Walton said later. His companions drove off in fear, leaving him there on the ground. He was found five days later near the town of Heber, Arizona. (Walton, T., *Fire in the Sky.* 1996; Flying Saucer Review 21, no. 5, February 1976; see also Chapter 8.)

Case R. 76-09-03 It was 9:00 P.M. when two elderly ladies were walking home after visiting a friend in their village of Fencehouses, England. The evening was cool and a slight breeze could be felt. As they passed an open field, they saw a strange oblong, asymmetrical object about five feet long and three feet tall sitting on a mound of earth. They first stopped and then walked towards it; they said they felt somehow "attracted towards it." The object stood off the ground supported by chrome or steel runners—like a sled—and its central section was like a "glass compartment." Its top was an orange color.

When they reached the edge of the object they noticed that *wind and traffic noises stopped.* One of the elderly ladies touched the surface of the object and it *felt warm.* Suddenly two "strange looking beings" were seen inside the glass-enclosed area. The remaining details are not cited here except that the ladies left the scene in a great hurry out of fright and, when they did, they both could hear the ambient noises once again. The UFO then took off at high speed emitting a humming sound. (*Flying Saucer Review* 22, no. 6, April 1977.)

Case S. 76-11-14(?) Mr. Eero Lammi, a schoolboy in Uleaborg, Finland, watched a luminous object cross the Finland-Sweden border and then land in a nearby field. As he was walking toward the object to look at it, *a light ray suddenly shot out of the object hitting him in the chest.* He felt a "searing" pain and then he passed out. Doctors subsequently found burn marks on his back and chest (*MUFON UFO Journal,* nos. 110 & 284, January 1977; *Flying Saucer Review,* 22, no. 6.)

Case T. 82-?-? (exact date unknown) An article in the *Herald Express* (Torquay, England; 7 October 1982) stated that a man who was camping in a field one night spotted bright lights and went to find out what they were. He neared a stream and saw a strange luminous object on the ground on the other side about 25 feet away. Almost immediately, *he was thrown to the ground by "an invisible force." About two hours later he was able to break free from the force and he felt cold and a pain in his back.* There were puncture marks on his ankle."

Case U. 96-06-26 Two Lithuanian police officers sighted a UFO near Vilnius according to the Russian news agency Itar-Tass. At 12:30 A.M., the round, shining object was seen illuminating the highway between Vilnius and Miadininkai, near the village of Nemejis. The object flashed its lights as it hovered from twenty to thirty meters above the ground. From time to time, the police officers heard a crackling sound like electricity. After watching it for almost thirty minutes, they began to approach it. When they were about fifty meters from it, the *UFO started to move upwards and away from them,* accelerating in the direction of Vilnius.

51

The policemen radioed an alert and soon van-loads of rapid reaction police and tracking dogs arrived, but the object had disappeared. Various tests were conducted on the ground beneath the location where the object had hovered, radiation measurements were made of the air, and sound recordings were made. Grass within about a ten meter area "around where the UFO was sighted was visibly flattened." (*Agence France Presse* 26 June 1996)

Preliminary Overview

Humans experienced sensation in nineteen of the twenty-one preliminary cases listed above. Of these:

Electrical shock was felt in three cases (15.8%)
Unconsciousness/fainting occurred in two cases (10.5%)
Pain was felt in two cases (10.5%)
Paralysis occurred in two cases (10.5%)
Sense of calm was experienced in one case (5.3%)
Complete silence was experienced in one case (5.3%)
Invisible physical barrier or wall felt in one case (5.3%)
Odor was smelled in one case (5.3%)
No particular sensations were felt in one case (5.3%)
No data provided in five cases (26.3%)

Of the nineteen cases in which a sound was heard, it consisted of:
A hum/vibration in three cases (15.8%)
A buzzing/bee sound in two cases (10.5%)
A whine in two cases (10.5%)
A cyclic throbbing or engine sound in two cases (10.5%)
An electrical crackling sound in one case (5.2%)
An "unspecified" noise in one case (5.2%)
No specific sounds described in eight cases (42.1%)

Seven different names were used to describe the physical form of the UFO in these nineteen cases; "domed disc" and "object" occurred most frequently. Two encounters occurred during daylight hours, thirteen during darkness, and six were not specified.

"Friendly" from whose point of view?

Of course, what is considered friendly to one race of people may be judged as hostile by another. We must be as careful as possible in our acceptance of the claimed events, allowing the widest possible latitude in our interpretation. Nevertheless, we must have some convenient starting point in our consideration of these events. If the witness' motives were claimed to have been friendly and no sharp objects or bullets were sent crashing into the side of the UFO, perhaps we can accept their behavior as being friendly.

Should the above cases be considered to be examples of friendly, hostile, or another kind of human behavior from the visitor's point of view? If it is difficult for us to classify them, one can imagine how hard it would be for our alleged visitors to do so! Can these alleged extraterrestrial beings accurately assess our weaponry, our feelings, our motives?

It is likely that the physical injury which humans report after venturing too close to a UFO may be not at all deliberate on the part of the phenomenon but, rather, some form of unavoidable energy transmission such as infrared (heat), microwaves (heat), ultraviolet, or electrostatic discharge. This may also be true in some cases involving a visible ray of light which could be nothing more mysterious than the transient ionization path of the transmitted energy bolt.

An event timeline

In the preceding cases, it is important to know when each of the UFO and human behaviors occurred, i.e., which came first (see Figure 8). For instance, referring to Part A, did the human signal to the aerial object before the UFO became visible (point A), or after (point B)? Did the UFO increase in brightness only after the human's first signal (A or B)? Or did the human signal only after seeing two or more light flashes (C)? Part B also illustrates two other possibilities, i.e., the UFO responds first and elicits some response from the witness (D) and where the UFO responds in some manner but the human does not reply (Y). The correct interpretation of the narrative will largely depend on knowing these details.

53

Figure 8

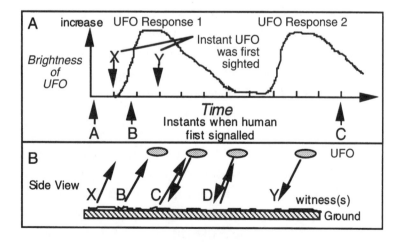

Theoretical event timeline

Without complete information we can speculate as to the course of events, but we can never be certain. This illustrates the importance of accurately recording the time of each event whenever possible.

The evidence

Equivocal Evidence? Most of the above cases are equivocal in terms of whether the UFO's response was at all caused by human behavior. But each of these cases is only suggestive that this was indeed the case. When taken at face value, each case includes anecdotal evidence that these objects were, at least, under some degree of intelligent, or perhaps, pre-planned control.

In one instance, Mr. Woody Polston tried to attract a UFO using his flashlight while in Yellowstone National Park. Polston had no success the first two nights. On the third night, he claims that two UFOs arrived near him. One looked like a "combined" disk and cylinder and the other was large, with lights on one side and an indentation on the other. Polston's recollection is poor after this point, claiming to have gone back to his car and either fallen asleep or unconscious for twelve hours. Later, he said that his watch was ten hours slow and he suffered from nightmares. ("Wyoming—Another Abduction," *APRO Bulletin* 28, no. 9, March 1980.)

There are also many published accounts of humans trying, but failing, to evoke a response from a UFO at close range. One of many such events took place on December 27, 1954, in Cobalt, Ontario, Canada. Mr. John Hunt, reporter for the *North Bay Daily Nugget,* waved his flashlight toward an angularly small UFO which had only just before cavorted around the sky at low altitude above the Agaunica Cobalt mine. Neither Hunt nor another eye witness present saw any response from the now stationary light in the night sky. (Woodley, M., The Grating UFO of Cobalt. *Fate,* 27 December, 1954, 35)

Clearer Evidence? Each of the cases to follow was selected because they present not only more details than were available in the above cases but also because they suggest a definite link between human behavior and a subsequent response by the UFO. Each narrative is presented in a common format which is for the benefit of those who want to perform other analyses. The alleged CE-5 portion of the narrative is italicized. The cases are presented chronologically. The number of witnesses is noted directly with a plus (+) value indicating other possible witnesses.

A comprehensive explanation of the following key can be found in chapter 2.

DB: duration of human behavior
CB: complexity of human behavior
PB-1: premeditation of human behavior (visible UFO)
PB-2: premeditation of human behavior (invisible UFO)
UFOD: UFO response delay

Case Files

Case No. 1

Date: 48-11-18
Local Time: 2145
Duration: approximately 10 minutes
Location: Andrews Air Force Base, Maryland
No. Witnesses: 1
DB: cc   CB: e   PB-1: a   PB-2: x   UFOD: b

55

Abstract: Air Force Lt. Henry G. Combs was about to land at Andrews Air Force Base in a T-6 airplane when he saw an "odd light" above the base. He terminated his landing approach in order to try to pursue it. The light *"began to take violent evasive action,"* he said. *The light always evaded his approaches by turning in a smaller radius and at speeds as high as 600 mph.* Its flight path was erratic as well, making it even harder to intercept. After about ten minutes of unsuccessful interception attempts, Combs turned his landing lights on during a close pass at the light. In the bright white illumination he made out a dark gray, oval-shaped object which was smaller than his own airplane (42-ft. wing span; 29-ft. length).

Comments: Such aerial interceptions are very common in UFO literature, and strongly suggest both advanced technology and intelligent guidance when compared with the relative state of aeroscience and technology. It also must be recognized that pilots are subject to various motion-related illusions in flight (Haines and Flatau 1994). However, there are no such illusions that will produce these particular UFO motions.

References: Haines, R.F., and C.L. Flatau. 1994. *Night Flying*, Blue Ridge Summit, Pa.: Tab Books.

Hall, R. 1964. *The UFO Evidence*. Washington, D.C.: NICAP, 23.

Cross references: Case No. 3 (Chapter 3); 9 (3)

Case No. 2

Date: 49-08-19 & 9 other dates
Local Time: night
Duration: various
Location: Norwood, Ohio
No. Witnesses: 20 + hundreds
DB: dd   CB: a   PB-1: x   PB-2: x   UFOD: d

Abstract: This controversial and complex set of sightings may qualify as a CE-5 only because the aerial phenomenon seemed to reappear at least nine times over a two-month period. The last time being March 10, 1950. Each time it was seen within the beam of an eight-million-candlepower, collimated, carbon-arc searchlight.[1] Newspaper articles (some with photographs) were published over the approximate period September 11, 1949, to April 6, 1950.[2] At the location where the UFO was seen on eight different nights, the grounds of the Saints Peter and Paul Catholic Church in Norwood, Ohio, a searchlight was installed. It was operated periodically (not on consecutive nights) after dark during their carnival celebration of a special church festival. The searchlight was also used at St. Gertrude Church in Madeira, Ohio, when the object was also seen in its beam once (49-09-11) and in Milford, Ohio (49-09-17).

The many witnesses of the strange aerial objects caught in the light of the searchlight included a church priest, a newspaper publisher, a U.S. Army sergeant, the town mayor, two scientists, dozens of other military witnesses, and hundreds of civilians. Mr. Robert Linn, editor of the *Cincinnati Post*, saw the unidentified "but definite object" illuminated by the beam of bright white light. The searchlight operator, Army Sgt. Donald R. Berger, also kept a nightly log of events, including many references to the aerial object.[3] Father Gregory A. Miller at the church also took a 16-millimeter movie of the UFO in 1952[4] and Norwood Police Sgt. Leo Davidson obtained still photos as well [5] (see Figure 9). Davidson used a Speed-Graphic camera with 15-inch focal length Wollensak Telephoto lens.

According to Stringfield's account, which includes Berger's searchlight operating notebook entries, the ten dates on which the aerial object was seen in the searchlight beam and what was seen were: 49-08-19 (stationary glowing disc); 49-09-11 (object disappeared vertically in several seconds); 49-09-17 (object which reflected searchlight beam); 49-10-23 (main disc from which emerged two separate groups of five triangular objects each; all smaller objects descended down beam some distance then flew away horizontally); 49-10-24 (object); 49-11-19 (object); 49-12-20 (small, faint object which brightened over time and enlarged to beam diameter); 50-01-11 (object—smaller objects also seen passing through the beam); 50-03-09 (object—two smaller objects emerged from it, displacing object laterally out of the beam); and 50-03-10 (Stringfield 1977). The object, which moved up and across the beam, then disappeared. It reappeared thirty minutes later in its original position.

Figure 9

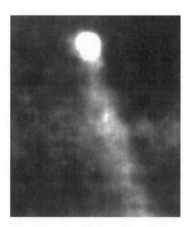

Photograph of searchlight beam illuminating
UFO during Winter 1949 at Norwood, Ohio
(Courtesy of Kenny Young)

Some eye witnesses sighted two balls of fire hanging over Cincinnati on August 20 in the early morning hours. They "seemed to grow dim and then get brighter again" (*Cincinnati Post* 20 August 1949). Hundreds of other people telephoned local radio stations, newspapers, and the weather bureau about two bright phosphorescent lights, appearing as a cluster of stars to some (*Cincinnati Enquirer* 20 August 1949). Around October 24 witnesses reported seeing a "chain of flashes" within the searchlight's beam "just below the saucer." Other shape names given to the phenomena were "slightly conical," "circular," and "triangular."

Norwood mayor, Ed Tepe said that two scientists (Dr. Dare A. Wells; Prof. Paul Herget) studied the object as did two persons from the USAF Office of Scientific Investigations on December 20, 1949. Prof. Herget allegedly claimed that "it's not a fake;" what was seen could have been, "the illumination of gas in the atmosphere." Dr. Wells dismissed the UFO as an "optical illusion." Some naturalists said that the lights were only birds, while an official of the U.S. Weather Bureau said that there were "two comets…poised over the downtown area, perhaps with the evil intention of blotting out the so-called Cincinnati area" (*Cincinnati Enquirer* 20 August 1949). Such a ludicrous explanation was not only irresponsible but patently impossible.

Comments: What kind of natural phenomenon reappears in the same location in the sky over such a prolonged period of time, almost as if it wanted to be seen? Narratives provided by different witnesses are not clear whether it was only reflecting the searchlight's beam or emitting its own light. A large proportion of cases in which a bright searchlight is used to signal to—or merely illuminate—a UFO result in sudden departure of the object. I am indebted to Mr. Kenny Young who provided most of this information.

> Notes: 1. I am personally familiar with this searchlight, having used one in high-luminance vision research studies over the years. The diameter of the beam at the searchlight is sixty inches. The beam diverges only slightly with distance; the circle of illumination measures (in most models) about twelve feet across at a distance of 5,000 feet. The beam cannot be converged to a point or diverged without extensive mechanical adjustments (which most advertisers do not do). 2. According to one newspaper article, the object was first seen on August 29, 1949. 3. Sgt. Berger was serving in the U.S. Army and was stationed at the University of Cincinnati in the Reserve Officer Training Corps office there. 4. The noted UFO investigator, Leonard Stringfield was the first to look into this interesting series of events and allegedly viewed the 16-mm movie taken by the parish priest. Stringfield obtained one photograph from *Time Magazine* staffer Harry Mayo in the 1950s. 5. Mr. Young has tried to locate these photos but they are now missing. Mr. Charles Geraci, president of the Norwood Chamber of Commerce, claims to have seen the police photos in the early 1980s.

References: Stringfield, L. H. 1977. *Situation Red—the UFO Siege!* New York: Doubleday.

Young, Kenny. 1966. Interview by author. 19 August and 19 December

Cross references: 62 (3); 93 (4)

Case No. 3

Date: 50-03-27
Local Time: night
Duration: estimated 4+ minutes
Location: near Goshen, Indiana
No. Witnesses: many
DB: v CB: a PB-1: a PB-2: x UFOD: a

Abstract: The cockpit of a Trans World Airways DC-3, manned by Capt. Robert Adickes and First Officer Robert F. Manning, was the venue for this interesting high-altitude encounter. The flight crew sighted a bright orange-red disc approaching and then slowing to keep pace with them. It appeared like "a big red wheel rolling along."

Capt. Adickes gently banked toward it several times and, each time, *the UFO moved away* "as if controlled by repulse radar." It dove downward at one point, permitting the crew to see it from the side.

Comments: Why did the UFO fly away from the airplane each time it was approached? While unusual, the shape of this UFO is not unfamiliar. A similar shaped UFO was reported during the Korean war in the Spring of 1951 (Haines 1990).

59

References: Hall, R.H. 1964. *The UFO Evidence*. Washington, D.C.: NICAP, 38.

Keyhoe, D. E. 1953. *Flying Saucers from Outer Space*. New York: Henry Holt, 145-148.

Cross references: 4 (3); 9 (3); 86 (3)

Case No. 4

Date: 51-09
Local Time: 1800
Duration: estimated 5 minutes
Location: near Memphis, Tennessee
No. Witnesses: 30+
DB: o CB: a PB-1: a PB-2: x UFOD: a

Abstract: Two U.S. Navy radar-equipped airplanes were flying in the general vicinity of one another during training exercises conducted in close coordination with the ground radar tracking facility at the Naval Air Station, Memphis. Each airplane had a dozen students on board and a regular flight crew. The pilot of the lead airplane radioed the second airplane's pilot about a silver object which was flying off his left side about 1,200 to 1,500 feet away. He was at 5,000 feet altitude at the time. The lead airplane pilot made a gentle left-hand banked turn in the direction of the UFO and, *"the object moved off immediately at right angles to its original direction of flight,* the ground radar tracking at a speed in excess of 5,000 kilometers per hour."

Comments: It is impossible to determine how the UFO interpreted the approach of the Navy airplane. Was it considered a friendly act or potentially hostile? This incident is included in this chapter because of the airplane's gentle turn toward the silver object and because no weapons were fired at it. Apart from the UFO's response, another interesting aspect of this case has to do with what happened after the two airplanes landed. The air station's public information officer (PIO) met the witnesses immediately upon landing and interviewed every man. He then called the *Commercial Appeal*, Memphis' morning newspaper, and gave them many of the details which appeared the next morning on the front page. By noon of the same day, the station's commanding officer received a dispatch from the Department of Defense in Washington, D.C. It instructed the commanding officer that "under no circumstances was he to release such UFO information for public consumption without first clearing with the Department of Defense." One can only imagine what the commanding officer said to his PIO. One also can ask how Washington, D.C., learned about the newspaper article this rapidly.

References: W. Brown, Jr. to J. Allen Hynek dtd., 28 August 1974, Northwestern University, Evanston, Ill.

Cross reference: 1 (3)

Case No. 5

Date: 51-10-09
Local Time: 1345
Duration: estimated 3 minutes
Location: Near Paris, Illinois

No. Witnesses: 3+

DB: o   CB: a   PB-1: a   PB-2: x   UFOD: a

Abstract: This incident involved two separate witnesses. The first was a private pilot flying toward the west and approaching the town of Paris, Illinois, at 1:45 P.M. He noticed a "flattened sphere" flying near him and banked his airplane toward it to get a better view. To his amazement, he saw it suddenly *accelerate away toward the northeast.*

The other sighting occurred at about 1:43 P.M., two minutes earlier. Mr. R. L. Messmore, a CAA Airways Operations Specialist, saw (from the ground at Terre Haute, Indiana—nineteen miles from Paris) an unusual flying object approach from the southeast. He quickly called a colleague to observe it as well. The flattened-round object continued to fly above them toward the northwest, disappearing in about fifteen seconds. The two men made a rough estimate of the velocity of the object using nearby references. They said that it must have been moving at over 2,800 mph (if at treetop level) or 18,000 mph if at 3,000 feet altitude.

61

Comments: There is little doubt that both incidents involved the same aerial object considering the proximity of the witnesses, the similarity of the object's shape, and its common direction of flight. Of course, there is no way to determine the actual altitude of this object without radar data or simultaneous tracking by multiple theodolites, neither of which were available. There are four reasons why it is very unlikely that the object was actually traveling at 18,000 mph at 3,000 feet altitude. First, an object possessing any appreciable mass and cross section would soon heat up in the air. Second, both reported velocities would produce a sonic boom. Nevertheless, there are scores of well documented cases of supersonic objects which do not produce a sonic boom. Third, the object would become nearly invisible; and fourth, the distance between these two locations and time of each sighting does not allow for such extreme speeds. Nevertheless, the witnesses did see something highly unusual and had the courage to report it.

Reference: Hall, R.H. 1964. *The UFO Evidence.* Washington, D.C.: NICAP, 44.

Cross references: 3 (3); 9 (3)

Case No. 6

Date: 52-07-13
Local Time: 0300
Duration: estimated 4 minutes
Location: 60 miles southwest of Washington, D.C.
No. Witnesses: 2+
DB: v   CB: a   PB-1: a   PB-2: x   UFOD: a

Abstract: The flight crew of a National Airlines passenger plane sighted a blue-white light below them which *seemed to be rising in altitude and approaching their position.* They radioed the air route traffic control to report the light and identify it. *The light "came up to altitude of aircraft, hovered two miles to the left of northbound aircraft.* (The) pilot turned on all lights. *(The) ball of light took off, going up and away."*

Comments: Of course it's problematic whether the radio message to ARTC initiated the UFO's response.

Reference: Hall, R. 1964. *The UFO Evidence.* Washington, D.C.: NICAP, 158. Original report from Civil Aeronautics Authority.

Cross references: 18 (3); 86 (3)

Case No. 7

Date: 54-09-07
Local Time: night
Duration: brief
Location: Origny, France
No. Witnesses: 2
DB: i   CB: a   PB-1: a   PB-2: x   UFOD: b

Abstract: Many French motorists saw a luminous, disc-shaped object moving up and down in the night sky. At one point they were able to turn their headlights in its direction. *The UFO took off at high acceleration, disappearing into the distance.*

Reference: Hall, R. 1964. *The UFO Evidence.* Washington, D.C.: NICAP, 11.

Cross references: 14 (3); 36 (3); 40 (3); 45 (3); 47 (3); 65 (3)

Case No. 8

Date: 54-11-26
Local Time: night
Duration: unknown
Location: Millville, New Jersey
No. Witnesses: several
DB: h   CB: a   PB-1: x   PB-2: x   UFOD: a

Abstract: A large, disc-shaped object with four lights located at the corners of a rectangle was seen circling in the sky by Mrs. Lois Barbour and several other town residents. Its lights were a red-yellow hue. They all watched it *fly away* when a searchlight beam struck it.

Comments: Once again, the UFO departed when illuminated. The long series of events which occurred in and near Norwood, Ohio, (1949-1950) involved a UFO which did not always fly away when illuminated by a searchlight. It reappeared at night for hours at a time over a period of many weeks.

References: Hall, R. 1964. *The UFO Evidence.* Washington, D.C.: NICAP, 11.

Keyhoe, D. 1955. *Flying Saucer Conspiracy.* New York: Henry Holt, 233.

Cross references: 2 (3); 20 (3); 22 (3); 23 (3); 25 (3); 27 (3); 46 (3); 50 (3); 62 (3); 71 (3); 79 (3)

63

Case No. 9

Date: 55-02-02
Local Time: 1115
Duration: estimated 5 minutes
Location: Between Merida and Maiquetia, Venezuela
No. Witnesses: 2+
DB: i   CB: a   PB-1: a   PB-2: x   UFOD: a

Abstract: Capt. Dario Celis and First Officer B.J. Cortes were flying an Aeropost Airlines airplane in western Venezuela when they saw a green, glowing sphere or "apparatus." As it approached them, it seemed to rotate counter-clockwise. It had a reddish ring around its middle that gave off flashes of bright light. The men observed markings like "portholes" above and below the center ring. When they tried to contact the authorities on the ground by radio, their communication was cut off for unknown reasons. Celis banked his airplane in the direction of the strange object and the UFO *rapidly whirled downward, leveled off, and sped away.*

Comments: Here is another aerial encounter where the pilot's behavior led to a rapid departure of the strange object. A three-man flight crew was typical in the 1950s.

References: Hall, R. 1964. *The UFO Evidence.* Washington, D.C.: NICAP, 121.

Keyhoe, D. 1955. *Flying Saucer Conspiracy.* New York: Henry Holt, 249.

Cross references: 3 (3); 4 (3); 9 (3); 86 (3)

Case No. 10

Date: 55-09-03
Local Time: 2115
Duration: estimated 3 to 4 minutes
Location: Cincinnati, Ohio
No. Witnesses: 2
DB: h   CB: a   PB-1: a   PB-2: x   UFOD: a

64

Abstract: Mr. Frank Flaig and his wife were driving on Boomer Road on the west side of Cincinnati. Moonlight reflected off a small—estimated four foot diameter—aerial object, causing the object to become visible as it dropped downward out of the night sky. It passed behind the roofline of a nearby building and disappeared from sight. The couple stopped their car, thinking that it must have crashed into the ground only about forty yards away. When Mr. Flaig got out and walked to the approximate impact point he was astonished to see a smooth-surfaced ball floating just above the ground. As he approached it, "his wife, who was still in the car and was concerned about her husband, called out to him. Mr. Flaig received the definite impression that the sound of his wife's voice alarmed the sphere because the instant she yelled, *the object shot skyward on a 45 degree angle.*"

Comments: Is this another case of coincidence or was Mr. Flaig's interpretation correct? It is problematic whether the woman's voice caused the UFO to respond as it did. If it did it implies that the UFO possessed not only an auditory sensor system of very great sensitivity and intelligence, but also a pre-programmed evasion propulsion system. Terrestrial technology has not yet developed either.

We don't know how close Mr. Flaig was to the object or whether he experienced any physical sensations.

References: *C.R.I.F.O. Newsletter 2*, no. 2 (7 October 1955): 2.

Flying Saucer Review 1, no. 5 (November-December 1955): 28.

Gross, L.E. 1992. *The Fifth Horseman of the Apocalypse. UFOs: A History, July–September 15th.*, Published privately, 82.

Case No. 11

Date: 55-11-14
Local Time: night
Duration: estimated 5 minutes
Location: San Bernardino Mountains, Southern California
No. Witnesses: 2
DB: v   CB: e   PB-1: a   PB-2: x   UFOD: a

Abstract: Mr. Gene Miller was piloting a private airplane from Phoenix, Arizona, to Banning, California. Dr. Leslie Ward, a doctor living in Redlands, California, at the time, was his passenger. They both sighted a "globe of white light" ahead of them. Miller blinked his landing lights twice, believing that they had sighted a commercial airplane at their altitude. This is normal procedure under these "see-and-be-seen" flight conditions. *The object disappeared from view twice and then reappeared.* Nevertheless, the small plane seemed to be gaining on the unknown object; it seemed to become larger. Becoming more concerned, the pilot flashed his landing lights three more times in succession and, to his amazement, *the globe of light blinked three times and "suddenly backed up in mid-air."*

65

Comments: The UFO's light signals were obviously in response to the pilot's light signals. The pilot followed correct procedures here since he kept his airplane under control and also reported the incident to the authorities. I suggest that neither of these would have occurred had this been a deliberate hoax nor would the pilot have allowed his name to be released.

References: Hall, R. 1964. *The UFO Evidence.* Washington, D.C.: NICAP, 41.

Los Angeles Times, 26 November 1955.

Cross reference: 54 (3)

Case No. 12

Date: 56-01-08
Local Time: sunset
Duration: 48 hours
Location: Robertson Island, Antarctica
No. Witnesses: 4
DB: b   CB: a   PB-1: a   PB-2: x   UFOD: a

Abstract: This event took place during the Second International Geophysical Year (1956-1958) at Weddell Sea, Antarctica. Four Chilean scientists were camped at a remote site in tents. One man stepped outside a tent at sunset and was surprised to see "two metallic, cigar-shaped objects in a vertical position, perfectly still and silent, and flashing vividly the reflected rays of the sun. One of the objects was almost at the mid-heaven, and the other at a distance of some 30 degrees from the first...the things looked utterly solid, with smooth, polished, seemingly metallic surfaces." He called a second man to come outside to look at the "spindles" as he called them.

The two men then walked about 100 yards to see if some sort of atmospheric optical effect could be causing them; the objects did not change appearance. After going back to their tent, two more men at the camp also saw the hovering objects.

The two UFOs remained motionless until 1:00 P.M. when the upper-most object tilted into the horizontal position and started to brighten and emit various unspecified colors. Then it began to move and executed a number of maneuvers (90 degree turns, zig-zag flight, instantaneous starts and stops, etc.). It came to a stop and hovered again. The second object then began to fly around for three minutes like the first object had just finished doing, and then it too stopped and hovered. The four witnesses were afraid, yet nothing untoward happened. Then, late in the afternoon of the second day, clouds approached, the altitude of which were known approximately. Using a theodolite, the altitude of the objects was determined to be about 24,000 feet, with each cigar about 450 feet long and 75 feet in diameter. One of the men aimed a polarized spotlight at one of the objects *which unexpectedly flashed a bright light in return and descended in altitude.*

After a "long pause," it rose again to its original position and stopped. Then, it conducted "another fantastic sky dance." Its velocity was measured at about

40,000 km per hour starting from a dead stop.[1] The objects eventually disappeared from view because of clouds that came in.

Comments: The credibility of all witnesses is very high, and the event is not at all brief. The capability of the witnesses to obtain accurate measurements lends further credibility to this case. Their use of a theodolite to measure the altitude of the objects was pragmatic and effective. As occurs so often in such cases, the humans waited a long time before attempting to signal the UFO—in this case over twenty-four hours. One of the objects flashed a bright light after a scientist shone a polarized spotlight at it. It descended, but it is not clear that the object's motion was also caused by the use of the spotlight. The fact that the objects had moved around prior to this CE-5 event suggests that it wasn't.

> Note. 1. This is equivalent to 24,800 mph which, if the object started at rest, represents a very high acceleration.

Reference: Creighton, G. 1968. *Flying Saucer Review* 14, no. 2 (March-April): 20.

Cross references: 22 (3); 50 (3); 71 (3); 76 (3); 87 (3)

Case No. 13

67

    Date: 56-09-02
    Local Time: 0430
    Duration: estimated 5 minutes
    Location: Dayton, Ohio
    No. Witnesses: 1
    DB: j   CB: a   PB-1: a   PB-2: x   UFOD: a

Abstract: This early morning encounter took place on the grounds of the Dayton Country Club as a 19-year-old security guard was making his final check. A green, self-luminous object was seen hovering five to six feet over the ground as it traveled slowly toward the clubhouse where he was standing. Not knowing what it was, he flashed his powerful flashlight directly at it when it was at the "far end of the grounds, but nothing happened." As he began walking towards it, he neither heard nor smelled anything unusual. The oval-shaped object was estimated to be about 15 feet to 20 feet wide and 8 feet to 10 feet thick.

"This object came to 150 feet or less of me when I flashed my light on it again," he said. *"It disappeared from sight in an instance without moving any."*[1] Before it disappeared, the object illuminated its entire surroundings.

Comments: Vallee's interpretation is that the object probably was a balloon drifting along the ground (1965). But how did it become self-luminous, and why did it suddenly disappear upon being illuminated by the watchman's flashlight? This event is very similar to others presented.

Note: 1. A dog inside a stable nearby began howling "something terrible" just then.

References: Gross, L.E. 1994. *The Fifth Horseman of the Apocalypse. UFOs: A History 1956 September—October.* Privately published, 1.

USAF Project Blue Book files, 2 September 1956.

Vallee, J. 1965. *Anatomy of a Phenomenon.* New York: Ace Books, 190.

Case No. 14

Date: 56-11-14
Local Time: 0115
Duration: 5 hours[1]
Location: between Ortonville and Graceville, Minnesota
No. Witnesses: 1+
DB: i   CB: a   PB-1: a   PB-2: x   UFOD: b

68

Abstract: Mr. Marlen Hewitt, a young truck driver, was near Graceville en route to Ortonville when he first noticed what he thought was just a very bright star in the dark night sky. It appeared to change colors several times. Hewitt stopped his truck and got out to see it better. The UFO swooped downward in his direction, stopping about 1,000 feet in the air to the east of him.[2] Hewitt was terrified. The slate-grey object appeared to be metal and was estimated to be about twelve feet thick and as large in diameter as a town block (about 1,500 feet). Figure 10 illustrates what the UFO may have looked like. The edge of the silent object was outlined in light and its body changed from a blue-white to orange or red. The object was not stable but seemed to tilt at different angles and gain 3,000 to 4,000 feet "in the flick of an eye."[3] From an opening in the object's bottom came an intense beam.

Hewitt said later, "I wanted to get out of there, so I jumped back into the truck and turned on the lights. *When my lights went on, the thing turned a bright cherry red and shot up into the high clouds."* He said that he had turned his headlights off at some point and that when he did so the UFO again "darted away" with some smoke coming from it. As Hewitt continued driving, the UFO came back down toward the Earth to the southwest, stopped, and hovered. It finally disappeared about five hours after it had first been spotted. Hewitt traveled a distance of about 150 miles during the encounter.

Comments: Whether or not turning on one's headlights qualifies as a friendly act is debatable. Once again, why did the UFO react to headlights being turned on? This incident includes such familiar details as object color changes, erratic flight, projected beam of light, and utter silence.

> Notes: 1. This aerial object remained in sight over five hours and about 150 miles of travel. Was it putting on a display for the benefit of the witness? What natural phenomenon could the witness have misidentified? How could such a large object move so fast without causing a sonic boom? 2. Its distance was estimated to be about $\frac{1}{2}$ mile. If the width of the object was 1,500 feet and its slant range was 2,500 feet, its width would subtend an angle of 31 degrees, truly huge. 3. The reported high mobility of UFOs obviously place the human at a great disadvantage should he try to escape. Indeed, there is no reasonable possibility of doing so. This fact certainly contributes to a witness's sense of helplessness and fear.

Figure 10

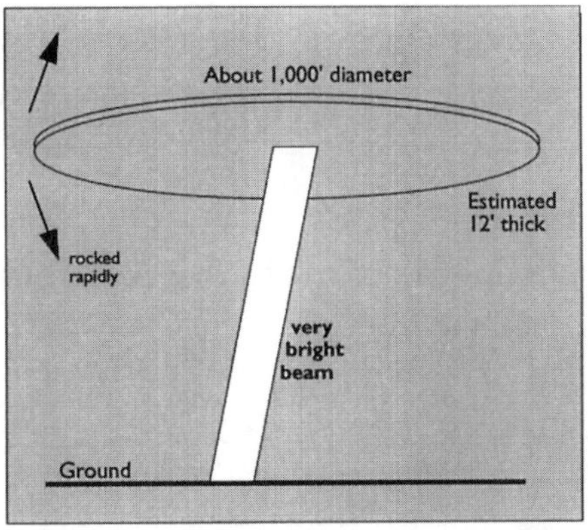

Sketch of object with beam seen on November 14, 1956, near Graceville, Minnesota

Reference: *Flying Saucer Review* 2, no. 1 (January-February 1957): 3.

Gross, L.E. 1994. *The Fifth Horseman of the Apocalypse. UFOs: A History 1956 November–December,* 12. Published Privately.

Press (Redfield, S.D.), 22 November 1956.

Cross references: 36 (3); 45 (3); 87 (3) 183 (6); 193 (6)

Case No. 15

Date: 57-01-11
Local Time: sunset
Duration: estimated 5 minutes
Location: Brazil, Indiana
No. Witnesses: 2
DB: j   CB: a   PB-1: a   PB-2: x   UFOD: 0

Abstract: Four "glowing objects" flying in formation were noticed by Mr. Donald Hodden at sunset. The air was very cold (-10 degrees fahrenheit) and clear. Nevertheless, he went indoors and asked his mother to come outside to see them. They both watched as the luminous points disappeared in the darkening eastern sky. But soon the lights returned accompanied by a fifth light source, all flying slowly and coming from the same spot in the sky where they had disappeared. All were a red-yellow color and approximately as intense as a first-magnitude star. Donald ran inside his house for a Kodak Brownie camera and a flashlight. He aimed his flashlight at the lights just as the UFOs "maneuvered about" the sky, trying to signal to the "disc pilots." It was Hodden's impression *the UFOs glowed brighter when the flashlight beam was directed at them and turned on and off.*

70

Comments: Other witnesses reported seeing a similar formation of lights as well. Nothing is known about any photographs that might have been taken.

References: Barker, G. 1957. Chasing the Saucers. *Flying Saucers,* (June): 31-21.

Gross, L.E. 1995. *The Fifth Horseman of the Apocalypse. UFOs: A History 1957 January–March 22nd.* Published privately, 11.

Case No. 16

Date: 57-03-27
Local Time: 2035
Duration: 5 to 10 seconds
Location: Roswell, New Mexico
No. Witnesses: 1+
DB: i   CB: a   PB-1: a   PB-2: x   UFOD: a

Abstract: Lt. Sondheimer was piloting an Air Force C-45 cargo plane near Roswell, New Mexico, when he saw three UFOs approaching his position from the left. All three objects were very bright white and round; they maintained a fixed, small separation distance from one another, resembling airplane landing lights. When Sondheimer realized that the UFOs were on a collision course with him "he immediately flashed his taxi lights on. *One of the objects shot straight up in the air above him while the other two continued on and passed in front of his aircraft.* When the pilot flashed his taxi lights, the objects *immediately blacked themselves out,* thereby disappearing from sight."

Comments: It is not clear how the two UFOs, which allegedly passed in front of the airplane, could be seen if they immediately blacked out. Did the pilot flash his taxi lights more than once?

References: Air Force Teletype Message. 1957. COMDR 686th ACWRO WALKER AFB to RJEDEN/COMDF ATIC WP AFB OHIO. Project Blue Book files.

Gross, L.E. 1995. *The Fifth Horseman of the Apocalypse. UFOs: A History 1957 March 23rd.–May 25th.* Published privately, 10.

71

Case No. 17

Date: 57-04-07
Local Time: 2330
Duration: estimated 4 minutes
Location: New Plymouth (North Island), New Zealand
No. Witnesses: 2
DB: v   CB: a   PB-1: a   PB-2: x   UFOD: b

Abstract: The following letter to the editor of New Zealand's Taranaki Herald was received: "Sir, on Sunday night at approximately 11:30 P.M. when coming past Bell Block Airport, my friend and I saw an orange light about the size of a light bulb in the sky. It came from the direction of the city of Waitara[1] and seemed to circle around us. Whilst it was circling it was going up and down like a spring. When we put the car lights on and started to move off, *it turned to a bright red and shot off* in the direction of New Plymouth. It was only a matter of seconds before it disappeared completely. While we were watching it, we heard a sound for all the world like someone banging a piece of iron. (Signed) 'Very Scared.'"

Comments: The UFO was obviously very small (a point source). Its vertical bouncing motion is not uncommon in UFO literature and suggests a very low mass (low inertia). Does the fact that the UFO circled the automobile indicate a deliberate attempt to communicate with the witnesses? Clearly, the driver had stopped and perhaps even gotten out to watch this light show.

Notes: 1. Located 160 miles to the west-northwest of Wellington (North Island) on the Tasman Sea coast.

References: Anon. 1957. Quarterly J. Civilian Saucer Intelligence 5, no. 2., 12-13.

Gross, L.E. 1995. *The Fifth Horseman of the Apocalypse. UFOs: A History 1957 March 23rd.–May 25th.* Published privately, 10.

Case No. 18

Date: 57-06-29
Local Time: 1830
Duration: 40 minutes
Location: Belo Horizonte to Rio de Janeiro, Brazil
No. Witnesses: several
DB: cc   CB: d   PB-1: a   PB-2: x   UFOD: b

Abstract: This interesting incident involved the flight crew and all of the passengers of a Real-Aerovias Airlines DC-3 propeller-driven airplane flying toward Rio de Janeiro on a dark, winter night. Some scattered cloud lay above their altitude of 9,900 feet. The co-pilot, Jackson Lopes Correia, was the first to notice a red glow some distance behind and on the right side of the airplane at the apparent altitude of the horizon. The glow was catching up with them at about their altitude and he pointed it out to Capt. Gabriel Junqueira Giovanini and the radio operator, Aluisio Meneses de Araujo. The red light was very bright and definitely was not a wingtip light of another airplane.

The captain instructed the stewardess, Berenice de Oliveira, to ask all of the passengers to look out at the light. Both the captain and several passengers viewed the light through a passenger's binoculars. Its extreme brightness made it difficult to see any other detail.

After pacing the airplane for fifteen minutes, the light approached them and stopped to maintain a fixed, but much closer, separation distance. At this point everyone could see a "flattened disk...surrounded by a faint red glow." Then, it "rushed toward the plane and passed above it at tremendous speed." Passenger Germano da Silva got a close view of it, noting three orange-red lights on its bottom forming an isosceles triangle with the acute angle toward the front. The UFO maneuvered around the airplane "making several passes" at it, crossing above and below it, in front of and behind it.

Giovanini "decided to do something. He ordered the crew to switch off all the lights of the plane. Then, he turned on the landing lights. The powerful lights in the wings flashed on. A blinding beam of white light reached out through the darkness toward the place where the orange-red lights marked the location of the UAO.[1]" For several seconds, the crew and all of the passengers[2] clearly saw the object outlined against the sky. *The UAO didn't approve of the landing lights. It sped up to escape, turned on edge and veered sharply to one side, disappearing swiftly from sight. Then it appeared again, but now it was at some distance from the DC-3—some miles behind it.*" The UFO remained with the airplane for almost the remainder of its trip to the Rio de Janeiro airport, departing only twenty miles or so north of the airport.

73

Comments: Author and investigator Fontes systematically excluded all of the usual explanations for such lights (i.e., ground fire, reflection from a cloud, jet engine exhaust flame, an atmospheric reflection of the DC-3 off ice crystals, balloon, guided missile, meteorite, or mirage). As he states, "It performed controlled maneuvers and displayed curiosity about the airliner but its tactics suggest that it was toying with the plane—like a cat plays with a mouse." It also displayed the now-familiar light-avoidance response. No portholes or other surface detail could be seen.

Notes: 1. UAO stands for unidentified aerial object. 2. While this detail may represent an overstatement, it is possible that some passengers may have viewed the object through the open cabin doorway and others through cabin windows when the object was beside the airplane.

Reference: Fontes, O. 1958. The Shadow of the Unknown. *APRO Bulletin* (March): 4, 7.

Cross references: 6 (3); 9 (3); 86 (3)

Case No. 19

Date: 57-07-14
Local Time: 2030
Duration: 70 minutes
Location: Between Pampulha and Belo Horizonte, Brazil
No. Witnesses: estimated 3+
DB: cc   CB: a   PB-1: a   PB-2: x   UFOD: a

Abstract: An anonymous airport tower controller sighted a stationary light in the night sky on a bearing of 270 degrees magnetic from the Pico do Couto airport, Pampulha, Minas Gerais State. He radioed a request to a Brazilian Air Force B-26 in the region to investigate. The airplane's flight crew saw an unusual luminous object for about an hour as they were en route from Pico do Couto to Belo Horizonte. Later, the pilot said he "wanted to escape it" (NICAP 1959). As he radioed the control tower for permission to land, the tower operator noticed a light hovering in the sky on a magnetic bearing of 270 degrees West. The tower operator also contacted the Meteorological Institute to find out if it might be a weather balloon. He also instructed the pilot to change his heading to 270 degrees in order to try to see the light. The flight crew saw the light, reported it, and then landed at the airfield.

74

"Disappointed that the pilot had not investigated the light [further], the tower controller decided to fire some flares toward the UFOs. Just at this moment *the UFO began to change colors, from white to amber to an intense green. Then it turned white again and darted upwards, disappearing in the darkness.*

"Startled by this reaction, the controller phoned the pilot, who by this time was in the canteen, to question him about the UFO. The officer answered that he had no desire to get any closer to the unknown object" (NICAP 1959). Twenty minutes later, the pilot asked permission to take off without any lights on—even though the plane was armed—so the UFO would not follow him.

Comments: This event was entered into the official airport logbook and also transmitted to the Air Route Administration of the Federal Aviation Board. While the CE-5 events are not as positively related, the alleged rapid color changes of the UFO to the flares suggests a deliberate, elicited response. The color of the flares is not known but were probably red. The airplane carried a crew of five, however it is not known how many of them saw the UFO.

References: National Investigations Committee on Aerial Phenomena, UFO Investigator. Hall, R., ed. November 1959, 1.

The UFO Evidence. 1964. Washington D.C.: NICAP, 12, 120.

Gross, L.E. UFOs: *A History 1957, May 24th–July 31st.* Published privately.

UFO Critical Bulletin, Sao Paulo, Brazil.

Cross reference: 43 (3)

Case No. 20

Date: 57-11-05
Local Time: night
Duration: unknown
Location: Witwatersrand, Southern Transvaal, South Africa
No. Witnesses: several
DB: kk   CB: a   PB-1: a   PB-2: x   UFOD: a

Abstract: A huge UFO in the approximate shape of a cylinder was witnessed by hundreds of people. It hovered for some time. A searchlight at the South African Air Force base in Dunnottar illuminated the object, after which the UFO immediately *"shot behind cloud."*

75

Comments: Here is yet another light avoidance response.

References: Hall, R. 1964. *The UFO Evidence.* Washington, D.C.: NICAP, 124.

Herald (New Zealand), 7 November 1957; dateline Johannesburg.

Cross references: 8 (3); 22 (3); 23 (3); 25 (3); 27 (3); 62 (3); 63 (3)

Case No. 21

Date: 57-12-211
Local Time: 2230
Duration: 25+ minutes
Location: Ponta Pora, Brazil
No. Witnesses: 5
DB: cc   CB: e   PB-1: a   PB-2: x   UFOD: b

Abstract: "On the night of March 3, 1958, I decided to invite my girlfriend for a ride in the moonlight. [2]...The time was 10:30 P.M. I was talking with my sweetheart, when suddenly she looked as if she was going to get sick. She didn't talk, but I saw through her frightened eyes that something was wrong, very wrong. [Then]...I saw it too. Something very bright was hovering in mid-air over the thicket ahead of us. The boys had sighted it too, and started to run towards the Jeep. [3] They were shouting: 'The saucer, the saucer.' Their behavior apparently didn't please the UFO. It must have seen their shadows moving in the moonlight and...with unexpected readiness...*it shot straight toward them at high speed. It rapidly went after them, coming down from a westerly direction. It appeared to be 'wobbling in mid-air' as it dived to the ground,* flying slightly above their heads. Fortunately, the boys were not too far away and reached the vehicle in time—just before the arrival of that 'thing.' [4] Then everything lighted up, as if the sun had suddenly come out. But the light was as red as blood. My hair rose straight up the back of my head. Where was the UFO?

"We were directly under the 'thing.' It was something made of polished metal, of a silvery color. It was larger than a car and shaped like a round ball. It appeared to be so near that we could have hit it with a pole. The boys had jumped into the Jeep. Everyone was very scared. At any moment that thing might have landed on the car and we would have been smashed and killed. I decided to run away at once.

"I started the car and drove hurriedly to the town. *The UFO followed us, staying just behind the Jeep—flying at about 9 feet from the ground.* It was very low, indeed; the vehicle as well as its surroundings were illuminated by the bright glow emitted by the object. There was no noise, or heat coming from it. I began to drive more rapidly.

"The chase lasted for fifteen minutes, until we reached another thicket. The road passed through the woods, and I hoped the object couldn't fly under the small trees. I was right. We saw it pass behind the trees and disappear. We then relaxed, thinking it was gone....When we left the protection of the woods, we spotted it hovering over the thicket—higher in the sky. It was clear that it was waiting for us to appear....Then, the thing saw our car; *it dived at high speed,* but not toward the Jeep. It left the road and cut its way to the hill; it shot straight toward the top of the hill. *After that maneuver, it was ahead of us.*

"The UFO descended over the hill, stopping on the ground just in the middle

76

of the road...*the object had maneuvered to cut off our way to the town. In fact, our path was blocked, for it was now on the road ahead, hovering about a few feet above the ground*. The Jeep was starting to climb up the hill—the UFO was there, just waiting....I had only one alternative left: to drive the car on a collision course against the object. I couldn't stop; to come back, on the other side, was to do exactly what 'they' wanted....The Jeep picked up speed and we prepared for the crash.

"Then an odd thing happened. As the car was climbing up the hill, its headlights gradually were raised until their beams of light hit the UFO directly. To our amazement, *the object reacted strangely* as if its propulsion system had been disturbed by the light rays. *When the beam of light from the car reached out and found it, the UFO wobbled violently in mid-air and shot straight up*, I don't know why, but it was running away from the light....But it was good for us because I saw our chance and crossed the dangerous point rapidly. We passed just under the object, but there was no interference....Some minutes later, *the UFO began to follow us again*, but cautious, never coming close (as it did before). After ten minutes, it changed course for the last time. It climbed up vertically, switched off its lights at a height of about 300 feet, and vanished into the darkness."

Comments: This case involved two adults, one of whom, Marcio Goncalves, was already knowledgeable about UFO phenomena, and three teenagers. It is clear that the night this event took place the adults were not at all interested in seeing or contacting UFOs.

Dr. Fontes (1959) wrote, concerning a previous incident, "This incident is unique in the whole history of the UFOs: for the first time, a group of responsible and reliable people had left their town to meet the UFOs—and found them, but they got more than they wanted."

It is intriguing to discover that the UFO began to wobble violently when the car's headlights shown upon it. The UFO reported in Chorwon, Korea, in the Spring of 1951 also wobbled when a rifle's round struck its surface. What is the likelihood that light (or solid objects) interferes with some aspect of UFO propulsion? Does light act like an electrical discharge path or does it disturb force field lines? [5]

Notes: 1. A possible alternative date for this event is 58-03-03. 2. The two adults were on a date, however, in order to obtain permission from the girl's parents, the young man had to invite his teenage brother and some of his brother's friends along. He accomplished this by telling them they were going on another "flying saucer" hunt. 3. The three teenage boys had

77

flashlights and were running out across the fields flashing light signals prior to this close approach by the UFO. 4. Goncalves said prior to this experience that, "The experience with those eerie "things" was something that never could be forgotten....The UFOs I had sighted were controlled machines, certainly with a crew inside....We should discover what that something they were doing might be....The only way was to make other night trips to the spots where they had been sighted before. I gathered a group of friends for such a purpose and, night after night, we made many exploratory trips through the roads around the town. Our efforts failed, however" (Fontes 1959). These comments suggest that Goncalves' earlier motive to seek out a UFO encounter was based purely on curiosity and not for economic or ego gratification reasons. 5. Part I of this article was published in the July 1958 issue of the APRO Bulletin and included sightings in the region of Rio de Janeiro of 58-07-07, 58-07-13, and 58-07-14 (Belo Horizonte, Brazil). The location of the Ponta Pora case reviewed above was some 780 miles due west and some four months earlier.

Reference: Fontes, O.T. 1959. The shadow of the unknown. *APRO Bulletin,* Part II (May): 6-9.

Cross references: 44 (3); 79 (3); 85 (3); 194 (6)

Case No. 22

78

Date: 58-05-17
Local Time: night
Duration: estimated 2+ minutes
Location: Ft. Lauderdale, Florida
No. Witnesses: 2
DB: v   CB: a   PB-1: a   PB-2: x   UFOD: a

Abstract: A father and his son saw an orange light approaching them from the north at a low altitude. *The man turned his high-powered spotlight on the light and the unidentified light flared in intensity and then shot out of sight.*

Comments: Here is another instance of a light-avoidance response by a UFO.

Reference: Hall, R. 1964. *The UFO Evidence.* Washington, D.C.: NICAP, 121.

Cross references: 12 (3); 50 (3); 71 (3); 76 (3)

Case No. 23

Date: 58-10-03
Local Time: 0310

Duration: 70 minutes
Location: near Rossville (Clinton County), Indiana
No. Witnesses: approximately 5
DB: v   CB: a   PB-1: a   PB-2: x   UFOD: a

Abstract: The engineer, fireman, and other railroad crewmen on board a Monon Railroad freight train witnessed a formation of four "odd white lights" flying across the tracks ahead of the engine. The lights then changed direction and traveled parallel with the cars in the opposite direction (toward the caboose), traveling about one-half mile, and were seen by everyone on the train. Upon reaching the caboose, they turned east again and began following the train. Their exact shape was hard to see due to their brightness. However, the lights seemed like flattened objects as they flew abreast of one another "with coordinated motions."

At one point, the conductor aimed a bright searchlight at them. *"Immediately, the UFOs sped away, but returned quickly and continued to pace the train."* They eventually flew away toward the northeast and disappeared from sight.

Comments: Here is another account of UFO showing apparent "interest" in terrestrial vehicles. The train is clearly the focus of the UFO's flight path. The avoidance of light at night is also a common UFO response.

79

References: Edwards, F. 1967. *Flying Saucers-Serious Business,* 63.

Hall, R. 1964. *The UFO Evidence.* Washington, D.C.: NICAP, 9.

Cross reference: 8 (3)

Case No. 24

Date: 59-07-08
Local Time: night
Duration: 1+ minute
Location: North Vernon, Indiana
No. Witnesses: 2+
DB: cc   CB: e   PB-1: a   PB-2: x   UFOD: c

Abstract: The Robert F. Baker family was driving in their automobile from two miles north of North Vernon to about seven miles south of Columbus (13 miles total) when they sighted a formation of "strange blinking orange, soundless lights." At first, the witnesses thought it was an airplane in trouble, and they pulled off to the side of the road and blinked their headlights. *"Then the [UFOs] broke their triangular formation and came back together again—then he gave chase. He estimated their speed at 15 mph until he began blinking his lights."*

Comment: While this account does not say so, it implies that the speed of the UFO increased after the headlights were blinked.

Reference: Anon. 1959. *APRO Bulletin* (July): 6.

Case No. 25

Date: 60-08-13
Local Time: 2350
Duration: 2 hour 35 minutes
Location: Corning, California
No. Witnesses: 5+
DB: x   CB: d   PB-1: a   PB-2: x   UFOD: a

80

Abstract: Two officers of the California State Patrol were driving east of Corning, California, when they noticed what they first thought was "a huge airliner dropping from the sky." They pulled over and jumped out in order to get a good sight bearing on the object's point of impact, thinking that it was going to crash and they would need to respond. The first thing they noticed was the total silence in the air, no airplane engine sounds at all. The object continued to descend to within 100 to 200 feet of the ground and then rose rapidly to about 500 feet. Then it stopped. The two astonished witnesses could see it clearly from their vantage. The object was round or oblong and surrounded by a glow and a red light at each end. From time to time they also could make out about five white lights between each of the red lights. As they watched the UFO began to dance erratically in the sky. Radio interference was experienced each time the object was near their car.

The object approached the patrol car twice, sweeping the area with a "huge red light." One of the men got inside the car and turned their intense red light on, aiming it directly at the UFO. *"It immediately went away from us,"* one officer said.

The object began flying slowly toward the east. The officers followed it in their car. They watched as another object in the sky approached from the south and joined the first object. The two objects stopped, hovering near one another and flashing red beams. Finally, the two objects disappeared from sight over the hills to the east.

Comments: This case is particularly intriguing because it was reported by two highly credible witnesses. It occurred during a six-day period of many such sightings in the north-central Sacramento valley and included reports from fourteen police officers who also had seen the same object. During the above encounter the UFO used its red light to illuminate the ground and sky six or seven times, according to the officers' reports.

Interestingly, the second patrolman's report stated, "We made several attempts to follow it, or I should say get closer to it, but *the object seemed aware of us*. We were more successful remaining motionless and allowing it to approach us, which it did on several occasions."

Reference: Hall, R., ed. *The UFO Evidence*. Washington, D.C.: NICAP, 61.

Cross references: 46 (3); 62 (3)

Case No. 26   81

> Date: 61-09-?
> Local Time: 2300
> Duration: 16+ minutes
> Location: Long Lake, Quebec, Canada
> No. Witnesses: 2 + 2
> DB: w   CB: d   PB-1: a   PB-2: x   UFOD: a

Abstract: Two youth were walking together toward a fishing lodge after having finished chores; they walked between stands of cedar and pine trees, using a flashlight to see. It was a clear, crisp night. Suddenly they noticed that the entire open area ahead of them was illuminated with a reddish hue. They stopped and saw a "huge red and orange globe" hovering just above the tree line in front of the lodge. After ten or fifteen minutes, the twelve-year-old flashed a powerful flashlight directly at it *"and got an immediate reaction. The center became white, then yellow, then an orange red. Glowing red (concentric) rings emanated from it, like the ripples...in a calm pond."* The two boys ran into the lodge and pleaded with others to come and look at it. The object descended to the ground within approximately ten minutes. The witnesses don't recall any further details.

Comments: The expanding concentric light rings are particularly interesting since they seem to have been stimulated by the powerful flashlight beam. They were not visible before this.

Reference: Cameron, V. 1995. *Don't Tell Anyone but…: UFO Experiences in Canada*. Burnstown, Ontario, Canada: General Store Publishing House.

Case No. 27

Date: 62-09-24
Local Time: night
Duration: unknown
Location: Hawthorne, New Jersey
No. Witnesses: 2+
DB: kk   CB: a   PB-1: a   PB-2: x   UFOD: 0

Abstract: This event was preceded by others (62-09-18: greenish saucer seen over a house; 62-09-24: saucer seen hovering over a rock quarry). On this date the object was witnessed both by police officers and nearby residents. At an unknown hour, a bright source of light was seen hovering in the sky. *When police directed a bright spotlight on it, it "moved away."*

Comment: Here is another example of an apparent light-avoidance response.

Reference: Hall, R., ed. 1964. *The UFO Evidence*. Washington, D.C.: NICAP, 12.

Cross references: 25 (3); 46 (3); 62 (3); 63 (3)

Case No. 28

Date: 63-10-21
Local Time: 2130
Duration: 45 minutes
Location: Trancas, Argentina
No. Witnesses: 9+
DB: cc   CB: d   PB-1: a   PB-2: x   UFOD: b

Abstract: This little known but important case is filled with interesting facts which relate to other close encounters reported here. There were nine people in the Moreno house (and other neighbors) all of whom sighted a number of aerial objects which hovered and emitted colored, collimated beams of light. The terrifying event began when a servant, Ms. Dora Martin Guzman, glanced out a window about 9:30 P.M. toward the nearby Belgrano Railway Line tracks some 300 yards away. She saw what appeared to be a very bright "small train" on or near the tracks with men moving around it. Thinking there must have been an accident she told Mrs. Julia Moreno de Colotti, 21, about it; Julia went out into the garden and confirmed what Dora had seen. At one point, the string of intense lights seemed to split apart with one half suddenly shooting away from the rest horizontally about one-half mile. Three separate circular flying objects could be seen streaking away. Presumably, three objects remained behind by the tracks. Returning to the house to tell the others, Julia got a lantern and returned outside to find out what had happened. She set up the lantern near the front gate "presumably intending to venture out further....*As soon as the women had set up the lantern, and as though in response to it, one of the brilliant beams of light emanating from one of the machines on or above the railway track at once turned from white to violet, and was switched round so as to play upon the two women in the garden.* They immediately were overcome by suffocating heat, and prickling or tingling sensations in their bodies, and were obliged to run back into the house" (Creighton 1966, 23). Then they noticed that the beam had changed from violet to red and swept over the house for the next forty minutes causing everyone to become terrified. The air became hotter and hotter and an odor like sulfur could be smelled. Everyone felt a prickling-burning sensation. The matron of the house, Dona Teresa Kairus de Moreno, 63, bravely peeped out several windows in turn to discover at least five discs suspended in the sky nearby (two were only about 70 yards away and 30 to 40 feet up). She and her husband Antonio Moreno, 72, estimated their diameter to be 8 meters.[1]

One of the objects emitted a beam that continually swept the house and gardens. The light beam was white and collimated and had sharply defined edges. The second object emitted a reddish-violet collimated ray.[2] The interior of their house was now like a furnace. After forty-five fearful minutes, the family saw the UFO near the tracks begin to rise and depart. The light beams suddenly shut off and the objects also flew away in the direction of the Sierra de Medina mountains.[3]

Comments: It is obvious that the rapidity with which the UFOs reacted to the lantern becoming visible indicates a constant monitoring of the locale around the phenomena. The manner of their response is also not unfamiliar. These multiple objects were not any kind of known terrestrial flying objects for many reasons. The elements of instantaneous physiological sensations of burning-prickling associated with the light beams, their differential color, and rise in air temperature all point to the use of microwave radiation of a high-power density which impinged upon the clay tile roof. When a clay is used that contains metals, the tile can absorb microwave energy and radiate it downward throughout the entire house relatively rapidly. Whether or not the sensations experienced were caused by this thermal, infrared energy is problematic. I suggest that one of the two rays may have produced the heating effect and the other the burning-prickling feelings. It is possible that, since microwaves are not visible, the radiating source could have visible wavelengths added to aid in aiming. Could the sulphurous smell be due to vaporization of certain chemicals in the soil or house? A police investigation the following morning discovered that several neighbors (at least 2 kilometers away) also had seen an "extraordinary illumination on the railway embankment." Another neighbor, Francisco Tropiano had, at about 10:15 P.M. to 10:20 P.M., watched six disc-shaped objects passing across the sky.

Notes: 1. Mr. Moreno claimed he could see the brightly illuminated interior of one disc through six portholes near its center as well as "figures silhouetted at the portholes." 2. One witness described seeing a "whitish gas" and hearing a "howling noise" that accompanied these beams. 3. For at least a day after the encounter, a thick mist smelling like sulphur, hung over their garden and was observed by a journalist and police inspector from Tucuman. Certain unnamed people arrived the next day from the National University at Tucuman to collect and analyze rock and soil samples near the tracks. Nothing more is known of their findings, allegedly due to a "blanket of silence" from Air Force and Navy officials in Argentina.

References: *Accion (Monte Video)*. 24 October 1963.

Creighton, G. 1966. Argentina 1963-64 Part II. *Flying Saucer Review* 12, no. 1 (January-February):23-26.

Case No. 29

Date: 64-06-08
Local Time: night
Duration: estimated 10+ minutes
Location: Burlington, Wyoming
No. Witnesses: 5+
DB: q   CB: a   PB-1: a   PB-2: x   UFOD: b

Abstract: This event took place after dark in a deserted pasture where two youth were looking for cows. Gary Brown and Richard Briggs caught sight of three objects. One of the objects appeared to be in some kind of trouble though nothing is said about their shape or the kind of difficulty. The boys also heard a shrill sound. They reported the incident to Big Horn County Sheriff L.C. Brinkerhoff. The sheriff, three other men, and the boys drove back out to the site the following night. An object was seen moving over grazing land at a high rate of speed; the sheriff remarked, "No wheeled vehicle could have gone across the muddy ground that fast." One of the witnesses, Harvey Baliso, a local publisher said, *"when a strong spotlight was turned on the object, a strange bluish-green light came from it.* He said the light was ten to fifteen times stronger than the aircraft landing light Brinkerhoff uses for a spotlight." The group departed the scene when the light disappeared. The group returned the following day and found nothing.

Comments: It is obvious that certain facts were left out of these newspaper accounts, probably because publishing them would be embarrassing to the witnesses involved. Was the UFO stationary when the group arrived the second night? Did it begin to move before or after being illuminated by the officer's spotlight? What else happened and why did they leave the scene rather than try to investigate further?

85

References: *Flying Saucer Review 10*, no. 5, (September-October 1964): 21-22.

Sheridan Press (Wyoming). 10 June 1964.

Case No. 30

Date: 65-03-15
Local Time: 0100
Duration: estimated 45+ minutes
Location: Everglades (Big Cyprus
           Indian Reservation), Florida
No. Witnesses: 1
DB: q   CB: a   PB-1: a   PB-2: x   UFOD: c

Abstract: Mr. James W. Flynn, a former constable of Fort Myers, Florida, and a rancher, took his four dogs and camping supplies on a trip into the Everglades on March 12. Flynn was an experienced woodsman and very familiar with the

Everglades. Two nights later, he was riding his swamp buggy at a slow pace when his dogs spotted a deer and ran after it. They would not return to Flynn's whistles, continuing to bark and bay in the distance. Finally, one dog returned. Flynn then heard a noise like a gun shot. Afraid for his dogs, he rode the buggy in the direction from which the sound had come. About one hour later (1:00 A.M.) he saw "a huge light in the sky above the cypress trees about a mile away" (APRO 1980). The light seemed to be some 200 feet up.

According to the article published by the *Aerial Phenomena Research Organization*, Flynn watched it for some time as it "moved from east to west and back to its original position four times." Then it descended and hovered just above the ground for about five minutes. He watched it through binoculars and, thinking it might be a helicopter, he drove toward it with his lights off.

The metallic-appearing object was cone-shaped with a rounded top (see Figure 11). Its base was about sixty feet across and its height was estimated (relative to nearby cypress trees) to be about thirty feet. A horizontal row of windows was positioned approximately eight feet from the top. Two more rows were below the top row, each one emitted a "dull yellow light" and was approximately two feet by two feet in size.[1] Flynn saw an orangish-red glow on the ground beneath the UFO. He waited about a quarter mile away for forty minutes, watching it through his binoculars. Thinking that it must be some new device from Cape Kennedy in trouble, he decided to move closer with his lights on.

Figure 11

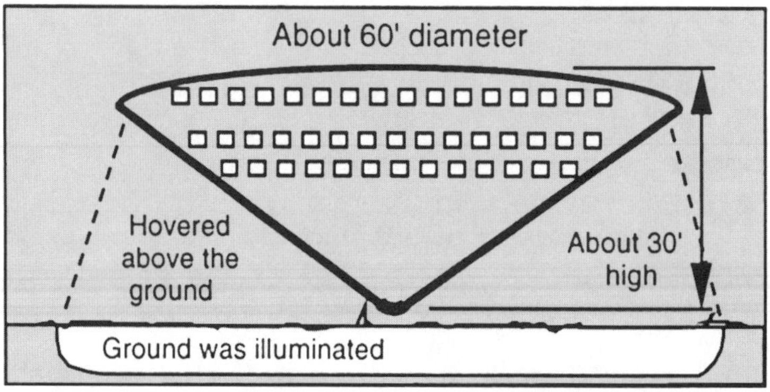

Reconstruction of object shape and details from March 15, 1965, at Everglades, Florida

He drove to about 200 yards from the UFO and stopped. The object was not supported physically; it hovered about four feet off the ground. A sound like that of a diesel generator came from it.[2] Apparently, the noise caused one of the dogs in a cage in the buggy to become very agitated and the dog tried to "tear out of its cage." As the APRO report states, "Flynn walked to the edge of the lighted area, raised his arm and waved. He got no response, and after waiting an estimated one half minute, he walked about six feet into the lighted area, raised his arm, and waved again" (1980, 133). He said later that *he heard a jet-like noise and felt a strong wind coming from it which almost knocked him over.*

A "short beam" of light erupted from just under the bottom of the windows and struck Flynn on the forehead. It was like a "welder's torch." He lost consciousness," awoke briefly, and blacked out again.

Flynn awoke twenty-four hours later, still flat on the ground. He got up and noticed a perfect circle of burned ground below where the object had hovered. The tops of nearby cypress trees were also burned. He made his way to the home of a friend on the Seminole reservation with difficulty and finally arrived home on Wednesday afternoon.

Comments: Flynn's physician of twenty-five years examined him after this encounter and attested to his reliability and emotional stability. His eyes were examined by an ophthalmologist at Lee Memorial Hospital. He was totally blind in his right eye and had partial vision in his left eye. His right eye was hemorrhaging into the anterior chamber of the eye. Other abnormal findings were neurological. "No paralysis was noted, but the deep tendon reflexes of biceps, triceps, patellas, and achilles were absent. Plantars and abdominals were absent, but cremasterics were present." These deficiencies returned gradually over the following five to seven days. The cause of these neurologic symptoms is not known. This particular array raises some clinical questions: 1) Was the cremaster reflex actually tested? 2) How was it tested? 3) Why would it not be affected (its neural innervation is at the first and second lumbar spinal level) when an abdominal reflex, whose innervation derives at nearly the same level, was not found? 4) If these reflex inhibitions were hormonally triggered at the synaptic level by a propranolol-like agent—which binds the beta receptor sites and which is at least 90 percent bound in the plasma—why did the recovery period last from five to seven days rather than 3.4 to 6.0 hours which is the measured half-life of propranolol

(Evans and Shand 1973, 487-93; Hughes 1977, 321-28). 5) How can such agents be stimulated in vitro or otherwise quickly introduced into the human being?

Flynn's physician accompanied him back to the site and noted the fresh burn marks and scorched tree tops. In his report he wrote, "there was no mark of any kind on the soft dried marsh underneath." Interestingly, the NICAP version of this event indicated that when the witness and four others returned to the site they found that the sawgrass was burned in a circle 72 feet in diameter and the earth "turned up." This event contains several similarities with the Michalak case of 67-05-20 (chapter 4).

Notes: 1. The NICAP account cites four rows of windows. 2. A very similar sound was described by Pfc. Francis Wall in the early Spring of 1951 (chapter 4) immediately after he fired a rifle bullet at the hovering craft.

References: Anon. 1965. Violent encounter in Florida. *Flying Saucer Review 11,* no. 4 (July-August): 14.

Aerial Phenomena Research Organization. 1980. Flynn encounter. In *The UFO Encyclopedia,* ed. R.D. Story, 132-134. Garden City, New York; Doubleday & Co.

Evans and Shand. 1973. *Clin. Pharmacol.* Ther. 14.

Hugues, F.C., et al. 1977. Determination de la demi-vie pharmacologique du propranolol chez 'Homme. *Therapi 32.*

Keyhoe, D.E., and G. Lore 1969. *Strange Effects from UFOs: A NICAP Special Report.* Washington, D.C.: NICAP, 12-16.

McDonald, J. E. dtd. letter. 31 October 1966.

Cross references: 81 (3); 121 (4); 131 (4); 231 (8)

Case No. 31

Date: 65-3-31
Local Time: 1916
Duration: 19 minutes
Location: Fujinomia City, (west of Mt. Fuji) Japan
No. Witnesses: 6
DB: p   CB: a   PB-1: a   PB-2: x   UFOD: a

Abstract: Six boy scouts were going on a camping trip to the western base of Mt. Fuji. But fifteen-year-old Inoue and the others became lost. A truck stopped and the driver offered his assistance. Since it was becoming dark and much colder, and their back packs were heavy, they made arrangements for the truck to carry the packs to their destination and then return for the boys. It was then that Inoue and the others saw a "bright orange object" in the shape of a hat hovering just to the left of the top of the mountain. It was on a compass heading of 110 degrees and an elevation angle of about 20 degrees. *He used his flashlight to signal S.O.S. toward the light over a period of eight to ten seconds and was surprised to see it begin to move "almost simultaneously."* Figure 12 (drawn by Inoue) shows the shape of the UFO (A) and its various movements (B) allegedly made to the flashlight signal.

After about fifteen minutes, the flatbed truck returned and the excited boys were transported to their campsite. They continued to watch the UFO during their ride and noticed it disappear briefly, reappear for a several seconds, and finally disappear for good.

Figure 12

(A)   (B)

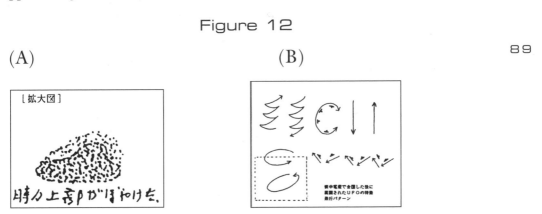

Drawing of UFO shape and motions in the night sky on 65-03-31 in Japan

Comments: A chart of this region shows that there are very few towns near where the boys were. While this might suggest that the flashlight beam would not compete with other ground lights from the vantage of the UFO, there is no way to be certain that the object's motions were actually responding to the S.O.S. signal. The five types of object motion shown here are very familiar to researchers as is the hat shape of the light source.

References: Amamiya, K. 1995. *Ufology.* Japan, 102-103.

Amamiya, K. 1998. Letter to author. 26 February.

Case No. 32

Date: 66-08
Local Time: 2230
Duration: estimated 1+ minute
Location: Montsoreau (Maine-et-Loire), France
No. Witnesses: 2
DB: p   CB: a   PB-1: a   PB-2: x   UFOD: a

Abstract: Mr. and Mrs. M. Lacoste witnessed a "dazzlingly luminous ball rise up." It then flew in circles above a nearby wheat field for forty seconds and then dropped downward again into the field. *The couple ran to their car in fear and Mr. Lacoste "put on the headlights, whereupon the luminous ball at once shot up from the field and vanished at a staggering speed."* A ten-square-meter area in the field was found to be crushed and oil marks were also found.

Comments: The claimed effect of the headlights is well-known while the reason for it is not.

References: Anon. 1967. World Round-up, France. Landing at Montsoreau. *Flying Saucer Review* 13, no. 2 (March-April): 31.

Paris-Jour. 13 August 1966.

Case No. 33

Date: 67-04-17
Local Time: 2045
Duration: estimated 5 minutes
Location: New Haven, West Virginia
No. Witnesses: several
DB: v   CB: a   PB-1: a   PB-2: x   UFOD: 0

Abstract: Mr. Lewis Summers used his car to chase a huge UFO as "big as a C-45 airplane." The object emitted two shafts of light from its bottom surface and flew at an altitude of about 500 feet. He followed it on Route 33 until the object crossed the highway a half-mile ahead of him and disappeared behind the hills. At one point, Summers pulled over and stopped and "flashed the headlights off and on several times." He said that, "The spotlights (on the craft) also were turned off and on several times, but I do not know whether they were answering my signals or not."

Comments: Here is an account of an alleged mimicking of a human behavior by the UFO. Is this response deliberate or merely some pre-programmed activity? The NICAP report includes details of three other eye witnesses to the same object. Hundreds of calls allegedly came into radio station WMPO about the aerial object.

References: Gribble, R. 1992. Looking Back. *MUFON UFO Journal,* no. 288 (April): 18.

Keyhoe, D.E., and G. Lore, Jr. 1969. Strange *Effects from UFOs: A NICAP Special Report.* Washington, D.C.: NICAP, 41.

Cross references: 87 (3)

Case No. 34

Date: 67-07-26, 27
Local Time: 2300-0130
Duration: 30+ minutes
Location: Newton, New Hampshire (and other towns)
No. Witnesses: 2
DB: cc   CB: e   PB-1: a   PB-2: x   UFOD: c

Abstract: Mr. Gary Storey was visiting his sister and brother-in-law, Mr. and Mrs. Francis Frappier, in Newton, New Hampshire, and had brought his telescope with him to view the stars. He set his telescope up in a nearby field. It was a clear, moonlight night; the moon was nearing last quarter and rose at 10:55 P.M. At about 11:00 P.M., they all noticed a very bright star located near the pointer stars of Ursa Major but didn't pay very much attention to it. They didn't even look at it through the telescope. By about 1:00 A.M. on July 27, the light began to move. Storey changed the eyepiece from 350x to 75x power and adjusted the mount and focus. Initially white, the source of light changed to orange "as soon as it began to move. When it apparently came closer each of them could see two white lights forming the base of a triangle and a red light forming a point just above and in the center of them."

As the witnesses watched, the object approached from the north-northeast and flew to the east-northeast, then retraced its path to where it had originally started. It repeated this same flight path seven or more times, each time, Storey was able to keep it in view in his telescope. A drawing of the object is presented here based on the witness's description (see Figure 13).

The five dimmer white lights located in a straight line along the tubular section of the baton flashed in a regular sequenced manner forwards and backwards in a 1-2-3-4-5-4-3-2-1 sequence.

Figure 13

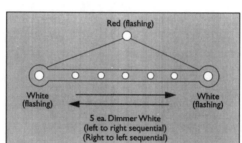

Reconstruction of UFO seen on July 26, 1967, in Newton, New Hampshire

"As it was moving away from them in an easterly direction at a point about east-northeast from them, Francis flashed his flashlight at it with three one-second-interval flashes. *The object abruptly changed directions seemingly without turning and started to retrace its path backwards. When it reached mid-point of its flight, all of its lights dimmed*—Gary could still see its dim outline in the telescope when it dimmed—*and its two end lights flashed back three flashes seemingly in response.* During its estimated seven passes back-and-forth, it appeared to get closer each time as its apparent size became larger. Francis flashed the light at it a number of times with various types of signals and *the object would dim its lights and flash back replica flashes in the same manner as described above. Only its two end lights would flash.* The red light remained dimmed. There was an estimated time lag of ten seconds between the flashlight signals and a response. When a jet air-craft...was heard, they looked for it and easily recognized its running lights. The jet headed directly toward the lighted object and *the object immediately shut off all its lights.* The jet continued on out of sight but still could be heard for awhile. The object was still nowhere in sight and it was thought it had gone away when *suddenly it appeared in the same spot where it had shut off its lights but as a bright flare-like white oval ten to fifteen times brighter than it had been before.* This flare effect only lasted a few seconds and then dimmed back to reveal the previous lighting configuration of two white lights and one red light with the object and 'flashing sequence' of lights seen through the telescope. On its final pass as it was disappearing into the east, *it flashed twice and disappeared."*

Comments: The first of the two CE-5 segments of this multiple-adult-witness sighting consisted of the repeated light flashes by the UFO seemingly to that of Frappier's flashlight. However, the delay was estimated to be about ten seconds between the two. The second part was the UFO's disappearance as the jet airplane approached its location and its increase in intensity after the jet was out of sight. One must ask what kind of natural phenomena, if any, respond in these ways if, in fact, it was a response?

In such cases as this we must consider the visibility of a flashlight as seen from high altitude. Newton, New Hampshire, is located within a heavily populated region with many hundreds of thousands of lights on all night. Many of them will flash on and off (e.g., advertising lights, house lights blocked by swaying tree limbs, vehicles traveling under bridges, etc.). How is it that three relatively dim, one-second flashes from a flashlight were detected by this aerial object? Either the UFO's "response" was coincidental, it possessed an extremely sensitive and intelligent sensing system, or a totally different communication mechanism was involved.

Other features of this case are interesting as well. Why did the object pass back and forth so many times, following the same flight path, while acknowledging its presence with sequenced strobe lights? Did it want to be seen? The constant angular rate of travel of the UFO was equivalent to that of a conventional helicopter seen two or three miles distance yet no sound was ever heard from the UFO. The sounds of the jet airplane were heard, however. The UFO made abrupt reversals of direction and never passed out of sight behind anything. Investigator checks failed to uncover any explanation for the UFO. The two men present were both former radar operators.

Reference: Fowler, R. 1967. Field report dated 11 August.[1]

Fowler, R. 1974. *UFOs—Interplanetary Visitors.* Jericho, New York: Exposition Press, 151-55.

Cross references: 38 (3); 69 (3); 75 (3)

Note: 1. I gratefully acknowledge the outstanding field research and clear reporting accomplished here.

93

Case No. 35

Date: 69-07-04
Local Time: 2000
Duration: estimated 20+ minutes
Location: Anolaima (40 miles northwest of Bogota),
    Columbia
No. Witnesses: 9+
DB: w   CB: d   PB-1: a   PB-2: x   UFOD: a

Abstract: Only the CE-5 portion of this complex case is reviewed here. Mauricio Gnecco, 13, and his friend Enrique Osorio, 12, noticed a red-yellow light traveling across the sky from east to west. Gnecco yelled for other children and adults to come outside to see the "flying saucer." They eventually came outside and everyone saw the light, at an estimated distance of about six hundred feet. *"Mauricio obtained a flashlight (there is no electricity in the area) and began to send signals in imitation of Morse Code. At that moment, the light source approached the house at a considerable speed and remained suspended between two tall trees about 150 feet from the farmhouse, where it hovered for about five seconds"* (see Figure 14). Brookesmith comments that the UFO responded immediately to the flashlight signals by increasing its speed and approaching the house (1984).

94

"While this occurred, [another witness] shouted to Mauricio: 'That thing is coming down upon us—turn that flashlight off, Mauricio!'" The completely silent object was from four to six feet tall with an arc of light surrounding it. It had "two luminous legs—blue with green tips." The UFO suddenly flew beside the farmhouse, appearing to fly at low altitude over a nearby hill.

Figure 14

Eye witness drawing of UFO seen on July 4, 1969, at Anolaima, Columbia

Comments: The report also provides fascinating details of physical injury leading to the death of Mr. Arcesio Bermudez who took the flashlight from Mauricio and ran off in the direction of the object. He claimed to have gotten within about 20 feet of the landed UFO (which had unexpectedly blinked out). He shone the flashlight directly at the object and said that he saw a "person inside." The being's upper half was normal but his lower half "appeared to be like the letter 'A' [and] was luminous."[1] Two days after this close encounter, Bermudez became very ill—reduced body temperature, black vomit, diarrhea with blood in stool. After he died, the cause of death was listed as gastroenteritis.[2]

Notes: 1. The interested reader should consult Mueller (1995) for several other accounts of beings taking the form of the letter "A." 2. The wrist watch and clothing of Bermudez were sent to the Columbian Institute of Nuclear Affairs for analysis (no data received), this institute said that the reported physical symptoms were similar to those caused by a "lethal does of gamma" radiation.

References: *APRO Bulletin*, July-August 1969; 1, 5.

Brookesmith, P. ed. 1984. *The UFO Casebook*. London: Orbis Publishing, 85-86.

Mueller, R. 1995. *The Fundamental*. Port St. Lucie, Florida: Arcturus Books.

Cross references: 38 (3); 69 (3); 133 (4)

Case No. 36

Date: 70-12-07
Local Time: 0330-0515
Duration: 1 hour 45 minutes
Location: Thadura Copper Mines, Western Australia
No. Witnesses: 18+
DB: kk CB: a PB-1: a PB-2: x UFOD: b

Abstract: At least eighteen workmen at the copper mines saw what looked like a very intense white star which was hovering silently one-half mile away from them above the southeast corner of the property. Mr. Ted Murphy, their supervisor, was awakened to see the oval light. He was told that a scraper operator had been the first to see the UFO and had *"flickered his headlights at it." As he did so, the object shot off to its position southeast of the camp.* The UFO had an orange-red tint at one end. The light disappeared from sight about two hours later, while several men were looking at it.

Comments: This event took place at an open pit copper mine, about 128 miles northeast of Meekatharra. The Civil Aviation Authority suggested that the witnesses had seen a light airplane operated by local farmers. However, this explanation disregards the reported silent hovering behavior of the light over such a long period of time.

Reference: *APRO Bulletin.* March-April 1971, 4-5.

Cross references: 14 (3); 47 (3); 65 (3)

Case No. 37

Date: 72-05-13
Local Time: 2140
Duration: estimated 4 minutes
Location: Canterbury, New Hampshire
No. Witnesses: 4
DB: v   CB: d   PB-1: a   PB-2: x   UFOD: b

96

Abstract: Four young boys were planning to sleep outside next to a barn on the edge of an open field. As one of the boys was looking at the stars, he noticed a fiery point of light that seemed to be moving toward them. All four observed the light descending into the field nearby. Thinking it was a helicopter, the boys crawled out of their sleeping bags and ran to the edge of the field to watch. "One of the youngsters used his flashlight and aimed it at the approaching object."

The object suddenly, yet smoothly, began to fall downward in a zig-zag manner like a "floating balloon." They were illuminated by a bright white light coming from the front of the object. Each boy described the object in the same way—i.e., an eight-sided craft which was tilted downward at a 45 degree angle as it moved downward. Four thin, pipe-like legs with pads at their ends protruded downward. From its bottom side came a ten-foot-long fiery exhaust that changed colors during the fly-over at about 150 feet altitude. From the object's top extended several bright, silver, inverted U-shaped, cable-like figures.

Comments: The witnesses estimated the object to be about twice the length and height of an average automobile. We do not know whether the flashlight signal played any role in the subsequent response by the UFO. Since it was already approaching toward the field it is likely that it did not.

References: Gribble, R. 1992. Looking Back. *MUFON UFO Journal,* no. 289 (May): 20.

UFO Investigator. July 1972.

Case No. 38

Date: 72-08
Local Time: 2200
Duration: estimated 20+ minutes
Location: near Karlstad, Minnesota
No. Witnesses: 6
DB: w   CB: e   PB-1: a   PB-2: x   UFOD: b

Abstract: A bright aerial light paced a family of six in their car while returning from their vacation in Canada on Route 59. They were between Tolstoi, Manitoba, and Thief River Falls, Minnesota. Suddenly, the light zoomed directly toward the right side of the stationwagon; the interior of the vehicle was illuminated "brighter than daylight" by its light. The light awoke the four children asleep in the back. *The father told his ten-year-old son, Wayne, to "use the flashlight and flash four times—short flashes—and immediately the object flashed back four times with short flashes.* I must explain that the object completely extinguished all light four times," he said.

97

"*My wife then took the flashlight and sent a series of long flashes and short ones and immediately thereafter, the object did exactly the same, this time coming even closer to the car.*"[1] The interior of the vehicle began to become very hot and the engine finally stopped altogether.

As the driver pulled off the road and got out, he saw a round object hovering above him.[2] He began to feel a prickly sensation all over his body "like small electric shocks." *The UFO suddenly shot away into the sky when another vehicle approached; the second vehicle did not stop.* The first vehicle's engine started immediately thereafter and the family drove away.

With the UFO now only a tiny speck in the dark sky, the family felt more relieved. Soon, however, *the object approached them again* but not as close as before. "*My wife flashed our flashlight at the object again and we received the same signal back.* The UFO seemed at a much safer distance now, so I stopped the car to get a better look at it. Then, everyone noticed three smaller discs come out of the larger one (see Figure 15). One immediately flew to the north, one to the south, and the third to the east.[3] The main object "sort of wobbled" and "left at an incredible speed, diminishing to nothing in perhaps not even a couple of seconds."

Comments: This narrative includes many interesting details which typify a close encounter: approach by a large, self-luminous, structured object; temporary electro-magnetic effects; ejection of several smaller luminous objects which fly independently from one another; and most importantly for this study, three separate CE-5 events.

Figure 15

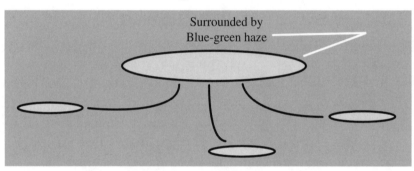

Drawing of UFO with three smaller objects seen in August 1972 in Minnesota

Notes: 1. The car's radio began to malfunction and the engine began to cough and sputter at this point. 2. He also opened the hood of the car and asked his wife to try to start the engine. It did turn over but electrical sparks jumped from the spark plug wires, across the coil, to the metal body of the car and back again. 3. The departure of the three smaller objects from one another in orthogonal directions is similar to the 78-08-28 pilot sighting over Nattenheim, Germany, reported in my book *Project Delta* (1994, 160-61).

References: Anon. 1976. Object emits three discs. *APRO Bulletin* 25, no. 2 (August): 1, 3.

Rodeghier, M. 1981. *UFO Reports Involving Vehicle Interference*. J. Allen Hynek Center for UFO Studies.

Cross references: 11 (3); 69 (3): 75 (3)

Case No. 39

Date: 72-08-13
Local Time: 0200
Duration: estimated 2 hours 40 minutes
Location: Taize, (eastern) France
No. Witnesses: approx. 35
DB: dd   CB: a   PB-1: a   PB-2: x   UFOD: 0

Abstract: The aerial light show which occurred during this close encounter event is so complex as to be almost impossible to describe adequately. Nevertheless, the recorder of the event, Mr. Tyrode, a schoolmaster, did an admirable job. The appearance of the various light sources are recounted here only because they may be relevant to an understanding of the CE-5 aspects of this case.

The group of eye witnesses were together at an open-air theater of a Protestant retreat center on the crest of a hill. They were facing west overlooking a low valley; higher hills lay beyond. Miss Renata and others saw a tiny point of white light falling downward. It landed about 1,700 meters away upon or very near the ground across the valley at the same elevation. While it seemed to illuminate the fields of grain around it, its body was dark.[1] Later, witnesses estimated its diameter to be about 30 meters. Suddenly another object lit up. It was hovering in the air just above the top of the hill beyond the valley. Seven individual yellow lights were seen in a horizontal line with another two orange lights grouped near one another to their left. Situated just above the two left-most yellow lights were "cupolas." Sometime later, five of the seven yellow lights emitted rays vertically [2] that slowly extended downward to the ground. [3] During the transit downward of the rays, a "train of red sparks" was seen near the right side of the object. Soon, the sparks had disappeared only to be replaced by three red disks, each having two individual red points of light on them. These disks flew erratically around the larger "object" until 4:40 A.M. when the object departed in the direction of the village of Cluny. It wasn't until about 3:00 A.M. that the CE-5 event took place.

Four of the witnesses—Miss Renata, Mr. F. Tantot, a young man from Dijon, and a student from Italy—found flashlights and agreed to walk together toward the fantastic display of lights. The remaining thirty witnesses watched from the relative safety provided by their distance. When they were about half-way across the field, the vertical beam second from the left end of the object suddenly

99

intensified with "showers of red particles." The particles filled the air all around the four curious youth. The tiny particles seemed to cover the ground around them as well. Then, the four sighted a "dark, haystack-shaped mass" on the ground ahead of them. Its height was at least 6 meters and a unitary, tiny red spot of light buzzed around it like a mosquito.[4] One of the youth aimed his flashlight directly at the haystack and was amazed to see *its beam turn ninety degrees upward,* about 1.5 feet in front of the "hedge." When the other three did the same, *their flashlight beams also performed the same unexplainable feat. Then the lights on the hovering object overhead went out and then came back on again; the three red disks quickly flew "into the big object."* Tantot then flashed his light directly at the hovering object. "As if in response, the largest beam from the UFO rose so that it shown directly at the witnesses."[5] The UFO began to rise into the night sky and then accelerated away.

Comments: This fascinating series of events include several which violate known laws of physics—i.e., apparently solid light beams which bend, rotate, and extend slowly from separate spots of light, and red particles which continue to glow for a relatively long period of time.

Notes: 1. It must have been seen by silhouette. 2. Several witnesses had the impression that the larger object was resting upon these five vertical beams. A tingling sensation like that from an electrical stimulation was experienced at this time by several witnesses. 3. Many witnesses claimed that the vertical beams rotated about their longitudinal axes in a vortex-like tornado. Others saw a row of portholes giving off white light; they went out after about twenty minutes. 4. Multiple points of red light are commonly associated with larger UFO. The four explorers also thought they saw a dark, low, horizontal hedge ten feet in front of them which (they knew from prior experience, was not actually there). 5. All four felt a surge of heat and were dazzled at this moment.

References: *Flying Saucer Review* 19, no. 4.

Garreau, C. 1973. *Soucoupes Volantes, Vingt-cinq Ans d'Enquetes.* 3rd. ed. Maison Mame, France, 186.

Case No. 40

Date: 72-09-20
Local Time: night
Duration: estimated 10 minutes
Location: near Mount Rouge, Quebec, Canada
No. Witnesses: 1
DB: kk   CB: a   PB-1: a   PB-2: x   UFOD: O

Abstract: A man driving a car watched as an unidentified object was settling down on Mt. Rouge near Montreal. *He deliberately drove to the mountain and then flashed his headlights. He said that the object then flew directly towards him.* As it passed only about thirty feet overhead, it caused his engine, radio, and lights to fail and produced elevated air temperature inside the car. He estimated that it was from 100 to 150 feet across.

Comments: It is problematic whether the UFO's behavior was in direct response to the driver's behavior or whether it was coincidental. See Cameron (1995, 60-61) for a similar incident in Buckingham, Quebec, only eighty-five miles west of this spot, also in 1992. In this second case, car fog lights were turned on toward the round UFO, which had descended out of sight behind some trees. It rose into sight, flew almost instantaneously toward the frightened witness hovering above the car for about ten seconds. It then zig-zagged its way toward the west out of sight.

References: Cameron, V. 1995. *Don't Tell Anyone, but…:UFO Experiences in Canada*. Burnstown, Ontario, Canada: General Store Publishing House.

Canadian UFO Report 2, no. 1.

Rodeghier, M. 1981. *UFO Reports Involving Vehicle Interference*. Evanston, Ill.: J. Allen Hynek Center for UFO Studies.

Cross references: 33 (3); 67 (3); 68 (3); 87 (3)

101

Case No. 41

Date: 73-10-18
Local Time: 2110
Duration: 16 minutes
Location: Maubeuge, Nord, France
No. Witnesses: 2
DB: o   CB: d   PB-1: a   PB-2: x   UFOD: a

Abstract: Two young men, age 17, had ridden their bicycles along a dark road out of town in order to see the stars more clearly. At 9:10 P.M., they saw a diamond-shaped object pass smoothly across the sky from southwest to north-east. It seemed to flash orange, blue, and red as if rotating to display different colored facets. It disappeared after about two minutes. Continuing to watch, they then noticed a "large star" moving along the same path at 9:20 P.M., however, it had a cigar shape (as viewed through binoculars). It too was red, orange, and blue with eight separate lights spaced around its periphery. Its small red lights blinked on and off. One of the witnesses aimed his flashlight at it and flashed several times. *It suddenly stopped and enlarged in size "enormously,"*[1] *remaining that way for fifteen to twenty seconds. Then it shrunk in size and continued on its original course* to the northeast. It displayed three red lights on top, three on its bottom, and one small blinking light at its rear end.

Then, at 9:25 p.m., they saw a third strange object come into sight also from the southwest horizon. It was round with four equally spaced, red, blinking lights located on its circumference. It was a bright orange. One of the witnesses flashed the flashlight at it and it stopped and became angularly larger. They had the impression that it was approaching them directly. *After about twenty more seconds it shrunk back to its original size and started flying to the northeast on the same path as the others had followed.* No sounds were heard at any time. The astonished young men felt that all three objects were at a relatively low altitude—2,000 to 2,500 meters—although there is no way to verify this. No other anomalous objects were reported by them.

Comments: It may be significant to mention that other witnesses in the village of Gommegnies, Nord, saw a large red star passing across the sky from the northeast to southwest at 9:25 P.M. that night which suddenly stopped and fell downward to earth, disappearing behind a stand of trees. It isn't known whether it was the same light.

Note: 1. A drawing of this cigar was 30-mm long and 8-mm thick with a large red light in front of its bluish front third, orange mid-section, and reddish rear third. It traveled in a horizontal orientation.

Reference: Bigorne, J.M. 1974. Enquetes Diverses. *Lumieres dans la Nuit*, no. 138 (October): 17.

Cross references: 53 (3); 66 (3); 68 (3); 92(3)

Case No. 42

Date: 73-11-02
Local Time: 0400

Duration: estimated 30+ minutes
Location: just west of Manchester, New Hampshire
No. Witnesses: 5+
DB: jj   CB: a   PB-1: a   PB-2: x   UFOD: a

Abstract: This interesting case involved Mrs. Lyndia Morel, who lived in New Boston, New Hampshire. She left work to return home at 2:45 A.M. in her Corvair, stopped at a coffee shop with a friend until about 4:14 A.M., and then started the twenty-mile drive home. Driving on Mast Street, she sighted a small, star-like orange source of light which was very bright. As she continued glancing at it, other colors—red, blue, and green—could be seen. "It was like a spotlight shining down," she said. As she rounded curve after curve in the road she realized that it was following her; it stayed to her right side, yet remained above the top of the nearby trees. The light went out as she reached the intersection of Route 114 and 114A and then popped back on again. By now it was seen as having a round shape.

The light disappeared from sight as she passed through Boston, but reappeared again as she neared her home. The object was "huge," about the size of a large car. She said, "I saw somebody standing in it looking down at me and I felt like an ant in a jar." She felt paralyzed. The being's eyes "were as though they came out, staring at me as if looking right through me. The craft looked like a sphere, not like a saucer, more round and it was like honeycombed texture. I could see through it; it was bright orange" (Holzer 1976, 161).

Driving beside a cemetery, she claims to have been in an altered state of consciousness,[1] not fully in control of the car. To get away from the object, she covered her eyes and yanked the steering wheel to one side, ending up on the lawn of a house nearby. She ran to the front door and knocked and screamed until Mr. and Mrs. Beaudoin, the home-owners, answered it. The UFO continued to hover above the area.

Mr. Beaudoin called the local police immediately. Officer Jubinville listened to Morel's story and spoke to four other people who also saw the orange UFO between pine trees. "The policeman then took out his flashlight and put it up between the trees to point it out saying, 'Is that what you're talking about?' *Just then the light went out again.* 'I don't believe it!' the policeman said. *Just then, the light went back on again.*" Soon, Morel's husband arrived at the scene where he too saw the UFO in the sky. The remainder of this incident is left to the interested reader to review further.

Note: 1. She said, "Then the eyes seemed to come to me as if I were maybe six feet away from it, right through my windshield and I heard in my mind, through my ears, someone said, 'please don't be afraid.' My whole body felt numb at that." She also heard a high pitched whining hum inside of her head.

Reference: Fowler, R.E. 1979. *UFOs—Interplanetary Visitors.* New York: Bantam Books, 57.

Holzer, H. 1976. *The Ufonauts.* Greenwich, Conn.: Fawcett Publishing.

Cross references: 34 (3); 38 (3); 69 (3); 75 (3)

Case No. 43

Date: 73-12-10
Local Time: 2030
Duration: 3 hours 15 minutes
Location: Border of Austria and Bavaria
(southeast of Munich)
No. Witnesses: 2+
DB: dd   CB: d   PB-1: a   PB-2: x   UFOD: O

104

Abstract: This fascinating and well researched event involved two separate UFO responses to a red emergency aerial flare fired in its direction. Only the main features relevant to our subject are presented here. Friedrich Lennartz, 33, and Peter Zettel, 29, were having dinner in an otherwise unoccupied mountain-top cabin in the Alps at 8:30 P.M. The extra-clear air and full Moon provided them a beautiful view in all directions. They suddenly noticed a tiny "fiery spot" some distance away just north of Mt. Geigelstein to the southeast of their position and near the snow-covered horizon. Thinking that it was a crashed airplane on fire they viewed it through binoculars and saw an egg-shaped object standing on end and measuring about 12 or 14 meters high and nine meters wide (as compared with a nearby cross of known size). As the accompanying drawing shows (see Figure 16), the top one-third appeared like a yellowish dome.

At 8:40 P.M., Lennartz, an experienced ham radio operator, then tried to contact others with his 11-meter band, 2.8 watt set (frequency was probably between 27.215 and 27.275 MHz).[1] He successfully reached the Munich-Rosenheim regions 70 kilometers away without interference. At 8:50 P.M., Lennartz went outside and fired a red aerial signal rocket in the direction of the UFO.[2]

Figure 16

Eye witness sketch of UFO seen on December 10, 1973 in Bavarian Alps
(Adapted from *Flying Saucer Review*)

As he went back inside the hut to listen to his radio, *"the object became enveloped in a red glow that hid the rotating light…and started to rise slowly.…*It was a dazzling red, with only the cupola [at the top] remaining yellowish." *It rose so slowly that it required about five minutes to clear the top of a nearby mountain,* an altitude of about 200 meters from where it began. Then, according to the witness, *"it dashed towards my location…* I noticed static on my carrier wave, but at the same time my transmitting power was rising."[3] The egg traveled between five and six kilometers in about ten seconds at clearly supersonic speed.

The object came to a sudden stop about two kilometers distance from the astonished young men who continued to observe its fascinating details. After about thirty seconds more it made a sharp 90 degree turn to their right toward an uninhabited cabin where it stopped until 11:40 P.M. When Lennartz fired a second red flare into the air toward the UFO, *"The red mantle enveloped the 'egg,' it accelerated from zero velocity to an extreme rate and ascended with a slight western deviation."* Within thirty seconds, it had disappeared into the stars and the radio began operating normally again. The many other interesting details are left to the reader to discover.

Comments: The UFO in this case responded in the same way to both red flares. Would it have done so if a different colored flare had been used? How did the UFO "interpret" the sudden appearance of the flare, as a friendly gesture or as some impending military operation? As with many other cases (Hall 1964), the object changed hue just before accelerating. The radio transmission interference was related to the distance to the UFO and, if the UFO was radiating energy in the wave bands cited here, would have canceled or otherwise blocked the transmission.

Notes. 1. The "HAM" band extends from 28.0 to 30 MHz. It is more likely that Lennartz used a so-called "Citizen's" band radio having a frequency of from 27.0 to 28.0 MHz. 2. These rockets travel from about 300 to 400 m. 3. Increase in signal strength was deduced because both Munich and Rosenheim listeners said they were receiving his voice "extra loud." It is known that large swings in signal strength are common at 28.0 MHz.

Reference: Hall, R. H., ed. 1964. *The UFO Evidence.* Washington, D.C.: NICAP.

Schneider, A., and E. Berger. 1975. UFOs invade the Bavarian Alps—Part I. *Flying Saucer Review* 21, no. 1, 22-25.

Cross references: 19 (3)

Case No. 44

Date: 74-01-14
Local Time: night
Duration: estimated 4 minutes
Location: Moor Side, Lancashire, England
No. Witnesses: 3
DB: p   CB: a   PB-1: a   PB-2: x   UFOD: b

Abstract: Two teenage girls were walking back home after stabling their horses at a nearby farm. One of the girls had a powerful flashlight to light their way. While walking on a path toward home, they noticed a very bright object hovering above some cottages. It looked like an inverted saucer (see Figure 17), and had a flashing red light and a steady green light on it. It appeared somewhat hazy with a whitish bottom. *The girl aimed the flashlight directly at the object and "it appeared to react, rocking to and fro in a gentle motion."* The girls were very frightened and ran home as fast as they could.

Comments: Subsequent investigations found the girls to be considered very stable but also very disturbed by their encounter. A third eye witness from another part of town saw the same shaped object that night.

Perhaps this is as good a time as any to comment on the red and green lights that are often reported on UFO. Most terrestrial airplanes use these same colors for purposes of helping other pilots quickly and accurately determine the orientation and distance to another airplane at night (Haines and Flatau 1992). If UFOs are piloted by extraterrestrials, it may be asked why they deliberately choose these colors? Is it to help them become less conspicuous, to communicate their understanding of our color coding system through simple duplication, or for some other reason?

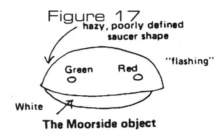

The Moorside object

Eye witness drawing of UFO seen on January 14, 1974, at Moor Side, England

Airplanes use a red light on the left wing tip and a green on the right. In level flight, an *approaching* airplane will thus show a green on the left and red to its right. This was also the case here, possibly suggesting that the UFO was some kind of terrestrial vehicle. However, a great many UFOs have one colored light on top and the other fixed or moving around its circumference. Neither placement would contribute to a human discrimination of UFO orientation because humans never know the shape of the UFO nor its colored light placement for sure. Perhaps colored lights on a UFO are not at all deliberately placed but are artifacts of their propulsion system. More analysis is called for on this important feature.

Reference: Haines, R.F., and C. Flatau. 1992. *Night Flying*. Blue Ridge Summit, Penn.: TAB Books.

Randles, J., 1976. Lancashire Round-up. *Flying Saucer Review* 21, no. 6, (April): 27.

Cross references: 79 (3)

Case No. 45

Date: 74-02-27
Local Time: 2145
Duration: 3 minutes

CE-5

Location: Latrobe, Tasmania
No. Witnesses: 3
DB: v   CB: a   PB-1: a   PB-2: x   UFOD: 0

Abstract: Greg Thornton and Sally Lamprey were driving toward Latrobe when they sighted a bright orange dot of light in the sky. The lights disappeared from view behind intervening trees for 150 yards but, when it reappeared, Greg pulled off the roadway and turned off his headlights. The UFO was about a mile west of them hovering above the Mercy Hospital area and two or three hundred feet in the air. *"Once he had turned off the car lights, the UFO approached rapidly but slowed as it neared the witnesses.* No sound could be heard at any stage." Lamprey said that it looked like a triangle with rounded corners (see Figure 18). Becoming more afraid, Greg started the car and then switched on his headlights. *"As if in response, the UFO turned onto its side.* "It appeared as a straight orange line, like a pencil at a 45 degree angle," said Lamprey. It then accelerated up to a very high velocity and disappeared behind nearby trees to the north.

Apparently, the same orange object paced another automobile only minutes later driven by Mr. J. about five miles to the east on the same road as above. He also stopped his car to get a clearer view of the silent, orange light. He estimated its size to be about twenty feet in diameter and round. The object also seemed to pulse in luminance. It finally just disappeared upon nearing the Bass Highway at Latrobe.

108

Figure 18

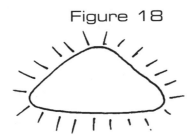

Eye witness drawing of UFO Seen on February 27, 1974, at Latrobe, Tasmania

Reference: Roberts, W.K. et al. 1975. From the Tasmanian "Flap" of 1974—Part I. *Flying Saucer Review* 21, no. 3/4 (November): 47-48.

Cross references: 40 (3); 87 (3)

Case No. 46

Date: 75-03-02
Local Time: night
Duration: 5 minutes
Location: Phillips, Wisconsin
No. Witnesses: 1
DB: q   CB: a   PB-1: a   PB-2: x   UFOD: b

Abstract: A local police officer began hearing strange noises on his patrol car's radio, "it went wild;...the noises were short weird responses." Then, he saw a 30-foot-diameter, disc-shaped UFO with a rounded hump on its top and bottom. It also had red and orange lights on its bottom edge. There were bright, blue-white areas of illumination on the ground about 1,000 feet away. *"He put [his] spotlight on [the] object, it rapidly ascended"* into the sky.

References: Spiegel, L., D. Flickinger. *UFO Central.* (case report). Evanston, Ill.: Center for UFO Studies.

Cross references: 8 (3); 12 (3); 20 (3); 22 (3); 23 (3); 25 (3); 27 (3); 50 (3); 62 (3); 63 (3); 71 (3)

Case No. 47

Date: 75-04-19
Local Time: 0100
Duration: 15-20 minutes
Location: Valle de Abdalajis, Spain
No. Witnesses: estimated 5+
DB: w   CB: a   PB-1: a   PB-2: x   UFOD: b

Abstract: This event took place during a UFO field investigation by members of a local study group in the region. Newspaper articles at that time indicate that there had been numerous UFO landings on nearby hills for over two months. Strange orange nocturnal lights had been seen moving above the El Cepillar plain, the Cerillo Conejo, and the Molina hill; some witnesses estimated their size to be about 1.5 meters.

The study group arrived at 10:00 P.M. on Saturday, April 19 to interview witnesses. At 1:00 A.M. the following Sunday morning, they saw an object giving off a very intense blue light "like a halo." It had a red light on one end and a yellowish glow seemed to envelope its whole surface. Then, two white lights came on from the body of the object and seemed to aim in particular directions.[1] *"One member of the party, who was sitting in a car, finally switched on the car headlights, whereupon the object made off at high speed."*

Comments: It's interesting to note the very long response time of the UFO investigators, a particularly common detail around the world. What other useful data might be obtained if there were rapid response teams properly trained and equipped?

Note: 1. Some members could not look at the UFO through their binoculars "for more than a few seconds" due to its extremely high luminance.

References: Anon. 1976. *Flying Saucer Review.* World round-up, no. 6 (April): 32. *El Chronista Commercial.* 19 May 1975.

Cross references: 7 (3); 14 (3); 36 (3); 65 (3); 84 (3)

Case No. 48

110
    Date: 75-09-14,15?
    Local Time: night
    Duration: estimated 5 minutes
    Location: Holybrook, Thames Valley, England
    No. Witnesses: 5+
    DB: v   CB: a   PB-1: a   PB-2: x   UFOD: 0

Abstract: Fifteen-year-old William Crowley (of Southcote) was the first in a group of five boys to see an unusual light in the darkened sky. He pointed it out to his buddies, and all watched as it hovered for a moment and then dove vertically toward the Earth. It rose upward again at an angle and flashed a red light. "Suddenly there was a group of them....Then a plane came and *they all disappeared*. Although it was very quick, I think they went towards the motorway," Crowley said.

Comments: A group of eight lights hovering in the sky was seen earlier that evening by Mrs. Carol Ward of Ufton Nervet.

Reference: *Reading Evening Post.* 15 September 1975.

Cross references: 11 (3)

Case No. 49

Date: 75-11
Local Time: night
Duration: estimated 6 minutes
Location: Wesendorf (Hammerstein Army Base),
           northern Germany
No. Witnesses: 2+
DB: bb   CB: a   PB-1: a   PB-2: x   UFOD: a

Abstract: This interesting CE-5 event took place during German tank training maneuvers on a large Army base in the Schleswig-Holstein province of Germany. The main witnesses were in a "Marder" tank on a very clear, cold night with some approaching clouds in the east. Mr. Ingo D. noticed an intense, white, self-luminous sphere the angular size of the full moon flying slowly under the clouds by their position from the east. It trailed a white vapor behind it. While it appeared extremely bright, it didn't cause the clouds above it or the ground beneath it to become brighter.[1] According to the primary witness, "More as a joke, our tank driver turned on the searchlights, and to the surprise of all of us, *the sphere answered in her way: It extended and contracted again. We repeated the experiment, and whenever the tank driver flashed with the searchlights, the sphere replied.*"[2] Then the light source diminished in diameter (to a tennis ball) and flew out of sight, leaving a very thin, reddish-orange trail in the sky visible for two more hours.

111

Notes: 1. One explanation for this rather commonly cited visual effect is that the witnesses were looking directly up a relatively narrow beam of light from the UFO which would not spread out to illuminate surrounding surfaces. 2. It isn't known how many times the tank crew repeated this signal nor how long the entire episode lasted.

Reference: Hesemann, M. 1994. *Geheimsache UFO*. Neuwied, 92.

Case No. 50

Date: 76-summer
Local Time: night
Duration: estimated 2 minutes
Location: Dartmoor (Devonshire), England
No. Witnesses: 1
DB: bb   CB: a   PB-1: a   PB-2: x   UFOD: a

Abstract: Mr. Peter Paget sighted a round, golden-hued object flying at about 3,000 feet altitude which seemed to be rotating. It was perfectly silent. *"When signaled by a 150-watt mercury quartz ultraviolet lamp, [the UFO] suddenly increased in brilliance, changed course by 30 degrees, and climbed away until it could no longer be seen in the starlit sky....*I can only state that the object was directly overhead when the light was kindled. The UFO's maneuver was instantaneous and for all the world looked to me like a high-speed evasive procedure away from our position, gaining both air speed and height combined with a change of heading."

Reference: Paget, P. 1979. *The Welsh Triangle.* Granada, London: Panther Books, 75.

Cross references: 12 (3); 22 (3); 71 (3); 75 (3)

Case No. 51

Date: 76-06-22
Local Time: 2130
Duration: estimated 2 minutes
Location: Las Rosas, Grand Canaria Island
No. Witnesses: 3+
DB: v CB: a PB-1: a PB-2: x UFOD: a

Abstract: Dr. Don Francisco-Julio Padron Leon, a medical doctor, was riding in a taxi to the village of Agaete located in the northwest part of the island. With him was Mr. Santiago del Pinto and the car's driver, Francisco Estevez. Del Pinto had asked Padron Leon to check on his sick mother. As the men drove around a bend in the road, they saw a huge spherical object hanging in the air several meters off the ground. Its outline was a pale grey-blue. Almost immediately, the car's radio stopped working and the three men felt a "surging wave of cold." The various visual details of the sphere and the two human-like figures inside it will not be recounted here.

The taxi driver impulsively turned on his spotlight. "At that instant the sphere began to rise until the watchers could see a transparent tube inside it that emitted a blue gas or liquid. This gradually filled the sphere, which expanded until it was as big as a thirty-story building although the beings and their console panels remained the same size."

The men sought refuge in a nearby house, learning from the family living there that their television had stopped working. Then, everyone saw the UFO still hovering in mid-air, the bluish gas inside it continued to swirl. The object gave off a "high-pitched-whistle" sound and accelerated in the direction of Tenerife.[1] Hundreds of others also sighted a UFO in Puerto de la Cruz as it flew above Tenerife Island. Investigators believe it was the same object.

The Spanish Air Ministry commanded everyone to not give out information on the event, yet some information leaked to the press. The case file was later released to the noted Spanish investigator Benitez.

Comments: The CE-5 response evoked was very likely that of light avoidance. The bizarre events reported here have all the characteristics of a Hollywood science fiction thriller created through advanced computer graphics. Of course, the technical apparatus that would be needed to produce such a virtual aerial image was not available in 1976. I want to speculate on the possible purpose of allowing humans to perceive such a complex display. Perhaps the E.T.s involved wanted to impress the witnesses with their technological prowess. Maybe they hoped to communicate some particular graphic message to the humans who would understand and re-transmit it to other humans. Or maybe the beings were oblivious to their presence and didn't care whether or not they were seen.

113

If the object was surrounded by an electromagnetic field of sufficient electrostatic intensity it could explain the radio failure reported and the visual images seen, but not how all three men saw the same thing at the same time. This would require a deliberate optical and/or holographic projection of some kind. A projection upon a planar, gaseous surface, or screen, would be far easier to generate than creating an actual three-dimensional object with transparent walls and highly detailed interior details that moved.

Note: 1. The witnesses saw the UFO change shape from a sphere into a spindle with a halo around it as it departed.

References: Brookesmith, P., ed. 1984. *The UFO Casebook*. London: Orbis Publishing, 30-31.

Flying Saucer Review 23, no. 3.

MUFON Journal, no. 278.

Case No. 52

Date: 76-08
Local Time: estimated 0100
Duration: estimated 6 minutes
Location: Mudbrook campsite, northern Maine
No. Witnesses: 4+
DB: v CB: a PB-1: a PB-2: x UFOD: O

Abstract: Four young men, all art students at the time, were on a camping and fishing trip to northern Maine when at dusk, with about twelve other campers also at the site, a large, round, whitish-yellow light with a reddish tinge was seen moving against the wind. It hovered above a small lake and then "imploded" into nothingness. Two days later, the four witnesses were at another nearby lake, Smith pond, to go night fishing. They first built a large beach fire to serve as a beacon to help them return to their campsite in the dark. Then they paddled some distance out in an aluminum canoe.

114　Charley Foltz was in the front of the canoe facing aft, Chuck Rak was in the rear, and twins Jack and Jim Weiner were in the middle. Foltz exclaimed, "what is that?" Everyone turned to look behind them where a huge, round light was rising up out of the trees on the far side of the pond. Jim said that it was perfectly silent and "changing in a very fluid, liquid way." Foltz recalled, "When it was nearest us, about 100 yards away, it paused." He flashed the flashlight to see what would happen. Jim Weiner remembered that *"the instant [Foltz] flashed his flashlight, this thing sent this beam of light out to us.* And we were in a sixteen-foot Grummond Canoe, which in the pitch black of night, lit up like a roman candle. We must have looked like [an] object just waiting to be approached on this lake." The remainder of the claimed events will not be recounted here, however, it does appear that extremely bizarre psychic events took place, leaving lasting psychological traces for many years.

Comments: The almost instantaneous UFO response to a flashlight beam is not at all uncommon. Surrounded by a pitch-black lake, the flashlight would have been particularly prominent, as would the beach fire nearby. Investigator Fowler has written a masterful account of these events which present the many strange elements which allegedly occurred. The two twins shared identical nightmares about this event over the years and underwent intensive study.

Reference: Fowler, R. 1993. *The Allagash Abductions.* Tigard, Oregon: Wild Flower Press.

Case No. 53

Date: 76-08-15
Local Time: 0300
Duration: estimated 5 minutes
Location: Real de la Jara (Province-Seville), Spain
No. Witnesses: 20+
DB: v   CB: a   PB-1: a   PB-2: x   UFOD: 0

Abstract: The drivers and occupants of about twenty automobiles traveling on the highway from El Real de la Jara to Sierra sighted a UFO in the sky at a distance estimated to be only about five kilometers away. Mr. Pablo Garcia Garcia, traveling with his wife and children, stopped his taxi by the side of the road, as did all of the other cars; they were all following his taxi. Garcia started to walk in the direction of the strange, very intense light. He then returned to the taxi and began flashing his headlights at the object. *It began to approach the line of cars* and everyone started yelling at Garcia to stop signaling. Garcia stopped, and the cars accelerated on toward Sierra.

115

Comments: It is unfortunate that nothing else is known about this case. Why did Garcia walk toward the light? Had he ever seen a UFO before? Were any electromagnetic effects experienced by any of the drivers?

Reference: Anon. 1977. Some recent Spanish reports. *Flying Saucer Review 22,* no. 6 (April): 28-29.

Case No. 54

Date: 77-01-21
Local Time: 2110
Duration: estimated 5 minutes
Location: near Ibaque, Columbia
No. Witnesses: 5+
DB: v CB: d PB-1: a PB-2: x UFOD: a

Abstract: Avianca flight 132 had just taken off from Eldorado Airport, Bogota, and climbed to its assigned cruise altitude of 20,000 feet. Shortly thereafter, Captain Gustavo Ferreira noticed an "extremely bright light heading straight towards them from the opposite direction." The first officer, flight engineer, and two cabin attendants also saw it from the cockpit.[1]

After discussing the strange light with his first officer and calling air traffic control at Bogota about other air traffic, Ferreira ordered that the airplane's landing lights be turned on in order to be seen by the oncoming object.[2] *Suddenly, the UFO changed from vivid white to red.* The captain then turned off all of his airplane's lights, leaving it in total darkness. He then turned his landing lights back on a second time *"and the UFO at once replied by flashing green lights.* This exchange of signals had taken about three minutes, after which the UFO moved away southwards and was rapidly out of sight" (Creighton 1977, 25). At the moment the UFO disappeared from sight, the radar operator cried, "I've lost it!" It had performed a ninety degree turn at a "fantastic speed" and departed off his scope.

Comments: This now classic case involves simultaneous visual-radar contact with a single high-intensity, multi-colored source of light. The temporal correspondence between the flight crew's behavior[2] and the object's change in appearance is striking, however, the exact delay between them is not specified.

116

Notes: 1. The onboard radar operator, Jorge Jiminez, also detected the UFO moving at a velocity of approximately 44,000 kilometers per hour. Its screen size was "three times the size of a normal aircraft," and moved in a pronounced zig-zag manner. The captain estimated it to be about 25 miles away. 2. Later, the pilot said that "although I was quite calm, I felt a weird sensation, of mingled surprise and bewilderment." The captain had flown for over twenty years and this was his first UFO sighting; he is now certain UFO exist and are not "nuclear weapons of a foreign power" or "the chariot of Elijah mentioned in the Bible."

References: Creighton, G. 1977. UFO Answers Signals over Columbia. *Flying Saucer Review* 23, no. 2.

El Tiempo (Bogota, Columbia). 18 February 1977.

IUR 2, no. 7.

Case No. 55

Date: 77-02-09
Local Time: 2030

Duration: estimated 4 minutes
Location: Flora (Madison County), Mississippi
No. Witnesses: 8
DB: v   CB: a   PB-1: a   PB-2: x   UFOD: O

Abstract: A strange aerial light was reported to the local sheriff's office by Mr. Fred Grant, and two officers were dispatched immediately.[1] Deputy Sheriff Ken Creel and Deputy James Luke drove along Coxferry Road, about five miles west of Flora, to Grant's house, when a UFO quickly flew to a position directly above their cruiser. It hovered at from twenty to fifty feet altitude, keeping pace with their automobile. It illuminated the vehicle with an intense light and their radio communication experienced interference.

Officer Hubert Roberts of the Flora Police Department was in his own cruiser parked on a gravel road off the main highway near Bogue Chitto Creek when he sighted the same UFO.[2] It was "sort of round," very bright, and had portholes in its sides. *"The officer blinked his headlights as a signal, and the UFO reportedly blinked two of its lights in apparent response."* The object shot away into the dark sky "with great speed."

117

Comments: Interestingly, nothing more is known about what happened to the two officers in the first patrol car.

Notes: 1. Louisiana and areas east of it had experienced numerous UFO sightings during January and February of 1977. 2. Four other eye witnesses also saw the object from different vantage points.

References: *MUFON UFO* Journal, no. 286.

Peters, T. 1977. Low lights in Louisiana, *APRO Bulletin* 25, no. 9(March): 1-2.

Cross references: 33 (3); 87 (3)

Case No. 56

Date: 77-04-?
Local Time: night
Duration: estimated 4+ minutes
Location: Roslin Glen, Scotland
No. Witnesses: 1
DB: v   CB: a   PB-1: a   PB-2: x   UFOD: a

Abstract: Mr. Derek Scott Lauder, 25, was outside his cabin in a remote glen near the ancient Roslin Castle. Because of a prior sighting, a keen interest in the night sky, and a prior suggestion by Arthur Shuttlewood that he should try to send light signals into the sky, Lauder took a flashlight with him (Lauder 1966). He said, "I went outside with my torch and began to signal with a couple of quick flashes of torchlight at various parts of the sky. I did not know what to expect. I signaled at the area of Leo Major and Cancer. Immediately afterwards, I saw two flashes of *bright white light from something which was 'covering' the star Acubens in the constellation of Cancer* [see Figure 19]. *I signaled again, and again I received replies of bright flashes of light....*This was my first attempt at using torchlight."

Subsequent attempts to obtain some response from the heavens did not always achieve success. However, Lauder has had so many alleged responses that he no longer keeps records of them. The celestial response was also witnessed by a UFO investigator who traveled to Scotland to "see it for himself."

Figure 19

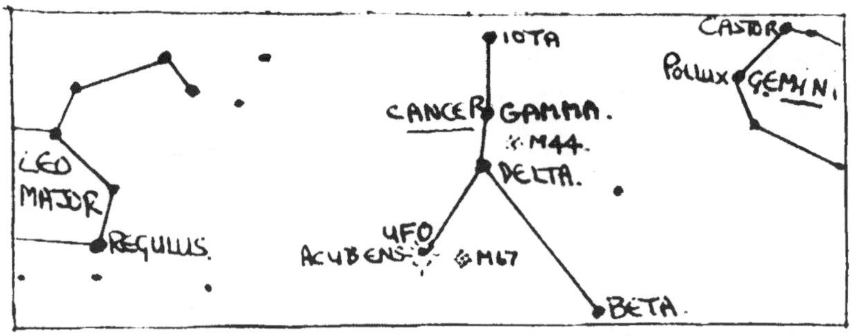

Location of flashing UFO seen in April 1977 (Sketch by Derek Scott Lauder)

Comments: It is clear from the witness's sketch that he has more than a casual knowledge of the constellations. However, what is very difficult to understand is how a moving object in space, or even the atmosphere, could come to a stop at a point exactly superimposed over a star? The odds of this happening by chance are truly astronomical. Correspondence with the witness as recent as March 1997 suggests that these CE-5 events continue.

Reference: Lauder, Derek S. 1996. Personal correspondence, 30 December.

Case No. 57

Date: 77-04 to 07
Local Time: night
Duration: various
Location: Pinheiro (Maranhao State), Brazil
No. Witnesses: 26+
DB: jj   CB: a   PB-1: a   PB-2: x   UFOD: 0

Abstract: A group of twenty-six workers were building fences about six kilometers from the village of Pinheiro. One of the group was selected to go fishing for the others and, while he was fishing, a UFO suddenly appeared just above him. Very afraid, he ran back to the others with the "fireball" following close behind. "Then everybody in the camp saw it too." After about thirty minutes, it departed into the sky. The following day when they had moved their work camp to a different location, they decided to construct a scarecrow out of sticks and some clothes. They placed a kerosene lamp on top and then hid in the nearby woods to watch what might happen.

119

"Later that night, *the object suddenly just appeared, very, very close to the kerosene lamp.*" It was so intense that no definite object or body could be made out. After about forty-five minutes the UFO rose up into the night sky as before. "After this, a lot of men left and went back home" (Pratt 1996, 136).

Comments: This CE-5 event involves a rather passive human behavior which, nonetheless, evoked a close approach by a UFO. How can something with apparent mass suddenly appear without first traveling to the site? Does it materialize or move silently to the site in one band of wavelengths and then modulate its wavelengths to the visible band for some reason? If the latter, why would it want to be seen at all, why not remain invisible? The strange features of all good UFO sightings are truly a challenge to science. The event took place during an awesome aerial display which occurred almost every night over four months (April-July) in the vicinity of Pineiro. Town Mayor, Manoel Paiva, estimated that as many as fifty thousand people had seen them!

Reference: Pratt, B. 1996. *UFO Danger Zone.* Madison, Wis.: Horus House Press.

Case No. 58

Date: 77-07-10
Local Time: 0100
Duration: estimated 6+ hours
Location: Pinheiro (Maranhao State), Brazil
No. Witnesses: 1
DB: p   CB: b   PB-1: a   PB-2: x   UFOD: a

Abstract: One early morning while walking from his home southwest of Pinheiro to catch a bus, farmer Jose Benedito Bogea noticed a bright greenish light in the dark sky. He was using a flashlight at the time to see his way. He started running away from the source of light but it followed him for several hundred meters. The object then flew on ahead of him and stopped over a bush.[1] It hovered only three or four meters above the ground. Bogea could make out an object fifteen to twenty meters long in the form of a "V." It emitted an orange beam toward the ground.

"I (Bogea) raised my arm and shined my flashlight at it, and in an instant I saw a bright flash of light. It knocked me down, and I felt like I'd had an electric shock. Then I passed out" (Pratt 1991, 101). Bogea woke up about 8:30 A.M. the next day, but now he was near Sao Luis, some 70 miles from where he had encountered the UFO. Several hours later he began experiencing terribly intense pain in his right arm, kidneys, spine, right side. He lost his appetite for eight days as well. For various reasons, Bogea interpreted his close encounter as "wonderful and very beautiful."

Comments: This CE-5 event also may qualify as an abduction (CE-4). The witness claims to have dreamt of waking up in a "strange city" populated by many similar human-looking people, "all dressed in gray and brown clothes, a few in light blue."[2] Other details of how his visual nearsightedness allegedly was cured as a result of this encounter along with partial recovery of lost hearing will not be presented here. Did his flashlight and arm act as an electrical conduction path for high potential within or surrounding the UFO? No skin burns were found on him, however. The similarity between this case and that of Travis Walton near Snowflake, Arizona, (cf. 75-11-05, chapter 8) is notable.

Notes: 1. Bogea stated that the UFO remained over the bush "for just a fraction of a second." I suspect that if Bogea was abducted, it occurred at this point in time. Abrupt memory interruptions such as this are common in the abduction literature. 2. His description of the humanoids, strange environment, and long-distance transport are similar in almost every respect to that recalled by Kathy and Susan [cf. case 75-11 in chapter 9], which allegedly took place underground near Sonora, California.

References: Pratt, B. 1991. Disturbing Encounters in North-East Brazil. In *The UFO Report*. ed. T. Good. London: Sidgwick & Jackson, London, 1990.

Pratt, B. 1996. *UFO Danger Zone*. Madison, Wis.: Horus House Press.

Schuessler, J. F. 1996. *UFO-Related Human Physiological Effects*. Houston, Texas: Privately published.

Cross references: 237 (9)

Case No. 59

Date: 77-early Autumn
Local Time: 1720
Duration: estimated 8 minutes
Location: Isfield, near Lewes, Sussex, U.K.
No. Witnesses: 1
DB: q   CB: a   PB-1: a   PB-2: x   UFOD: a

Abstract: Denise, a policewoman trained in aircraft recognition in the Royal Observer Corps, was waiting alone at a bus stop. She sighted a strange, angularly large, clearly defined, black oval or circular object at an estimated 300 feet altitude in the sky nearby. It seemed to be rising vertically at a very slow rate—approximately one degree arc per minute—but made no noise at all.[1] The object seemed to spin slowly in a clockwise direction. "On impulse she waved in its direction. "I hope you won't think I'm crazy," she said later, "but I thought 'come down here and let me get a good look at you." She was not at all frightened and sensed no hostility or danger. *Immediately, the disc changed course and began to move in her direction. "It then very slowly moved to an elevation of 90 degrees"* over the next ten minutes.[2]

The woman suddenly became aware of time. She reasoned, based on the bus schedule, that at least twenty minutes had elaspsed. Then the UFO appeared to shorten in width and she saw a "type of dome" on its top because of reflected sunlight. The dome had a "greenish-blue" light on top and seemed to be made of a light greenish-gray metal. A drawing of the disc is shown below (see Figure 20). Centered on its bottom surface was a very dense black circle or oval.

Figure 20

Artist's sketch of UFO seen in early Autumn 1977 in Isfield, England
(Courtesy of *Flying Saucer Review*)

Comments: The witness estimated that the object was no further than fifty feet away from her. She prefers to remain anonymous, partly due to official pressure. She experienced an acute headache from the time she boarded the bus until the next morning. At work, she was "nervous, shaky, and unaccountably clumsy in her movements." She also felt very thirsty all that evening. The many other physiological and psychological symptomatic details of this close encounter, including her conjunctivitis, are not recounted here.

Another close encounter with somewhat similar characteristics took place on December 12, 1978, at 9:00 P.M. in Ronneburg, Saxony, Germany (Hesemann 1996). It involved a lone 38-year-old woman who waved at a large (estimated 150-foot-diameter) bell-shaped object suspended about 75 feet in the air over nearby houses. Pulsating orange lights covered its bottom and a large, constant yellow light was located at its middle. Smaller red, green, and yellow lights were spaced around its periphery. As soon as she waved the *UFO tilted toward her* and displayed a "transparent screen behind which three huge men were standing near a kind of control panel." Other fascinating details including fifteen to twenty minutes of missing time are found in the original reference. Here we find yet more evidence either for an altered state of consciousness and/or direct distortion of time itself in the near vicinity of a UFO.

Notes: 1. She estimated its angular width to be four inches wide at arm's length or about 11 degrees. If it was a Harrier jet, which can hover, and was seen side-on, it would have been at a distance of only 237 feet and its engine noise would have been truly ear-deafening. It was not a Harrier, unless the woman became temporarily deaf and the airplane could change its shape. 2. The woman had the feeling that she was being observed by the UFO and of being surrounded by an "invisible ray" from the bottom of the object.

References: Good, T. 1988. *Above Top Secret: The Worldwide UFO Coverup.* New York: Quill, William Morrow, 115-116.

Grant, P.B. 1979. A very personal encounter "somewhere in Sussex." *Flying Saucer Review* 25, no. 2 (March-April): 18-21.

Hesemann, M. 1994. *Geheimsache UFO.* Neuwied, Germany, 463.

Hesemann, M. 1996. *UFOs uber Deutschland. Niedernhausen*, Germany, 120.

Case No. 60

Date: 77-03-08
Local Time: 2030
Duration: estimated 4 minutes
Location: Oswaldthistle, U.K.
No. Witnesses: 4
DB: kk   CB: a   PB-1: a   PB-2: x   UFOD: 0   123

Abstract: A lighted object was seen in the night sky by four night-shift workers. The object appeared to vary from round to triangular shape over time. One of the men aimed his flashlight upward at it. *It reacted by shining a beam back down.*

Comments: Interestingly, about fifteen minutes after this event, a bus load of forty people riding through Barnoldswick (East Lancashire) reported seeing a big light in the sky which changed to a triangular shape as it approached them.

Reference: Randles, J. 1984. The Pennine UFO Mystery. *APRO Bulletin* 32, no. 1 (March): London, 5.

Case No. 61

Date: 78-03-29
Local Time: 2130
Duration: estimated 5 minutes
Location: near Indianapolis, Indiana

No. Witnesses: approximately 6
DB: p   CB: d   PB-1: a   PB-2: x   UFOD: a

Abstract: A number of trucks with trailers were traveling together on Interstate 70 outside of Indianapolis[1] when all were illuminated by a "bright blue light" which lasted several seconds. During this period their diesel engines began to sputter, the Citizen's Band radios which had been in use stopped working, and a "complete silence" was reported. One of the drivers[2] shouted into his CB microphone, "'Hey UFO, if you have your ears on, I want to go with you!' *Suddenly, the blue light returned and again surrounded all the trucks for about fifteen seconds, resulting in the same [electromagnetic] effects as before.*"

Comments: It isn't clear the direction from which the blue light came nor other important details. Could the huge area of blue illumination have been some kind of natural phenomenon? If so, what caused it?

Notes: 1. Assuming five trucks (each 75 feet long) behind one another and 150 foot separation the total length of the caravan would have been about 975 feet. 2. The witness was in the rear-most truck. His comment indicates some interest in or knowledge of UFO.

Reference: Hall, R. 1988. *Uninvited Guests: A Documented History of UFO Sightings, Alien Encounters and Coverups.* Santa Fe, N.M.: Aurora Press, 53.

Case No. 62

Date: 78-06-11
Local Time: 2328
Duration: 8-10 seconds
Location: New Shrewsbury, New Jersey
No. Witnesses: 2
DB: o   CB: a   PB-1: a   PB-2: x   UFOD: a

Abstract: This incident took place at the U.S. Naval Ammunition Depot, Naval Weapons Station Earle in Eastern New Jersey. It was preceded by two other sightings (at 2:35 A.M. and 7:45 P.M., respectively) on the base which UFO investigators later judged to be not anomalous. This last event involved gunnery sergeant Brininger and Pfc. Johnny Johnson, 22, who were on duty outside an eighty-foot-tall tower. They sighted a "distinctly-outlined illuminated white ball with a short conical tail behind it." It was first seen near the horizon to the south, but within about five seconds reached a point due west of their position with about a 20 to 30 degrees arc above the horizon. Brininger thought the lighted object was only 300 feet away and about 200 feet altitude. He turned on a Navy spotlight nearby and swung it toward the object. *The light hit the object and "it abruptly changed course from its rapid horizontal flight, turning away from them and flying due west and climbing."*[1] The angular length of the tail behind the object seemed to become shorter after it turned west. At this point, its center seemed darker with only its rim brighter.

Comments: Most of the stated details may be questioned as to their accuracy, which is true for most of the incidents included in this book. Nevertheless, when these details are accepted as reasonably accurate, they contain insights into the core phenomenon. In this case the ice-cream-cone-shaped luminous source might have been a meteor, as suggested by the UFO investigator, except for its small radius turn and luminous appearance as it traveled away from the two witnesses. Its apparent climb during departure is harder to explain, since the eye witnesses were on the ground, making an illusory climb a far more difficult occurrence.

Note: 1. Johnson said that the UFO curved as it turned west and Brininger claimed it made a sharp 90 degree turn to the west.

Reference: *International UFO Reporter* 3, no. 7 (July): 4-5.

Cross references: 2 (3); 8 (3); 20 (3); 22 (3); 23 (3); 25 (3); 27 (3); 46 (3); 63 (3); 71 (3)

Case No. 63

Date: 78-08-11
Local Time: 2359
Duration: 3 minutes
Location: Great Mills, Maryland
No. Witnesses: 5
DB: v   CB: a   PB-1: a   PB-2: x   UFOD: c

Abstract: This encounter involved the use of a bright police spotlight to try to identify a "vaguely defined, silvery, blimp-shaped light form" above the deputy sheriff's car. A 38-year-old female deputy sheriff was returning from a benefit ballgame with four youngsters in her cruiser on Chancellors Run Road when they all saw a large UFO above some nearby trailers. It had several red glows "toward the bottom."

"The woman shone the cruiser's powerful spotlight toward it and scanned the length of the form, fully confident that she was close enough to light it up; yet the beam had no effect on the large shape. She moved in closer toward the hovering UFO, but *it began to get 'cloudy' on its underside, and started moving slowly down behind the trees.*" She continued on home and the children could still see it in the sky for awhile until it suddenly disappeared.

Comments: During an interview, the primary witness claimed no particular interest in UFOs nor had she read any books on the subject. She did go back to the location of the earlier sighting for possible traces but found none. The unusually slow departure of the UFO should also be noted.

Reference: Anon. 1978. *International UFO Reporter 3*, no. 9 (September): 9.

Cross references: 2 (3); 30 (3)

Case No. 64

Date: 80-07-13
Local Time: 0300
Duration: estimated 2 minutes
Location: Butzbach, Hessen, Germany
No. Witnesses: 4
DB: i   CB: a   PB-1: a   PB-2: x   UFOD: a

Abstract: Four men were camping together and had gone for a long late-night hike, returning to their camp at about 3:00 AM. Soon after climbing into their tents to sleep they heard a dog barking nearby. They looked outside to see what was the cause. Later, they learned that horses and sheep pastured in a field nearby were also "nervous" at the same time. Above them was a bright light in the clouds moving silently toward the north. It came to the top of a nearby mountain and stopped. "Although being completely afraid, one of the boys had the guts to flash a torch in the direction of the luminous mass. *The signal was answered by a bright spotlight-like beam which broke through the clouds and shone down on them.* They were too much afraid to repeat the experiment." Some (unknown) time later, the luminous object darkened until it was no longer visible.

Comments: This sighting is common with its multiple witnesses, apparently collimated beam of bright light from the UFO, and apparent human-initiated response from the unknown aerial object. Again, it raises the question of what would happen if more humans would show more initiative toward the phenomenon?

Reference: Hesemann, M. 1994. *Geheimsache UFO*. Neuwied, Germany, 117.

Case No. 65

Date: 80-09-05
Local Time: 0340-0415
Duration: 35+ minutes
Location: Tropy, Poland
No. Witnesses: 6
DB: jj   CB: a   PB-1: a   PB-2: x   UFOD: b

Abstract: Returning to a hospital in an ambulance from the small village of Zutawka, Poland, with a pregnant woman in labor, three medical personnel—a physician, a driver, and a stretcher bearer—all saw a "big red ball in the sky." It was about the angular size of the full moon, a dark crimson color, and was approaching them. By 3:40 A.M., they had passed the village of Kalwa and the UFO was still near them at tree-top level "swinging in gentle curves." Becoming more afraid, the driver sped up to 130 kilometers per hour *but the ball stayed exactly beside them "as if the ball were linked to them by cord."* The doctor said, "It seemed obvious to me that the object was under intelligent control." Then, between Kalwa and Sztum, near a railroad crossing, the UFO landed squarely in the road about 200 meters ahead of the car. It appeared to be hovering a few centimeters off the ground.

127

The UFO was so wide that its outer edges extended about 50 centimeters beyond each side of the road which was 6 meters wide. The object had curved bands and stripes with "a lot of black lines going up and down irregularly in each direction...like the veins inside a human body." Its basic surface was dullish but some surface areas changed color from moment to moment. Yellow-orange patches appeared on its deep crimson surface.

The physician got out of the ambulance to talk with two men on duty at a railroad crossing guard tower. They too were watching the unusual event. The UFO slowly moved away from the roadway, stopping behind nearby trees and brightly illuminating the ground. "No, I was not nervous," the physician claimed later. "Every moment I was aware of what I was doing, of thinking, of watching. I knew it was a strange object from the skies, and certainly not a natural phenomenon. I was a little bit frightened, but not scared. My mind was alert." The doctor was pressed to get her patient to the hospital, "surely it would know we have no time if we give it a signal?" *She asked the driver to flash the headlights, and he did so twice. The UFO disappeared* "like a TV set when switched off." The time was about 4:15 A.M. [1]

Comments: It isn't hard to agree with the chief witness concerning the apparent intelligence displayed in this case if these details are accurate. As with so many cases such as this, we can ask how the visitors interpret flashing headlights; as greetings, simple acknowledgment of one's presence, or as impending hostile action.

Note: 1. Happily, the woman delivered her fourth child at 6:10 A.M. A third party drove back to the site of the event soon afterward and found no marks or burned vegetation.

Reference: Popik, E. 1981. Under intelligent control? *Flying Saucer Review* 26, no. 6 (March): 2-4.

Case No. 66

128

Date: 80-10-23
Local Time: ?
Duration: estimated 5 minutes
Location: Morenci, Arizona
No. Witnesses: 5
DB: v   CB: a   PB-1: a   PB-2: x   UFOD: a

Abstract: A boomerang-shaped object approached smokestacks of a copper smelter in Morenci, Arizona. Five men in the area watched as the object emitted a brilliant light into each smokestack and then accelerated at a high rate to the southwest. Before it departed, the object release a small fireball. One witness, Joe Nevarez, allegedly said that he wished the UFO would return so that he could get a better look at it. According to the original article, *the UFO performed an instant reversal and returned to the slag dump area of the smelter after which it accelerated out of sight to the north.*

Comments: The validity of this case rests upon the credibility of the witnesses which cannot be verified. For this reason the overall reliability rating is considered to be low.

Reference: *The MUFON UFO Journal.* 1990. No. 270 (October).

Cross references: 21 (3); 82 (3); 135 (4)

Case No. 67

Date: 81-03-?
Local Time: early A.M.
Duration: 2 hours 20 minutes
Location: Tacuarembo, Uruguay
No. Witnesses: 4
DB: r   CB: a   PB-1: a   PB-2: x   UFOD: a

Abstract: Local chief of police Mr. Miguel Costa, his wife, and another couple were driving near Tacuarembo after midnight. They all sighted a very large aerial object in the dark sky. *Costa stopped the car and flashed his headlights on and off. "Immediately, the UFO stopped, and then zig-zagged in response."* As they started to drive forward, again the object seemed to follow them, so Costa stopped a second time and flashed his headlights. *The UFO allegedly stopped again and "wavered in reply."* Costa drove forward a third time and the object flew above them for about thirty minutes. During this time, the UFO came closer to the ground (estimated 50 to 100 yards) and a second object appeared as well. The object looked like a disc with a dome on its top. Costa drove for another ninety minutes with both UFOs *hovering above the car.*

129

Comments: It isn't known how this event ended and, as is relatively common, we cannot be certain that the responses of the UFO were in direct reply to the car headlights, although it would appear so.

Reference: Blundell, N., and R. Boar. 1989. *The World's Greatest UFO Mysteries.* London: Octopus Books, 140.

Cross references: 23 (3); 53 (3); 68 (3); 73 (3)

Case No. 68

Date: 82-02-mid
Local Time: 2100
Duration: estimated 4 minutes
Location: Bakersfield, Vermont
No. Witnesses: 1
DB: v   CB: a   PB-1: a   PB-2: x   UFOD: 0

Abstract: The woman who witnessed this event was about 1.5 miles from Bakersfield traveling by herself on Highway 36. Mrs. Aubre Brogden noticed a large white light moving slowly across the night sky. "At first I thought it was a plane about to land, so I flashed my car headlights to warn it off the road. *And then it started coming towards me*," she said.

Just before she lost sight of the craft, she noticed that it was triangular. As she drove into her gravel driveway, she was surprised to see the object hovering only twenty-five feet above her own backyard.[1] She parked, got out with two bags of groceries, and ran toward her house. "It was obviously watching me," she recalled. During her dash to the house she dropped both bags. She glanced down at them and then up at the object that had moved directly overhead without a sound. It was now as big as a football field. She ran inside and locked her door. The UFO flew away into the dark sky.

Note: 1. Like other similar cases, the UFO seems to know the destination of the human in advance.

References: *The Free Press (Burlington, Vt.).* 22 February 1983.

Gribble, R. 1992. UFOs over space and time. *MUFON UFO Journal.* No. 286 (February): 17.

Cross references: 35 (3); 45 (3); 53 (3); 73 (3); 74 (3); 78 (3); 84 (3); 88 (3)

Case No. 69

Date: 83-03-?
Local Time: night
Duration: 5+ minutes
Location: North Salem, New York
No. Witnesses: 2
DB: v   CB: d   PB-1: a   PB-2: x   UFOD: O

Abstract: The chief of the New Fairfield, Connecticut, fire department, David Athens, and his girlfriend saw a circle of separate lights in the dark sky about five miles southwest of Danbury. "I thought it was a jet far off in the distance with running lights,…but as the object got closer, I realized it had more than just two lights. There were four or five lights. As it got closer, even more lights became visible.

"I got out my flashlight and flashed it on and off. Then, it flashed at me. Then, it started to make a circle right in front of us. This went on for five minutes. I flashed my light three times and it would flash its lights three times. It was close enough that I could see it had eight lights and small red lights underneath. When it shut the big ones off there was just a circular pattern of red lights. The object then gained speed and passed out of sight beyond the trees."

Comments: We note the familiar "in-kind" response by the UFO. This is a logical response by someone who may want to communicate that the signal—e.g., a flashlight flash—has been received without indicating anything more.

Reference: Hynek, J.A., P. J. Imbrogno, and Bob Pratt. 1987. *Night Siege: The Hudson Valley UFO Sightings.* New York: Ballantine Books, 96.

Cross references: 11 (3); 34 (3); 38 (3); 75 (3)

Case No. 70

Date: 83-07-12
Local Time: 2130
Duration: estimated 5+ minutes
Location: Bethel, Connecticut
No. Witnesses: approximately 11
DB: q   CB: a   PB-1: a   PB-2: x   UFOD: b

Abstract: An officer of the Bethel police department answered a call at a location southeast of Danbury where he saw a number of people standing outside looking up into the sky at a circular pattern of lights flashing red, blue, and green. The silent object was estimated to be at a height of no more than five hundred feet and about three hundred feet in diameter. No stars could be seen inside the ring of lights, suggesting the presence of a dark opaque mass on which the lights were attached.

The officer turned his powerful spotlight on the aerial object. Suddenly, the object projected a brilliant flash of white light downward, covering everyone present. The object then began to accelerate toward the north and was lost behind nearby trees.

Comments: Here is another intriguing instance of an "in-kind" response by the UFO—i.e., a light-for-a-light. If the policeman's spotlight was hand-held, it is likely that his inability to aim it steadily at the UFO probably produced a brief light flash as "seen" by the object. This flash may have been merely repeated by the UFO. We don't know whether anyone felt unusual sensations from this light flash, as occurred in the Spring 1951 case near Chorwon, Korea, nor whether any of the witnesses became ill later. It is not likely that the flash seen from the UFO was from a surface reflection of the officer's spotlight.

Reference: Hynek, J.A., P. J. Imbrogno, and Bob Pratt. 1987. *Night Siege: The Hudson Valley UFO Sightings.* New York: Ballantine Books, 96.

Cross references: 12 (3); 76 (3); 78 (3); 83 (3); 149 (4)

Case No. 71

Date: 83-07-12
Local Time: 2255
Duration: estimated 4+ minutes
Location: southeast of Danbury, Connecticut
No. Witnesses: 4
DB: v   CB: e   PB-1: a   PB-2: x   UFOD: b

Abstract: This event took place in a small fishing boat with Chief Nelson Macedo, his brother-in-law Charles Yacuzzi, Yacuzzi's 15-year-old son Michael, and retired police officer Jim Lucksy on-board. They were in a cove called "Old Never Sink." As they talked, Michael yelled, "What's that behind us?"

Everyone turned and looked in the new direction. Macedo was quoted as saying, "We observed this very large, dull gray, circular object. Its altitude was high and very difficult to judge, even with a very clear night, a large harvest moon, and a quiet lake."

"It had twenty to thirty lights," Lucksy claimed. "Here's this thing in the air, all kinds of lights—blue, red, orange, green—spaced together. Bright. They looked like they were turning in a circular motion. While it was turning, the object was still motionless. It wasn't moving from that spot."

Macedo also pointed out that, "The object was noiseless. *Charley shut off the boats (sic) lights, and the object also shut its lights off, stopped, and hovered. It was very quiet for a few minutes. Charles turned the lights on and off, and the objects (sic) lights reacted again. The lights became brighter and brighter, almost a yellow.*

The object then moved behind the mountains and disappeared. My nephew Michael was so frightened that he almost jumped overboard."

Comments: This multiple-witness event details all the usual features of modern UFO sightings, i.e., a large, silent, hovering object with apparent mass and numerous colored lights. These lights appeared to be revolving in a circular pattern without corresponding object motion, i.e., the lights were probably fixed rigidly to the three-dimensional object but turned on and off in a temporal sequence which produces the illusion of motion resembling a theater marquee. It is interesting to note that the UFO's reaction to the boat's lights being turned on and off was in-kind, except that, at one point, its lights became increasingly bright—turning to a kind of yellow

Reference: Hynek, J.A., P. J. Imbrogno, and Bob Pratt. 1987. *Night Siege: The Hudson Valley UFO Sightings.* New York: Ballantine Books.

Case No. 72

Date: 91-04-14
Local Time: 2230
Duration: estimated 5 minutes
Location: Durant, Oklahoma
No. Witnesses: 2
DB: o   CB: e   PB-1: a   PB-2: x   UFOD: O

Abstract: This wonderful account clearly illustrates the burst of natural human emotions that often occurs during a close encounter with the unknown. A young mother was rocking her baby to sleep one night when she noticed an unusual red light through the curtained window. The baby's grandmother, also in the room, did not want to look at the light and closed the blinds. When the younger woman went outside to see it better, she noticed two objects in the sky. One was flying rapidly back and forth "along the southern sky" near their home. This "fireball" seemed small (about 18 inches in diameter). It then suddenly "swooshed" away to the northwest, disappearing from sight high in the dark sky.

The other object was huge, silver hued, and moved deliberately above a nearby field and stopped. It had a "nipple" on top along with an antenna and red light. A row of rectangular or square windows were situated alongside the nipple. A line of blue lights extended across its middle section and another line of red lights at the flat bottom. Also on the bottom was a "segment" that "protruded below the main hull." The woman saw this protrusion open and a white light flashed downward to the ground, almost touching her. The beam remained on continuously and tapered in diameter at the ground.

She was frightened, extremely happy, and crying. *She began jumping and, for some reason, jumped two jumps to the left. The object changed and it, too, seemed to respond by jerking two "steps" to her left. She thought this was wonderful so she jumped two steps to the right and the object responded by moving to the right. This was repeated several times.* Then the beam retracted into the object without varying in intensity.

Comments: This classic account involves a larger hovering object and a smaller self-luminous source which can execute very high velocities and high-g turns. The literature is replete with similar reports. What the function of smaller objects is remains to be seen, but they may act as remote sensor platforms for defense, scientific data collection, simple surveillance, or other functions. Once again, we see an almost simultaneous and broad array of human emotions displayed in the presence of the UFO phenomenon. Is this caused by the UFO in some way or an independent response?

Reference: Coyne, G. 1992. Current Case Log, *MUFON UFO Journal* 291 (July): 19.

Case No. 73

Date: 92-?-?
Local Time: night
Duration: estimated 2+ hours
Location: Florida
No. Witnesses: approximately 4
DB: bb   CB: a   PB-1: a   PB-2: x   UFOD: 0

Abstract: Four male witnesses were in an automobile on their way to hunt. It was a dark, clear night and the men were joking about the subject of UFOs.[1] For quite a while, the driver saw a bright white light in the distance which appeared to be sitting on top of a radio tower. He said, "Hey guys! See that light over there on top of the tower. I'll put on my emergency flashers and call it over here."

"At that point he put on his flashers and, to everyone's amazement, especially light started to...approach them and passed directly over them. It was, reported, the shape of a gigantic flying wing, grey in color and appeared to the mass[2] of ten to twenty Boeing 747s.[3] It passed directly over their heads about 500 feet in the air in total silence. Its light appeared to be from deep inside the craft, and he described it as shrouded. After the craft passed over them and was increasingly distant, the light appeared brighter and brighter. When it was about 10 miles away, the lights became very active and began to explode in blasts of beautiful bright white light. The light remained in the sky the whole night. As the sun rose and grew stronger, it began to look like an ordinary star."

Comments: This incident has all the bizarre elements that characterize a strange UFO sighting. First note the approach response made by the UFO to the human's behavior. Its seemingly gigantic size and subsequent light show also are not unfamiliar. Since the lights appeared to become more intense with increasing distance, it suggests a significant increase in radiated power by the UFO rather than a decrease or constant power output over this period of time.

The reference to "blasts of beautiful bright white light" is also quite familiar. Many minutes of video and motion picture film have captured very intense, blue-white, flickering lights suspended in the dark sky.

135

Notes. 1. The driver was known to have had a previous UFO sighting with a friend. 2. His use of the term "mass" is interesting since it is a precise, technical word that is usually poorly understood by the general public. Its use suggests that he is not particularly educated in the sciences and that he tried (but failed) to be scientifically credible. Indeed, the mass of a standard model B747 is about 750,000 lbs. Most people might use the word "size" or "length." See Figure 24 of America's B-2 stealth bomber. 3. A standard model Boeing 747 airplane is 231 feet 4 inches long with a 195' 8" wingspan. Thus, a conservative estimate of the largest dimension of the UFO would be about 2,310 feet long.

Reference: *The CESETI Newsletter* 1, no. 2 (April) 1992: 4.

Cross references: 35 (3); 45 (3); 67 (3); 68 (3); 78 (3); 84 (3); 88 (3)

Case No. 74

Date: 92-02-11
Local Time: 0045
Duration: estimated 2+ minutes
Location: Henri-Chapelle, Belgium

No. Witnesses: 4
DB: w   CB: b   PB-1: x   PB-2: b   UFOD: 0

Abstract: Four members of the Center for the Study of Extraterrestrial Intelligence (CSETI) field investigative team from America traveled to Belgium in early February 1992 to see if they could provoke a close encounter event or even a landing by a UFO. Just after midnight on February 11, the team found their way to a high ridge near the village of Henri-Chapelle. Using their specially devised contact protocol—projected thoughts, high intensity spotlight, special noises— *"they successfully vectored into the site a large triangular, silent craft which dipped below the cloud cover to reveal a brilliantly lit apex of the triangle.* The light on this part of the craft was about the size of a full moon. A few minutes later, the team heard a low vibrating rumbling directly above them that seemed to be turned on and off in rapid succession twice." This vibration sound seemed stationary within the clouds above the team.

Comments: In this case is found another claimed success to deliberately coax a UFO to approach humans on the ground using various means. The large triangular object with a thirty minute arc diameter light at its apex is interesting in light of the well-documented occurrence of a similarly shaped silent craft which was witnessed by thousands of Belgians on the ground and in the air during a period of weeks following November 29, 1989 (Anon. 1991).

136

References: Anon. 1991. *Vague d' OVNI sur la Belgique: Un Dossier Exceptionnel.* SOBEPS.

CSETI. 1992. Asheville, NC: Center for the Study of Extraterrestrial Intelligence, 33-36.

Cross references: 35 (3); 45 (3); 67 (3); 68 (3); 78 (3); 84 (3); 88 (3)

Case No. 75

Date: 92-03-14
Local Time: 2024
Duration: 10-12 minutes
Location: Santa Rosa Island, near Gulf Breeze, Florida
No. Witnesses: approximately 50
DB: cc   CB: e   PB-1: x   PB-2: b   UFOD: 0

Abstract: This interesting and complex event occurred during a Center for the Study of Extraterrestrial Intelligence (CSETI) training workshop in Gulf Breeze, Florida, a resort town that had experienced scores of previous UFO sightings seen by hundreds of witnesses and captured on film and video. At about 7:30 P.M., a group of thirty-nine people were present. "The sense of unity, peace and excitement in the group was very high and the atmosphere was filled with confidence that a significant event was about to unfold" (Greer 1992, 26).

The group used their coherent-thought sequencing (CTS) protocol to try to contact the "visitors." CTS involves focused mental imagery and "projecting" thoughts of peace, tranquillity, and welcome out into space. Soon afterward, witness Dr. Steven Greer and several others "sensed a definite consciousness 'lock-on' between our group and several E.T.s." Several people in the group had remotely viewed this same fact and Greer told witness Vicki Lyons that he felt "several—four or five—craft were en route or already in the area." Then, high candlepower spotlights were used to convey a signal, as seen from the air,[1] that there was intelligent activity at the group's location near the beach. Specifically, several angularly large triangular formations were "painted" in the sky using the light beams (see Figure 21), the corners of each triangle established by aiming the spotlight there for some seconds.[2]

At 8:24 P.M., after having repeated these projected light-beam triangles from twenty to thirty minutes, a squadron of five UFOs were seen in the west-northwest at 26 degrees above the horizon.[3] The UFO "appeared with a bright circle of white light and as they hovered this quickly evolved into a pulsating cherry red-orange energy/light source."[4] What follows is taken directly from the CSETI report of this event:

Immediately after regaining a group focus, Dr. Greer began signaling to the UFOs with a 500,000-candle power light in intelligent sequences. To everyone's astonishment, *when he flashed three times to the UFOs, the lead craft flashed back three times. Then after two flashes, it returned a signal of two flashes, then five, and so on.* There was clear and definite congruence of the human-initiated signaling and the return signals of the craft, a sort of "photon dialogue." After several minutes of this activity, the UFOs "winked out," but remained in the area. At this point, Dr. Greer began "drawing" a large equilateral triangle in the northwest sky with the light and *then three of the UFOs visibly returned and formed, in clear response, an equilateral triangle....The "light conversation" continued with a series of flashes, again with direct responses from the craft.* At the same time, the group was continuing coherent thought sequencing and directing the occupants to approach and, if possible, land on the beach behind us. Information concerning the peaceful intentions and motivations of the working group were similarly conveyed.

At this point, *the formation began to move directly towards the group* on the beach. Dr. Greer then took the light and flashed it in a continuous strobe-like manner, directing the formation of craft to the beach location. As he did this, *the lead craft rapidly responded to the signaling and began flashing in a similar strobe-like manner, and moved directly toward the group and into the zenith of the sky. It then flashed a bright ring of white light, and "winked out" as did the other craft.* The team sensed that these UFOs remained in the area without their lights on for some time after this, but no further signaling occurred" (Ibid., 27).

Another summary of this extraordinary event published by CSETI said, "this event was observed from six locations, and videotapes and photographic evidence were obtained. These objects were under intelligent control and moved towards the group while exchanging light signaling with the team."

138 **Comments:** Of course, extraordinary claims require particularly sound evidence before being accepted, particularly those claims which challenge the fundamental tenants of established science. So, is there anything that occurred on this beach that challenges current science? Other than the clear and documented presence of multiple point sources of white light in a geometric formation, which has been clearly documented hundreds of times before (Haines 1994; Hall 1964), my answer is simply no. Where can one find in any branch of established science any law, precept, theorem, or other verified description of how nature works which directly contradicts or makes impossible anything that took place on this particular night? Herein is a challenge to physical scientists to come forward with such evidence. The official account of this event (Greer 1992, 25) points out that while traveling to Florida on March 13, Greer had "subjectively received information indicating that a sighting of the Gulf Breeze UFO would occur at midnight on March 13. Were the subsequent events coincidental, fulfilled prophecy, or did he somehow manufacture this claim following the event?" There is no objective way to settle this important issue. However, knowing Greer personally, I can attest to his honesty. Other psychic and unexplainable events during this event included claimed remote viewing and various premonitions.

Notes: 1. A total of five video cameras were used at different times during this night's activities. Two separate video tapes show four of the five aerial lights and show some of the pulsed "signaling" by the UFO and also three white point sources which appear to travel to the corners of a triangle in the sky, seen simultaneously within the field of view of a camera. I have seen one of the video-tape recordings obtained during part of this event. It shows four white point sources against a dark sky background. Three are seen hovering at the corners of an non-equilateral triangle. Later, a fourth light appears to join them to form an irregular square. No video or other photographic evidence was been seen by the author of this event. 2. As Part A of Figure 21 illustrates, an equilateral triangle (shaded - X) may be outlined by the three temporarily stationary beams as viewed from the ground. However, as Part B illustrates, the only thing which will be seen from a point high in the sky will be three transitory linear rays of light, and, if the atmosphere is clear each beam will: (a) be almost invisible and (b) extend beyond the height of the UFO to an even higher altitude. An infinite number of triangular shapes will be produced by these three rays depending upon the inclination of the plane which intersects them all—i.e., there will be one and only one equilateral triangle formed as seen from the ground. Only if a separate UFO moves to and then stops within each stationary ray (and is illuminated by the ray) can it be said that they responded by a corresponding geometry as that projected from the ground. 3. It isn't clear how this particular angular value was arrived at. 4. Although many people had reported seeing UFOs in this area over the previous years there had never been five seen at the same time.

Figure 21

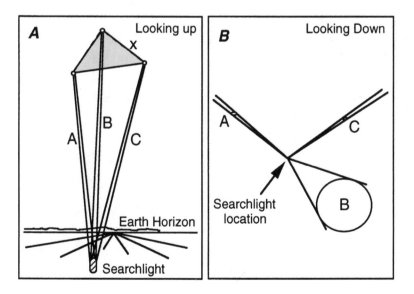

Geometry of single light beams as seen from ground (A)
and high in the sky (B) on March 14, 1992, near Gulf Breeze, Florida

References: Greer, S.M. 1992. CSETI. Privately published, 25-31.

Haines. R.F. 1994. *Project Delta: A Study of Multiple UFO.* Los Altos, Calif.: LDA Press.

Hall, R. 1964. *The UFO Evidence.* Washington, D.C.: NICAP.

Cross references: 11 (3); 34 (3); 38 (3); 69 (3); 75 (3)

Case No. 76

Date: 92-07-26
Local Time: 0110
Duration: estimated 2+ minutes
Location: Woodborough Hill, Alton Barnes, England
No. Witnesses: 5+
DB: kk   CB: b   PB-1: b   PB-2: x   UFOD: O

Abstract: This incident was witnessed by at least five American and British citizens who were taking part in field studies where crop circles were found. The Center for the Study of Extraterrestrial Intelligence (CSETI) affiliated group was standing together on the night of July 25 inside a circle in a grain field. At approximately 11:00 P.M., they sighted an intense yet perfectly silent "amber-orange colored object" traveling from north to south. Later, at 1:10 A.M., they shown a number of very intense spotlights vertically upward into a cloud overcast and were surprised to see from above the clouds, *"a bright light was shining down from the sky into the cloud bank, mimicking our light formation....*Because of the cloud cover which had moved in at this time, we could not see any specific UFO, however, the beam of light shining from the sky down into the clouds was striking and could not be explained" (Greer 1992, 12).

Other people were in the immediate area that same night on the tops of Adam's Grave, Milk Hill, Nap Hill, and Goldenball Hill with their own high-powered lights, "and were flashing them down onto our group trying to detect our location."

Comments: It is interesting that the eye witnesses on the ground saw separate areas of illumination moving around relative to the areas of illumination they themselves were producing. Nevertheless, the possibility is high that the other people nearby them on the ground could have aimed their spotlights upward upon the bottom of the cloud layer? This possibility must be addressed before accepting the more difficult possibility that the lights originated from above the cloud layer.

Reference: Greer, S.M. 1992. *Close Encounters of the 5th Kind: Contact in Southern England.* Privately published.

Cross references: 71 (3); 87 (3)

Case No. 77

Date: 92-07-27
Local Time: 0020
Duration: 10-15 minutes
Location: Alton Barnes, Wilshire, England
No. Witnesses: 5
DB: kk   CB: h   PB-1: x   PB-2: b   UFOD: 0

Abstract: This CE-5 event involved four members of a CSETI field investigation team who had traveled to England specifically to research crop circles and to attract a UFO to their location. While waiting in a field near Alton Barnes, south-west of London, the CSETI group had previously used coherent-thought sequencing, a high-intensity spotlight, and specially selected auditory sounds to try to vector a UFO to their location. But on this night it was raining so hard that team members were in cars waiting for the downpour to stop.[1] Suddenly, Chris Mansel alerted the three others present to the approach of "a large, brilliantly lit, disc-shaped, domed craft, measuring eighty to 150 feet in diameter." It had numerous intense blue-green, red, amber, and white lights that rotated counter-clockwise along its base, seeming to "blend into each other;" there also were three or four amber lights at its apex. The craft was silent, and less than 400 yards away, and came to within 10 to 30 feet off the ground.[2] It was in fact, in the same wheat field in which CSETI's team was located. Using an intense hand-held spotlight, the group *"signaled to it, two bright flashes and pause, and it then flashed back to us in the same sequence. This sequence was repeated again with a similar response from the spaceship."* A small amber object at one point detached from the upper right-hand side of the object and went up into the clouds. While at its closest proximity, this craft caused magnetic disturbances to the compass so that the needle rotated to a new "north" each time it was inspected. This near-landing event

141

concluded with the craft receding into the mist and out of visible range after ten to fifteen minutes of signaling with the team." The other synchronistic events that took place will not be recounted here.

Comments: It isn't known whether the compass needle rotated continuously or only once. A micro-cassette voice recorder continued to function normally during this event. Everyone present also felt an electrical charge—a tingling sensation—during the close encounter.

> Notes: 1. Greer and a British couple nearby (Peter Davenport and Judy Young) felt a compulsion to remain at the site, "sensing that something significant might happen." 2. The object's angular width was about 1.5 inches at arm's length. An object this size at 400 yards distance would be about 95 feet across.

Reference: Greer, S.M. 1992. *Close Encounters of the 5th Kind: Contact in Southern England. July 1992: An Interpretive Report.* Asheville, N.C.: CSETI, November: 15-17.

Cross references: 135 (4)

142 Case No. 78

Date: 93-01-30 to 93-02-04
Local Time: 0030
Duration: 10+ minutes
Location: approximately 55 miles southeast of Mexico City
No. Witnesses: 5
DB: kk   CB: c   PB-1: x   PB-2: b   UFOD: 0

Abstract: Five members of a field investigative team[1] from the Center for the Study of Extraterrestrial Intelligence (CSETI) traveled to the town of Amecameca west of Mt. Popocatapetl, an active volcano forty-five miles southeast of Mexico City.[2] On the night of January 31, one of the team members, Jeff Baker, left the others at the observation site to go back to the parked cars for equipment. Dr. Barbie Taylor, a witness, later wrote that the other four persons were "briefly engulfed in a beam of amber light, which originated in the northwest sky. No conventional source for this one-to-two-second-duration beam was found." Baker said that the beam came from somewhere behind him.

One of the team members present, Dr. Joseph Burkes, wrote to me about this incident in a letter dated December 17, 1996, "Dr. [Steven] Greer [a witness] was suddenly hit by a powerful amber-colored beam of light that appeared to come from the sky, although we could see no identifiable source. We were very surprised by this flash of light and no ill effects were felt." [3]

On the night of February 1, the team found an observing site just north of Metepec (see Figure 22). They used several means to try to signal a UFO: coherent-thought sequencing, high-intensity spotlight flashes, and moderately loud prerecorded noises to attempt to encourage a UFO to approach them. Greer provided the following description of what took place.[4] "Then, at 11:45 P.M., a CE-5 of profound significance occurred. The entire group was performing CTS,[5] and I was in a state aware only of 'one mind.' Suddenly, I sensed—knew—to sit up and look to my right, and there it was: a large amber craft moving obliquely away from us in the northwest sky." Burkes' recollection was that "Greer directed our attention to a light that was moving obliquely near our position." Taylor recalls, "It was 11:45 P.M. There was a large amber light moving low on the horizon. It had come from the northwest and was passing through the valley between the two volcanoes, Mt. Popo on the south and Mt. Izzy just north of it....[We] flashed [our] halogen light at it, and it immediately turned toward us and began approaching." (See dashed line, Figure 22) The small "x" marks the group's location.

143

Figure 22

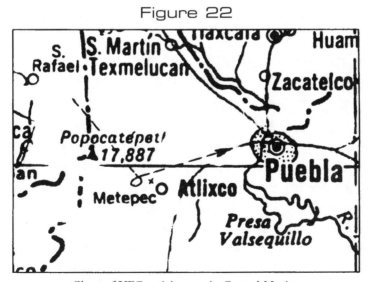

Chart of UFO activity area in Central Mexico

Greer wrote, "My immediate sense was that it was looking for us, that it was over the area of the pyramid-shaped mountain and the field of the eagle cloud. I immediately notified the team saying: 'This is it—this is the real thing!' With both CTS and powerful lights, we signaled for the spacecraft to come over to our location. *Immediately, it turned off it's present course, and moved directly towards us.*

"Initially, the spacecraft was two to five miles away, but as it silently glided towards us it came within 1500 to 3000 feet, and only 200 to 500 feet in the air![6] We immediately realized the historic significance of this event: a team of humans had consciously and intentionally vectored a spacecraft into a research area. *It was clearly responding by changing course and coming directly to us,* now in an apparent landing approach, or at least landing approach simulation."

Burkes' recollection of this event was that "It was totally silent, highly maneuverable and came towards us in a wide arc. By now the exact shape of the extraterrestrial spacecraft could be seen: it was a huge 300 to 900 foot diameter,[7] triangular-shaped structured craft, with a white light on the underside at each apex and a reddish glowing light in the very center on the underside....We noted that even though the wind was coming from it's direction, no sound was heard whatsoever!

"Jeff Baker moved the video camera to the adjacent field, and set out our portable-landing strobe lights—indicating to the E.T.s that we welcomed a landing, should this be safe and feasible for them....The rest of the team prepared for boarding and I sent signals with the high-powered light. *The spacecraft made a sweeping arc, approximately 180 degrees, and then came straight towards us in an apparent landing approach. As I signaled to it, it returned by illuminating large and powerful lights on the now 'front' Antero-grade of the triangular craft.*

"It descended to 200 to 300 feet, coming straight towards us, and as it did so, *it greatly increased the luminosity of the front lights, as if signaling to land.*[8] Greer and [witness Shari] Adamiak could also see a small 'scout ship,' red-orange in color, floating alongside and behind the large triangle....As Jeff Baker activated the high-powered camcorder, he noticed that no image was coming through. The camera, which had worked perfectly before the event, was now non-operational. Even his hand and flashlight in front of the camera lens were not visible in the viewfinder! Moreover, the still cameras (Canon EOS system with 70-210 lens), would not operate, nor would Shari Adamiak's little spring-powered Instamatic! It was clear to us that some interference prohibited the filming of this very close-range event. Was this intentional jamming of the equipment or an inadvertent electro-magnetic effect of the ship itself?

"As all this was happening, we suddenly realized that all our cameras...were not operational....It appeared as if someone or something did not want us to document this encounter with any type of photographs. We checked out all the equipment later, and it worked normally for the remainder of the trip."

Dr. Taylor's account is equivalent in every respect: *"At about the time [Baker] tried to film the spacecraft, it aborted its landing approach, turned to our left (east-northeast), and began moving away from the site...*the receding spacecraft illuminated its retrograde lights and clearly, unequivocally signaled with us as we flashed our lights to it....In one to three minutes, it dipped below a nearby ridge, and was not seen again at close range that night. The entire event lasted about ten minutes....After the spacecraft left, we were astonished, perhaps dismayed, to see that the powerful Canon LI Pro Camcorder was fully operational again!"

At its closest approach, the "ship tipped its nose up, and I saw the underside, it turned away from us and leveled out again," Taylor said. [We] flashed it again two more times (maybe three), and *it once again flashed back as it left—the same number of times as [our] flash....*The size of the object could be determined in comparison with this little church [near us on a ridge]. We estimated it to be about the size of a 747."

The remainder of this report will not be included here since it deals with other aspects of the group's experiences that night and their interpretations of what took place.

Shortly after midnight on February 2, "the same type craft returned, this time under 200 feet [altitude]," and again *signaled in an interactive manner with the team.* Burkes remembers it appearing "over the ridge on the horizon. As Greer started to reach for his light to signal the craft, it rapidly pulsed two flashes of light at us and then quickly turned to the north and descended behind a small hill and was out of view. It seemed to be more than a coincidence. It was as if it were looking for us."[9] During the last twenty-four hours the team was in Mexico, they encountered brilliant, shiny metallic discs during daylight hours which hovered or flew near the team members. Four such discs on four separate occasions were encountered by the team on February 3 and 4, 1993.

Comments: I have interviewed four of the five eye witnesses to this remarkable series of events and have found a high degree of consistency from account to account, a high degree of emotional response remaining as the psychological aftermath of what took place, and a reasonably detailed memory for what took place these nights. Minor discrepancies discovered in some details, as would be expected, add to the overall reliability of the reports. Nevertheless, it is possible that some amount of post hoc homogenization of facts occurred to arrive at a single narrative.

If this narrative is reliable, it, once again, raises fundamental questions which traditional science must confront immediately and with integrity. These questions are clear to anyone who is willing to think seriously about the implications of such contacts taking place with increasing frequency to the world's citizens.

Notes. 1. Drs. Barbie J. Taylor, Joseph Burkes, Steven Greer, Miss Shari Adamiak, and Mr. Jeff Baker. 2. Elevation is 17,883 feet. The most recent volcanic activity started in 1994 and continues into 1998. 3. Baker had gone back down the hill to the automobiles for something and did not see this light flash which came from behind him. Taylor wrote in a letter dated January 25, 1998, that this beam hit both Greer and herself "full in the face....We felt as though we had been 'scanned' by the E.T.s—that they knew we were there and just checked us out." An almost identical light flash involving Burkes occurred on September 5, 1993. 4. Burkes stated that local authorities were aware of UFO sightings in the area and that 80 percent of the local population had experienced a CE-1. 5. Coherent-thought sequencing is a behavioral procedure where everyone present follows the verbal guidance of a leader to make their thoughts and visual imagery more nearly coincident in content and "direction." The concept of focused light is somewhat apt here. By attempting to reduce cognitive "scatter" (dissonance, image intrusions, emotion-based thoughts, etc.), CTS is thought to project a more consistent or common idea into space than would otherwise occur. Such a projected idea may be received and responded to by extraterrestrial minds. 6. Taylor wrote, "We later estimated that it came within a mile of us by looking at topographical maps the next day." 7. Burkes estimated its size to be about 300 feet across. 8. Burkes stated, "The craft descended to less than 500 feet and turned on an incredibly powerful bank of forward-mounted lights." 9. If the light beam actually illuminated the CSETI group for any length of time (and not simply passing across them) it already "knew" where they were. The group's infrared "signature" would have been easy to detect given the surrounding background terrain's much colder signature.

References: Burkes, Joseph dtd. 1996. Letter to author. 17 December.

Greer, S.M. 1993. *Close Encounters of the Fifth Kind in Mexico. CSETI Special Report,* 30 April.

Taylor, Barbie dtd. 1998. Letter to author. 25 January.
http://volcano.und.nodak.edu/vwdocs/current_volcs/popo/mar5popo.html

Cross references: 74 (3); 83 (3)

Case No. 79

Date: 93-02-26
Local Time: 1930
Duration: 2+ minutes
Location: Louisville, Kentucky
No. Witnesses: 4
DB: p   CB: d   PB-1: a   PB-2: x   UFOD: b

Abstract: Two Jefferson County police officers in a police helicopter were on their way to investigate a possible burglary in progress near Appliance Park, General Electric's industrial center. Visibility was from five to six miles with scattered, broken cloud cover and the air temperature was about 20 degrees fahrenheit. Pilot Kenneth Graham, 39, saw the yellow-orange ball of light first. It was very near the ground "in the trees." Officer Kenneth Downs, 39, turned on their 1.5-million-candle-power searchlight and aimed it directly at the basketball-sized light source. *"When we hit it with the light, it seemed to begin swaying back and forth....It rocked back and forth in about a six-foot arc...then, it gained altitude, quickly."* It rose at a 45 degree angle to about 1,500 to 2,000 feet altitude and then stopped climbing!

147

During this aerial close encounter the pilot recalls changing his heading at least three separate times to keep the UFO directly in front of them for safety reasons. *Each time he did so the UFO flew to a position directly behind the helicopter.* The worried pilot then accelerated forward to about 120 knots, "and the thing passed us up; got ahead of us and did it pretty quickly," Downs explained. The remainder of the bizarre episode will not be recounted here.

Comments: The local police department received seventy-five telephone calls from residents who had seen this glowing aerial object. Despite the subsequent revelation of a local couple who said they had launched an internally illuminated hot air balloon (propelled only local winds and suspended lift from birthday-candle heat) that same night and had watched the helicopter shine its search light at it, this explanation does not explain most of the object's flight dynamics: (a) rapid ascent and stopping at the same altitude as the helicopter, (b) rapid flight to a position behind the helicopter three different times, and (c) horizontal flight which bypassed the helicopter that was traveling at 120 knots. Why the two officers could not correctly identify it as a homemade balloon, although it was clearly illuminated in their very bright beam of light, is also unclear. Also, UFO investigators discovered that the couple had a rather questionable relationship with the law and moved from their apartment within days after the story appeared in the newspapers and television.

Reference: Rutherford, G.O. 1993. Kentucky Chopper Case. *MUFON UFO Journal* 302 (June): 9-11.

Cross references: 44 (3)

Case No. 80

Date: 93-04-16
Local Time: 2120
Duration: 15 minutes
Location: Bayport, Florida
No. Witnesses: 1+
DB: w CB: a PB-1: a PB-2: x UFOD: b

148

Abstract: During this pacing of a sheriff's patrol car by an object with several lights, the deputy stopped briefly at an intersection and shown his spotlight up in the direction of the formation that was crossing the roadway ahead of him. He saw an object with the outline of a boomerang whose surface was completely matte black; interestingly, one wing was twice the length of the other.[1] Its belly had five "soft blue lights" and its trailing edge had five white lights.

As the deputy started forward again, the object began moving with him *"slowing and accelerating to match his varying speeds."* When he turned his spotlight on a second time, he was surprised to see the object's *"sudden, incredibly fast departure west by southwest."*

Comments: The witness subsequently located others who had also seen an "undefined array of lights for about ten seconds." A second group of people some thirteen miles to the south-southwest also claimed to have seen the same shaped object, but with six lights on it. It made a faint "droning sound" as it traveled toward the west out over the Gulf of Mexico.

Note: 1. This wing configuration would be impossible to fly if the normal laws of aerodynamic lift were applied. The fact that it hovered suggests a different means of lift was involved.

Reference: Spencer, T.D. 1996. Current Cases. *MUFON UFO Journal* 343 (November): 20.

Case No. 81

Date: 93-04-30
Local Time: 0205
Duration: estimated 2 minutes
Location: Millville, Massachusetts
No. Witnesses: 1
DB: p   CB: e   PB-1: a   PB-2: x   UFOD: a

Abstract: A 30-year-old man was walking from his house to his car when he first noticed a formation of lights approaching his general location. They were all red and in a triangular pattern. He made out a dark cigar-shaped craft having "at least nine red, blue, and white lights on it." Some blinked on and off. "He thought to wave his arm as it passed some 3000 feet from him at an angle of 70 degrees in the air. *Immediately, a white light moved from the rear of the object to the middle and beamed briefly right into his eyes, as if directed at him in response to his gesture.*" The UFO then continued flying in a straight line above Millville at an estimated 200 mph until it was no longer visible.

Comments: If the UFO's response to his wave was immediate, it implies an extremely advanced technology that is preprogrammed to take a particular action rather than an action that is preceded by some cognitive decision making or thought. If the white light which appeared from the larger object was an intense point source, it would have radiated light in all directions at the same time and would have appeared to shine in everyone's eyes regardless of their location on the ground or in the air. This explanation is much simpler than assuming, without any supporting evidence, that the white light emitted a narrow beam of light which only struck the witness.

149

Reference: Current Cases. *MUFON UFO Journal* 307 (November 1993): 17.

Cross references: 30 (3)

Case No. 82

Date: 93-07-10
Local Time: night
Duration: estimated 3+ minutes
Location: Asheville, North Carolina

No. Witnesses: 8
DB: j   CB: a   PB-1: b   PB-2: x   UFOD: c

Abstract: Eight members of the Asheville Center for the Study of Extraterrestrial Intelligence (CSETI) working group met one evening to attempt to attract a UFO and encourage it to approach the group. "Over six high flying satellite-like lights were observed following different trajectories, one of which changed directions.

"One member commented out loud that it would be nice if the lights would do a low fly-by. *Within the next few minutes, a brilliant golden-white, V-shaped-structured craft streaked across the sky and blinked out before reaching the western horizon.* It took almost three seconds to traverse the sky, leaving a long trail of blue-white light. It was the lowest altitude that the group had ever witnessed a craft of any sort flying. If it had flown any lower, in fact, it would have had to dip into the site, which is in a bowl-shaped area surrounded by mountains. There was little doubt that this was a [UFO] craft, and the member felt strongly that this was an acknowledgment of their efforts."

Comments: The verbal statement that it would be nice if the lights would do a low-level fly-by was obviously heard by others present. Whether these spoken words were also somehow "heard" at the distance of the UFO is highly unlikely, if current laws of atmospheric physics apply. Could the thought processes behind the spoken words have been the means of communication? Little scientific research has been performed on this important possibility. This issue is discussed in chapter 5.

Altitude and distance estimates are notoriously prone to large errors, particularly at night in the absence of references to size and distance. Such is probably the case here. Another difficulty is the rather long duration which occurred between the spoken request and the subsequent UFO response, if that is what it was.

Reference: *The CSETI Newsletter* 2, no. 3 (December 1993): 8.

Cross references: 66 (3)

Case No. 83

Date: 93-09-05
Local Time: 0145
Duration: approximately 15 minutes

Location: Santa Susana Pass, Chattsworth, California
No. Witnesses: 6
DB: v   CB: i   PB-1: b   PB-2: x   UFOD: a

Abstract: The CE-5 aspect of this Center for the Study of Extraterrestrial Intelligence (CSETI) sky watch is the primary focus here, other details are found elsewhere (Burkes 1996). A group of twelve adults[1] had spent from 11:00 P.M. on Friday night to 1:00 A.M. Saturday morning using an intense, hand-held halogen spotlight, coherent-thought sequencing, and other CSETI techniques to try to signal to an intelligence felt to be above them in the slightly overcast sky. At 1:00 A.M., six people left the site to go home and at 1:30 A.M., the rest of the group packed up and left the viewing location since it was getting colder and the wind was increasing. Their site overlooked the San Fernando Valley to the south with the town of Simi Valley over the taller hills to the west; Chattsworth was about three hundred feet lower elevation also to the south. While leaving, they first had to climb a ridge line separating them from the Santa Susana Pass to the north, and then descend a steep trail to a point located about a twenty-minute hike from where their cars were parked. There, the group of six witnesses saw "two powerful lights" located on a high ridge farther to the north, a part of Rocky Peak State Park. They were on the side of a ridge just beyond the San Fernandino Valley Freeway. The time was 1:45 A.M.

151

Each light source was round and estimated by Burkes to be from one to two feet in diameter, about 100 feet apart, about 400 feet higher elevation, and about 1,700 feet away.[2] The left-hand light appeared to be at a slightly higher elevation than the other. A narrow beam of white light came from the right-hand source and shone out across the San Fernando Valley roughly toward the southeast, while the left-hand source was coming on and then slowly fading out. Burkes used his spotlight to send one or two flashes toward the lights and both "fired in what appeared to be a random fashion. I couldn't seem to easily get their attention," Dr. Joseph Burkes wrote. "We (CSETI) use a predetermined sequence of flashes. If the object signals back in exactly the same fashion, then we know some kind of primitive communication has been established. I was hoping to get those strange lights to mimic my pattern of flashes" (Burkes 1996). Later, Burkes wrote that the left-hand light, "seemed to go on and off without any interactive component....By the start of the following events the beam coming from the left-hand source was aimed directly at the group.

A series of three light bursts was then tried by Burkes: one short flash, three-second pause, two short flashes, three-second pause, and finally three short flashes. *Suddenly, the left-hand light "fired back with a single bright flash*. Then suddenly it went silent.....I again sent out the pattern of light bursts a second time....*The left light signaled back by exactly mimicking my complex pattern....Each time I completed a phrase of "photon talk" an exact strobe reply flashed back from the hillside."* The reply was almost instantaneous and followed each set of flashes from the hand-held spotlight. The exchange continued three or four times over several minutes. Finally, Burkes decided to cut off the exercise[3] and everyone continued on down the trail to their cars and went home, cold, excited, and tired.

Comments: There are several reasons why it is hard to believe that someone had set up a deliberate hoax. First is the rugged terrain and the practical difficulties involved in moving all needed equipment. Second is the total lack of electrical power at the site. Third is the fact that no signs of any kind were found that humans had been where the lights were seen. Fourth, only a small number of people knew where the sky watch group was going that night. Fifth, the nearly instantaneous reply to each group of Burkes' spotlight flashes almost defied human abilities. Indeed, Burkes wrote later, "Even with considerable practice, my co-workers could not reproduce the precision signaling which transpired."

Everyone present also saw several very brief "lightning" like flashes at about 11:30 P.M. that night. Most participants had their eyes shut and yet still noticed these light flashes. The entire area around the group lit up briefly; the light source seemed to come from above the horizon from the southeast. No sound of any kind was heard during or after the flashes. Since closed Caucasian[4] eyelids transmit about four percent of incident illumination, and assuming an almost fully dark-adapted retina, the light flashes would have had to produce only about 10 foot-candles (or more)[5] at the eyes which is not particularly difficult to achieve. The possibility exists that what was seen was clear air lightning or earthquake-related luminous discharges. Indeed, the area surrounding the sighting location has several earthquake surface fault lines (Anon. 1997a); six tremors were recorded between 11:00 P.M. and midnight on September 4, 1992, five between midnight and 1:00 A.M. on September 5, and ten between 1:00 and 2:00 A.M. More details on them are found in Anon. (1997b). It is also possible that whomever was controlling the intense light sources above the group had passed the beam over the assembled group unexpectedly.

Notes: 1. The group included two medical doctors, two professional psychologists and others described as "solid middle-class people" by Dr. Joseph Burkes, the group's coordinator. 2. A team returned to the location of the two unidentified lights the next day and found that there was no local road access at all. The trail and locked gate in the vicinity had signs indicating "DAY USE ONLY." The group even had to rock climb to reach the ledge where the lights had been seen. No signs of human occupancy were found at the site. 3. Burkes concern was for the safety of his team members (Burkes 1997). 4. All the witnesses were Caucasian. 5. One ft-c is the amount of illuminance falling on a surface measuring one square foot in an area all points of which are one foot away from one whale sperm-lit candle.

Reference: Anon., 1997a.

http://www.scec.gps.caltech.edu/lafault.html#MAH.

Anon., 1997b. http://scec.gps.caltech.../catalogs/SCSN/1992.cat.

Burkes, J. 1996. LA-CSETI Files 92-III *Mystery Lights in the Santa Susana Pass.* http://www.cseti.org/freports/la921.htm.

Burkes, J. 1997. Letter to author. 20 December.

Cross references: 78 (3)

Case No. 84

Date: 93-11-22
Local Time: 2145
Duration: approximately 5 minutes
Location: Udell, Pennsylvania (Westmoreland County)
No. Witnesses: 4
DB: v   CB: a   PB-1: a   PB-2: x   UFOD: O

Abstract: This event took place on a deserted Brinkertown county road near Youngwood, Pennsylvania, on a cold, dark, partly cloudy night. Four men were looking for deer using a spotlight from their truck. Their engine was off. As one of the men trained a spotlight over the top of the cab into a nearby field to his left, the beam struck a distant object. He got back inside the cab. At first he thought the object was a tractor and told his companions so. The driver then rolled his window down but the men heard nothing. Just then, the object began to *rise from the ground and approached them in complete silence.* The length of the object was estimated to be about 60 feet and had an overall boomerang shape (see Figure 23). There was a white light at each end and a single red light on its bottom.

The object traveled toward the west for approximately three miles and then turned a sharp right-hand turn toward the northwest in a "fluttering motion." It disappeared from sight behind distant trees at an altitude estimated to be less than a mile.

Figure 23

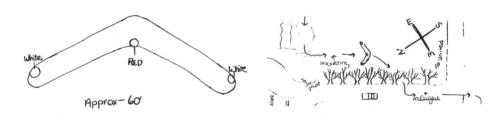

Eyewitness drawing of UFO seen on November 22, 1993, near Armbrust, Pennsylvania
(Courtesy of Stan Gordon)

154

Comments: No electromagnetic effects or physiological symptoms were reported. It is very unlikely that the object seen was the B-2 stealth bomber (see Figure 24) because of: (a) its vertical takeoff, (b) its complete silence, and (c) its small size. The wing span of the B-2 is 172 feet.

Reference: Anon. 1994. *The Anomalies Zone* 1, no. 1 (January-February): 3.

Cross references: 74 (3); 78 (3); 79 (3)

Figure 24

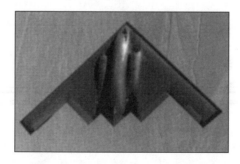

Photograph of the B-2 stealth bomber in flight as viewed from above

Case No. 85

Date: 93-12-01
Local Time: 1945
Duration: 15+ minutes
Location: Carp, Ontario, Canada
No. Witnesses: 2
DB: w   CB: d   PB-1: a   PB-2: x   UFOD: 0

Abstract: A mother and young son were returning home in her car after dark when they saw an intense light "like stadium lights" ahead of them hanging over the roadway. Moonlight reflected off the snow and the air was crystal clear. Intrigued at first, she drove past her driveway toward the UFO, realizing that the road ahead of her was a dead-end. Upon reaching the end, the lights disappeared behind trees to her left. She turned around and then noticed the same or similar lights several hundred yards directly ahead of her. Many trees lined this part of the road; *she drove on slowly and the UFO seemed to maintain a constant distance.* She thought it was oval in shape and had a strobing effect in its upper right side.

As they came to an opening in the nearby trees she "flashed her brights on and off. *The light started coming at us,* right down the middle of the road. That really frightened me." She remembered thinking at the time, "Oh God, I wasn't calling you." Then, she drove at high speed into her driveway, stopped, rolled her window down, and stuck her head out. The perfectly silent UFO was only forty feet up and a hundred yards away, hovering above a nearby house. It was half the size of the house.

155

Comments: During an interview with a UFO investigator the woman said, "I felt like it was locked onto my car, so I was controlling its movements. Whenever *I veered, it veered.* It stayed an exact distance ahead of me, traveling at the same speed. It only severed the connection when I panicked. This same type of comment has been made numerous times before, during Center for the Study of Extraterrestrial Intelligence (CSETI) field expeditions. Does human panic or fear somehow break the relationship that has been established?

Reference: Cameron, V. 1995. *Don't Tell Anyone, But...: UFO Experiences in Canada.* Burnstown, Ontario, Canada: General Store Publishing House, 128-32.

Case No. 86

Date: 94-02-27
Local Time: 0105Z
Duration: approximately 5 minutes
Location: northwest Ontario near Thunder Bay, Canada
No. Witnesses: 1
DB: v   CB: a   PB-1: a   PB-2: x   UFOD: b

Abstract: This night encounter took place at an altitude of about 7,500 feet when the witness was flying a Beech 18 twin engine airplane by himself on a magnetic heading of 73 degrees northeast. Approximately one hour after taking off, he noticed what appeared at first to be an aircraft one-half mile off his wing and below him. Soon, he realized that it wasn't an airplane but rather "something bowl shaped, brown and orange with darker spots in the center." Its top and bottom surfaces were convex and joined at a rim. It climbed to his level and then maintained a fixed position relative to his airplane. He banked away from it and, to his surprise, *it turned in the same direction to maintain a separation of about 300 yards.* Then he banked back to his original heading and "the object followed." The UFO slowed, "dived to the ground," and disappeared from sight.

156

Comment: The pilot reported the incident using CIRVIS procedures and said that he "will report if he sees it again but he doesn't want to see it again." He was shaken by the event. Visibility was fifteen miles with scattered clouds at 22,000 feet altitude. When contacted by investigator Mike McLarty, the pilot pointed out that he had two sightings on consecutive nights, one of which involved two luminous objects which appeared to pulse in response to each other and, that during one of the two sightings, he experienced "time distortion" lasting about fifteen minutes. It isn't clear what is meant by time distortion. Did he land at the correct time or otherwise have difficulty in processing flight information?

References: Unclassified report to National Research Council Ottawa, Herzberg Inst. of Astrophysics SD003, 1994.

CIRVIS (UFO) report received from the pilot by Thunder Bay Flight Service Station on February 27, 1994, 01:46:16.

Cross references: 3 (3); 6 (3); 9 (3); 18 (3)

Case No. 87

Date: 94-04-25
Local Time: approximately 2230
Duration: unknown
Location: 25 miles from Great Falls, Montana
No. Witnesses: several
DB: kk   CB: a   PB-1: a   PB-2: x   UFOD: 0

Abstract: A large, disc-shaped craft landed in a field some twenty-five miles from Great Falls. The field was owned by the eye witnesses. The object was "the size of a jeep, or import pick-up...."*It flashed red and blue lights and, when the witnesses tried flashing their headlamps at it, the craft responded in kind with a brilliant white light directed at the witnesses. It then flew straight up and disappeared."*

Comments: Here is another case involving a small unidentified aerial object which responds in-kind to the flashing of automobile headlights and then followed by the object's departure. Are the UFO lights but an artifact of the propulsion system, some deliberate signal prior to leaving, or something else?

Reference: *Bill Knell's UFO Newsletter.* 1994. Special summer issue, 1.

157

Cross references: 14 (3); 54 (3)

Case No. 88

Date: 94-08-01
Local Time: 0100
Duration: estimated 4 minutes
Location: Silbury Hill (Wiltshire area), southwest England
No. Witnesses: 3
DB: v   CB: a   PB-1: a   PB-2: x   UFOD: a

Abstract: According to an account by Adamiak and Russell (1994) describing Center for the Study of Extraterrestrial Intelligence (CSETI) field activities conducted in England in late July 1994, three people (Sonia, Paul, and Rob) were sitting together near the top of the northeast ridge of Silbury Hill looking out across the West Kennett Long Barrow fields. A thick fog had formed by 1:00 A.M., and visibility was low. Suddenly, they saw "two orangeish balls of light coming through the air from the Wadden Hill area to their left. The airborne lights floated smoothly and silently by the Hill."

The lights then changed their direction of flight, entered another wheat field, and seemed to become larger. Each light assumed the shape of a tetrahedron, "like small pyramids illuminated from within. They exclaimed to each other that this was better than [a movie by producer] Steven Spielberg. They could then see into the pyramids and discerned the shape of five or six beings. Sonia had the brilliant impulse to flash her torch (flashlight) at the tetrahedrons....*The craft immediately turned around to face the hill, flying swiftly towards the top.*" It came to a position within about fifty feet of the astonished witnesses. "*Some strong and sudden fear reactions arose. As if in response to these intense emotions, the ships moved back and down, going in to the same field from which they had come.*"

Reference: Adamiak, S., and R. Russell. 1994. *Energy Grids: The Koch and Kyborg Experiment with CSETI.* Unpublished paper, 1994.

Cross references: 35 (3); 74 (3); 78 (3); 84 (3)

158   Case No. 89

Date: 96-03-06
Local Time: 2300
Duration: approximately 90 minutes
Location: Friol (Lugo Province), Spain
No. Witnesses: 1+
DB: cc   CB: h   PB-1: a   PB-2: x   UFOD: a

Abstract: The owner of the Cantena de Retorta bar, Ms. Maria Se Pena Ceide, saw what looked like a globe in the night sky but angularly larger than the Moon. The object changed location over an hour's time. Just after midnight, Mr. Jose Immanuel Castro, a local farmer, also saw it shining through a window of his house. He went outside and observed a self-luminous sphere with the outline of three or four E.T.s in silhouette inside it. He quickly drove to his brother's house, but couldn't wake him. So he returned home; the sphere was still suspended over his house. Now quite curious, Castro flashed light signals at the object using a small flashlight and called out, "Baixade. Baixade." ("Come down. Come down!")[1] *Suddenly and almost immediately, the UFO descended, "plunging* *downward.*" Castro dashed into his house in fear and the ball of light landed some 100 feet away. Then, a luminous walkway somehow came out of the object and three small alien beings emerged. The witness was now "scared out of his wits" and quickly climbed under his bed where he remained the rest of the night, at least five hours. Later, he reported the incident to the Guardia Civil who arrived to make an inspection of his property. They found "strange footprints" in his field at the same location where he had seen the object land. The impressions were larger than a cow's hoof-print—i.e., at least six inches long.[2]

Notes: 1. A neighbor, Sera, heard him shouting but didn't look out. 2. Why the police report would refer to a cow's hoof as a size reference and not a human shoe or foot size isn't known. UFO investigator Manuel Carballal made plaster casts of the impressions.

Reference: Samizdat Special Report—Spain 1996: *The Galician UFO Wave of 1995-96,* April-May 1996.

Cross references: 88 (3)

Case No. 90

159

Date: 96-06-16
Local Time: 0205
Duration: estimated 5+ minutes
Location: Crestone, Colorado
No. Witnesses: 10
DB: v   CB: b   PB-1: a   PB-2: x   UFOD: a

Abstract: This event took place during a Center for the Study of Extraterrestrial Intelligence (CSETI) training weekend in the San Luis Valley of Colorado. As Sheri Adamiak (1996) stated in her report of this incident, "Instruction and practice in coherent-thought sequencing and remote viewing rounded out the afternoons, often outdoors under the crystalline, sapphire-blue skies. Each night was spent in the field, using the CSETI protocols of light, sound, and thought to invite extraterrestrial life forms to interact with us.

"About 1:40 A.M., Dr. [Steven] Greer dismissed the team; however, about ten of us remained behind. Greer had subjectively acquired information suggesting an event would occur about 2:00 A.M. Indeed, at 2:05 A.M., an orange-red elliptical pulsating object was seen moving towards Mt. Blanca. *Greer signaled to it with the high-powered light three times, whereupon the craft pulsed back three times. We sent thoughts of welcome, peace, and friendship to the craft. Immediately, the craft increased its illumination and pulsed brilliantly.* This was a definite CE-5-human-initiated communication—a photon dance, if you will.

"At 2:30 A.M., a craft glowing yellow-white with a small red light following closely behind was seen also moving towards Mt. Blanca. Seen through the night vision binoculars, Greer could discern a triangular shape to the craft. Moments later, a military jet was seen roaring towards *the craft which vanished behind the Sangre de Cristos mountains.*"

Comments: Events such as these call for detailed written reports completed independently by everyone present. Were the thoughts of welcome, peace, and friendship unanimous within the group? Were they transmitted simultaneously? Was any attempt made to obtain photographs or other records of the light source? Such evidence is absolutely necessary to convince the skeptically minded that a CE-5 took place.

Reference: Adamiak, S. 1996. Ambassadors to the Universe Training. Crestone, Colorado-June 1966. *CSETI Newsletter* 5, no. 2 (Autumn): 7-8.

Case No. 91

Date: 97-04-04
Local Time: 1938-2158
Duration: 1 hour 13 minutes
Location: north side of Lake Ontario, Canada
No. Witnesses: 3
DB: g   CB: a   PB-1: b   PB-2: x   UFOD: 0

Abstract: Three members of the Southern Ontario Center for the Study of Extraterrestrial Intelligence (CSETI) working group were continuing their sky watch sessions under cool and slightly hazy skies. Some stars were visible but the Hale-Bopp Comet could not be seen. They set up their camp at 7:30 P.M. and saw an angularly small orange light in the sky for about two minutes before it sank to the horizon. Seven minutes later, two orange lights appeared side by side at an eight degree arc elevation and then they both disappeared. These kinds of events continued until 8:35 P.M. when team leader J. Jarvis "surreptitiously" flashed a hand-held red laser pointer[1] several times at a "large orange light at 5 degrees above horizon. *Object rising up into the air at 8:36. Remaining stationary at same elevation around 10 degrees above horizon."* She repeated these brief laser flashes at 8:38 P.M. since the light source was becoming very bright. Witnesses watched for a response and, Jarvis wrote, "this could be coincidence, but object's light intensity increased considerably!" Again at 8:40, 8:57, 9:00, and 9:09, she aimed the laser pointer at the bright orange lights and several times noted an increase in intensity or [they] appeared to move apart. On the other occasions nothing unusual was seen to occur. When the two orange objects were angularly near each other, they seemed to move independently while pulsating in brightness. The group departed at 10:00 P.M. with the orange lights apparently still visible.

Comments: Given the extremely small diameter of a laser pointer, the apparently great distances to the luminous sources, normal atmospheric light absorption and scatter, and hand tremor while aiming the laser beam, the likelihood of the laser beam even striking the typically moving orange light sources is very small indeed. It is more likely that the lights' change in intensity, angular size, and position are not a direct result of the laser pointer. Yet, this still leaves other causal possibilities including the repeated presence of the sky watchers night after night in the same area and their other related behavior.

161

Note: 1. Most hand-held solid state laser (diode) pointers possess a very low power output for practical, economic, and eye safety reasons, emitting typically from two to five milliwatts of red light within a very narrow band of wavelengths (633 to 670 nanometers). The pencil diameter beam remains relatively parallel, reducing light loss over relatively long distances. Nevertheless, unless the laser beam is aimed directly at a distant object (a very difficult feat given normal hand tremors) and some sensor on the distant object which happens to be aimed directly back at the laser source no "contact" will be made.

Reference: CSETI report dtd. 4 April 1997 by J. Jarvis:
http: //www.cseti.org//freports/sreps/srep.htm)

Overview

By now, you must have gained a better idea about the kinds of responses UFO present to non-threatening human behavior, at least non-threatening from the human point of view. Whether our behavior is interpreted as non-hostile by the UFO is not yet known. Elsewhere, I have presented a set of simple procedures to follow in the event you find yourself confronted by a UFO (Haines 1997). These recommendations are meant to help you cope effectively with the entire event, recollections of which will probably last a lifetime.

Following are some general observations I have drawn from the ninety-one cases in this chapter. Other observations and statistics are presented in chapter 10.

UFO responses to bright lights

Of the sixty-three (69.2% of total) incidents in which some type of light was aimed or flashed at the UFO, the craft:

- departed in seventeen cases (27%)
- became brighter and/or changed color in sixteen cases (25.4%)
- approached the witness in eleven cases (17.5%)
- repeated flash sequence in ten cases (15.9%)
- disappeared or did something else in seven cases (11.1%)
- did not move or paced the vehicle in two cases (3.2%)

Of the sixteen cases in which the UFO became brighter or changed color, the UFO departed immediately thereafter in five cases (31.2%)

UFO responses to waving, shouts, or other gestures

Of the six incidents in which witnesses waved, shouted, or made gestures, the UFO:

- approached the witness in three cases (50%)
- departed in one case (16.7%)
- did something else in two cases (33.3 %)

UFO responses to airplane movement

Of the seven incidents in which a pilot flew his airplane toward the UFO, the UFO and/or its lights:

- accelerated away in six cases (85.7%)
- disappeared in one case (14.3%)

Fifteen other events occurred in which the human(s) behaved in some other manner which are not discussed here.

I echo the earlier statement by Hall that "the observed objects and patterns have no counterparts among atmospheric phenomena which could account for them" (1964, 9). This significant finding suggests that we are not dealing with natural phenomena. Indeed, lightning does not flash identical sequences when a searchlight is aimed in its direction and flashed on and off several times. And mirror-like reflections of one's searchlight beam do not wait for a second or two to respond. The likelihood that a meteorite just happens to be entering the earth's atmosphere at exactly the moment a pilot flashes his landing lights in its identical location is more difficult to accept than is the existence of UFOs themselves.

References

Haines, R.F. 1978a. UFO drawings by witnesses and non witnesses: Is there something in common? *UFO Phenomena* 2, no. 1. Bologna, Italy, 123-51.

Haines, R.F. 1978b. UFO drawings by eye witness and non-eye witnesses. Proceedings of Joint Symposium of the AIAA and the World Future Society. *Our Extraterrestrial Heritage—From UFOs to Space Colonies.* California Museum of Science and Industry, 28 January.

Haines, R.F. 1980. UFO Shapes. In *The Encyclopedia of UFOs.* ed. R.D. Story. New York: Doubleday & Co., 330-34

Haines, R.F., and B. Guenette. 1993. A large stationary object above Montreal. In *Alien Update: The Contact Continues.* ed. T. Good. London: Arrow Random House.

Haines, R.F. 1990. Advanced Aerial Devices Reported During the Korean War. Los Altox, California: LDA Press.

Haines, R.F. 1997. Preparing for the unknown. In *Making Contact.* ed. B. Fawcett. New York: Wm. Morrow and Co., 3-16.

Hall, R. H., ed. 1964. *The UFO Evidence.* Washington, D.C.: NICAP.

Keyhoe, D.E., and G. Lore. 1969. *Strange Effects from UFOs: A NICAP Special Report.* Washington, D.C.: NICAP.

Lade, J.M. 1960. Practical steps to encourage visitors from space. *Flying Saucer Review* 6, no. 5 (September-October): 13-15.

Pratt, B. 1996. *UFO Danger Zone.* Madison, Wis.: Horus House Press.

Rodeghier, M. 1981. *UFO Reports Involving Vehicle Interference: A Catalogue and Data Analysis.* Evanston, Ill.: Center for UFO Studies.

Walton, T. 1996. *Fire in the Sky.* New York: Marlowe & Co.

> "The early Sci-Fi pulp writers
> pushed war into the future
> with gory relish."
>
> —Harry Harrison

Chapter 4

UFO Responses to Overtly Hostile Human Behavior

(59 cases)

In the preceding chapter, many provocative reports were presented which showed that human actions can cause a UFO to approach and sometimes mimic or repeat human signaling behavior in some instances. Of course, human signaling can take many forms, most of which are harmless from the human point of view. But in this chapter we will review fifty-nine interesting events in which humans behave in an overtly aggressive manner toward the unidentified object or phenomenon. Of course, we don't know how any type behavior is interpreted by its recipients.

Possible reasons for aggressive human behavior

Why would someone behave aggressively? Perhaps the main cause is one's fear of the unknown and the threat that the unknown might harbor. Humans usually react in one of two ways to threat: fight or flight. These instinctive behaviors are thought to protect the survival of our species from extinction.

Movies, Magazines, Radio, Television. Immersed in today's society of science fiction movies and television thrillers with ever greater visual realism and powerful, wrap-around sound effects, those who experience such media may be more deeply affected by the sensory impact than we once thought. It is possible that this dynamically gripping imagery is subtly conditioning mankind to accept it as reality, particularly during the show. Yet, we don't know how long this effect lasts. What if there is a long-lasting trace of fear, curiosity, ambivalence, or some other negative emotion from watching movies? Could these reactions cause some people to react aggressively toward UFOs? The controversy continues to rage on this vitally important issue within the behavioral and social sciences. In the present context we can ask, has science fiction in itself provided a necessary and sufficient stimulus for subsequent CE-5 UFO reports?

This is an extremely difficult question. The answer is confounded by the many variables that are involved, such as what media is considered and who is its audience. What proportion of the population is subjected to this particular subject? What psychological impact does this genre actually have on its audience? Have there been any significant changes in the storylines of science fiction in America over the years toward greater warfare and violence? In his excellent and nearly comprehensive anthology of science fiction, B. Ash (1977) presents a listing of the magazine articles, books, films, and radio and television programs or serials presented each year from 1926 to 1976 that presented science fiction-based material. Each succeeding ten-year period showed a relatively small increase in all categories except radio, which peaked in the 1936 to 1945 period with two programs. Here is a compilation of these numbers (acknowledging that the database is not one hundred percent complete):

Table 2
Number of Magazine Stories, Books, Films, Radio Programs, and Television Programs on Science Fiction Themes
(Adapted from: Ash, 1977; Anon., 1992).

Year Period	Magazine Articles	Books	Films	Radio Programs	TV	U.S. Pop. (million)[1]
1926-1935	107	7	10	1	0	1.15829 (1.1%)[2]
1936-1945	137	6	4	2	0	1.27250 (1.0%)
1946-1955	188	31	10	1	0	1.32481 (1.2%)

Table 2 cont'd

Year Period	Magazine Articles	Books	Films	Radio Programs	TV	U.S. Pop. (million)[1]
1956-1965	181	36	10	0	7	1.64308 (1.2%)
1966-1975	182	44	9	0	5	1.93526 (1.1%)

Note. 1. Population figures are approximate and represent the years 1925,1935, 1945, 1955, and 1965. Percentage values in parentheses are increases from the previous value and year. 2. This value is based upon a population of 1.06461 million in 1920.

When the growing population of the United States is factored into these numbers and assuming that the number of copies printed of each article and book did not keep pace with the population increase, there very likely was a proportional decrease in the percentage of Americans who would have been influenced by these stories. Nevertheless, this dwindling percentage of Americans still could be responsible for some of the present CE-5 reports. Only by asking each witness whether or not they read or watched science-fiction material could we begin to gauge whether their CE-5 report had any roots there. And, even if they claimed a strong and abiding interest in science fiction, we still would need to prove a specific link between the details of each story and subsequent human and UFO behavior. The task is extremely difficult indeed. Nevertheless, it should be remembered that a large percentage of the CE-5 events presented in this book took place in foreign countries, sometimes involving illiterate witnesses—people who may not have been directly susceptible to American pop culture. A recent book by Rux (1997) sheds more light on this subject.

Innate Human Fear and Insecurity. What if some humans possess an irrational fear about the idea of extraterrestrial life forms? What if human beings harbor a subconscious societal memory of perhaps massive, earlier contact? This might underlie our fear of loss of self- or societal-control and even total domination by other races. As mentioned before, people often jump to the wrong conclusions during unexpected and psychologically stressful situations. Such individuals may make assumptions about the cause of their own stress response which are totally false. These assumptions may then lead to openly hostile behavior (see 126 (4)).

Researcher M. Moravec (1980) discovered that of the forty-six close-encounter cases in Australia that he reviewed, human fear was expressed in thirty-two of them (70%), curiosity in seven cases (15.2 %), amnesia in four cases (8.7%), and puzzlement, nightmares, or other dreams in another three cases (6.5%). There is little reason to expect different human responses elsewhere in the world.

Many of us believe that people possess a dark and destructive predisposition to do wrong rather than right, a morbid rather than a spiritually uplifting view of life, a fundamental behavioral "trait" referred to by some as "original sin." In fact, history has clearly illustrated a natural tendency to do wrong; it is very real and has survived through the ages despite the claim that humans are improving over time. Humanists are simply wrong in this matter.

168

This morbid psychological tendency to dwell on subjects that are dangerous or destructive seems more powerful than the tendency to think about peace, love, and tranquillity. Indeed, movies about war, murder, ghosts, mythological creatures from the past, for example, usually outsell their competition, at least in American box offices. And it is possible that this same morbid predisposition for the negative also leads some people to carry a weapon and fire it at UFOs. Such an event took place in July 1947 in Bluefield, West Virginia, according to an article in the *Sunset News*.

Case A. 47-07-12 Mrs. Johnny Johnson and some friends were returning home on Brush Fork road from a church meeting about 10:00 P.M. when they saw several disc-shaped objects at a low altitude. Upon arriving home, she woke her husband who grabbed his shotgun and ran outside, saw the objects, took aim, and shot at them several times. "However, the only [thing] that fell was Johnson's hope of capturing one of the nationally-known curiosities," said the article. The UFOs flew around for about thirty minutes and then departed. One might ask why Johnson behaved this way. What led to his apparent fear-based reaction?

Case B. 71-02-17 Pat McCollum was inside his trailer home located at Coma and Brim Creek Roads near Vader, Washington, when he saw two strangely lit objects outside fly past his window. Each had a blue light which glowed with an orange hue. McCollum also observed blue flashing lights against the night sky. They emitted a humming noise. He went outside with his gun and fired three shots at one of the objects. The first object was about six to eight feet above the ground while the second was about thirty feet up. (Anon. 1971. *Data-Net*. The UFO Amateur Radio Network 5, no. 6 (June); *Centralia-Chehalis Daily Chronicle (Washington)* 18 February 1971).

Case C. 81-08-30 While not strictly a CE-5 case, this incident illustrates how fear motivates one to perform an act of aggression. David J. Santoro, 27, a reserve police officer in Spring Valley, California, reported seeing a sixty-foot-wide, dark, disc-shaped object hovering only fifteen feet above the roadway ahead of him at about 10:30 P.M. His car came to an abrupt stop by itself—he did not apply the brakes. His field of view was filled by an intense luminosity. The next thing he recalls is being removed from his patrol car by two small beings; he was returned there in less than fifteen minutes and recalls seeing the dark object rising up into the night sky. But *as it rose it shone a ray of white light in circles repeatedly around his car.* "He fired four shots at the craft with his service revolver out the left window and drove home wildly at 90 mph, narrowly escaping an accident." The next day he noticed that the front end of his car was clean while the rear was still dirty with road dirt. *(The (San Diego) Tribune*, 1 July 1987; Gribble, R. 1991. Looking Back. *MUFON UFO Journal* 280 (August): 19.)

Curiosity. Perhaps there is another motive for showing aggressive behavior toward a UFO: simple curiosity. One may just be interested in finding out whether the object is solid, if makes a noise when hit, or how it will react to outside forces.

Case D. 88-early City councilman, rancher, and auto parts store owner, Antonio Fernandes Duarte, 43, admitted the following during an interview in 1991 in Mossoro, Brazil: "I was fascinated by it. I didn't know what to do. I had a revolver and I wanted to shoot [at the alien beings] to see their reaction but [I] didn't" (Pratt 1996, 221).

Nevertheless, it is irrational for someone to initiate hostility with something which very possibly could be far more militarily advanced and destructive than humans. We are taught from childhood to pick on someone our own size; to do otherwise invites a retribution we may not like. But, as we shall see in this chapter, humans don't always act rationally. Fortunately, there are instances where humans do act responsibly.

Attracting attention. In at least one case, a hunter fired his shotgun at a UFO which was flying past him in order to "attract it back." This failed! The hunter and his companion tried to follow the object in their truck for several miles, but the object simply flew out of sight, giving off a steady humming sound. (*International UFO Reporter* 2, no. 12 (December 1971): 2)

U.S. military responses

An early reference concerning official U.S. military orders to shoot at unidentified aerial phenomena—although the nation was at peace at the time—is found in a letter to the Aerial Phenomena Research Association in Seattle, signed by Commander R. W. Hendershott. It is about a radar contact with a UFO in the latter part of December 1944 or early 1945 near Pasco, Washington. Hendershott wrote, "Radar operators at the Naval Air Station, Pasco, Washington, reported unusual blips on their radar screens. These blips appeared out of nowhere and proceeded from northwest of the Air Station to the southeast and consequently off of the radar screens. A fighter pilot was made available with an armed F6F fighter and given orders to shoot down anything that appeared to be hostile. He was vectored out on two occasions that this writer remembers, but, in each instance, made no contact. The blip always acted much like a Piper Cub aircraft and at about the same speed." It is important to note that the propeller driven F6F combat airplane can easily out-fly a Piper Cub and also shoot one out of the sky, if necessary. Perhaps the orders to shoot at anything hostile in the sky were a result of the presence of the then top-secret atomic bomb manufacturing plant at Hanford, Washington, nearby.

It is one thing for a private citizen to fire a weapon at a UFO out of fear but it is quite another for a nation's military service to issue orders to fire willfully at UFO in an overtly aggressive manner. Indeed, many such incidents of military "action" exist. A headline in Fullerton California's *News Tribune* on July 26, 1956, stated, "Pacific Navy Fliers Ordered to Engage Saucers" (Honolulu, T.H., Overseas Central News Service). An article published in the British journal *Flying Saucer Review* said the following about this policy, "The United States Navy will not publicly admit that it believes in flying saucers, but it has officially ordered combat-ready pilots to 'shoot-to-kill' if saucers are encountered, OCNS has learned.

"The information was first learned when Navy pilots navigating trans-Pacific routes from the United States to Hawaii were ordered in a briefing session to engage and identify 'any unidentified flying objects.' If the UFOs appeared hostile, the briefing officer told pilots of Los Alimitos Naval Air Station reserve squadron VP 771, they are to be engaged in combat" (also cf. *Flying Saucer Review* 3, no. 1, 1957).

U.S. Navy pilots in Honolulu met to talk about this order. It was learned that extensive top-secret operational procedures for dealing with UFOs have been established, "including forms of combat." A Navy commander remarked, "It's gotten so we wouldn't dare say we've seen a UFO. If we did, every pilot in the Pacific would be ordered up. It would be pretty embarrassing if all we'd seen was a sunspot on the windshield." Another pilot said, "I believe there are such things but I think that Washington might be wrong in their 'shoot-to-kill' orders. The fact that saucers are in our atmosphere doesn't mean to me that there's any pending invasion, which is what Washington seems to believe. And if there were such an invasion, we'd do a lot better if we sent out a flight of priests and ministers, rather than a bunch of rockets and machine gun bullets. If anybody who could conceive a saucer wanted to invade us, there's no sense fighting them. They've got us licked from the start." Another pilot stated, "How do we know our bullets will work on a UFO? And if we do shoot, that's asking them to shoot back. And we don't know what they're going to shoot at us!" (Ibid.).

The *News Tribune* article also noted, "Operational procedures for a U.F.O. scramble apparently are highly classified. Most officers refused to discuss the Pentagon's plans or modes of saucer combat. However, it was learned that a concrete plan of action does exist. The plans reportedly can be swung into action within seconds."

U.S. Air Force Captain Wallace W. Elwood, Assistant Adjutant, Air Technical Intelligence Center at Wright-Patterson Air Force Base, Ohio, admitted, in a letter dated July 12, 1957, "Air Force interceptors still pursue unidentified flying objects as a matter of security to this country and to determine technical aspects involved. To date, the flying objects have imposed no threat to the security of the United States and its possessions." Elwood wrote in response to a letter of inquiry from U.S. Representative Lee Metcalf (D.—Montana) about the position of the air force regarding UFO encounters. The air force reply went on, "In a few cases, Air Force pilots have officially reported firing on flying objects which they could not identify, but which were later determined to be conventional objects. *The orders to pilots are to fire on an unidentified object only if it commits an act which is hostile, menacing, or constituting a danger to the United States.*"

One explanation given for the official shoot-to-kill order is that this might be one way of bringing one down and providing government officials with tangible evidence of UFOs. Author Peter Paget (1979) and others have argued a slightly different view. Paget wrote, "To the military-minded generals, the fruit that the superior UFOs held out to them was irresistible. They were concerned only with thoughts of capturing such a machine in order than they might duplicate its mechanics and thereby insure world security or at least their own nation's security....Needless to say, none of the interception attempts were successful, although it is thought that at least one UFO was destroyed by the direct hit of a missile. This apparently was launched from a United States vessel in the South Atlantic in 1963, a panic measure when the Navy Department took the situation into its own hands, fearing that the ship was in danger,...the missile...was fired from a range of four miles...and the result was a terrific explosion; but it produced only thousands of pieces which promptly sank to the ocean floor and were never recovered" (Ibid.).

Perhaps another indirect impetus for the above order to shoot to kill came from a rather profound statement made by General Douglas MacArthur on May 12, 1962, to cadets at the U.S. Military Academy. McArthur was an extremely powerful and respected national leader at the time and his speech would have carried a good deal of weight within the Pentagon. Did his words carry a deeper hidden meaning known only to those with a need to know?

His speech, entitled "Duty, Honor, Country," lasted about forty minutes, according to Dennis Stacy, editor of the *MUFON UFO Journal* (no. 320, December 1994). McArthur stated the following:

> You now face a new world, a world of change. The thrust into outer space of the satellites, spheres, and missiles marks a beginning of another epoch in the long history of mankind. We deal now, not with things of this world alone, but with the illimitable distances and as yet unfathomed mysteries of the universe. We are reaching out for a new and boundless frontier. We speak in strange terms: of harnessing the cosmic energy; of making winds and tides work for us; of creating unheard-of synthetic to supplement or even replace our old standard basics....of spaceships to the moon; of the primary target in war, no longer limited to the armed forces of an enemy, but instead to include his civil population; of ultimate conflict between a united human race and the sinister forces of some other planetary galaxy; of such dreams and fantasies as to make life the most exciting of all times.

While he made no direct mention of UFO nor did he disclose any basis for his comments, we may wonder if McArthur knew of the many scores of ground, sea, and air sightings which military intelligence had collected and analyzed during the Korean conflict (Haines 1990).

The former head of the U.S. Air Force's Project Blue Book study of UFO, Edward J. Ruppelt, wrote the following in the foreword to his outstanding book *The Report on Unidentified Flying Objects:* "Is it proof when a jet pilot fires at a UFO and sticks to his story even under the threat of court martial?" Such a statement suggests that pilots have actually fired at UFO and Ruppelt knew it. Airplane gun camera film, known to exist in military files, which shows smooth, symmetrical objects with no visible means of propulsion, provide further reason to believe that our military air crews have engaged UFO in combat.

It is interesting to note the 1977 edition of *Arms Control and Disarmament Agreements—Texts and History of Negotiations,* published by the U.S. Arms Control and Disarmament Agency, contains a chapter titled "Agreement on Measures to Reduce the Risk of Outbreak of Nuclear War Between the United States and the Union of Soviet Socialist Republics." The following statement is found there:

> Arrangements for immediate notification should a risk of nuclear war arise from such incidents, from detection of unidentified objects on early warning systems or from any accidental, unauthorized, or other unexplained incident involving a possible detonation of a nuclear weapon [should be made].

173

This "agreement" between America and the Soviet Union was signed on September 30, 1971. It is not known when the above clause was added to the agreement nor exactly what is meant by "unidentified objects."

An Air Force statement on UFO issued in 1980 stated, "There has been no evidence submitted to or discovered by the Air Force that sightings categorized as 'unidentified' represent technological developments or principles beyond the range of present-day scientific knowledge." Nevertheless, a more recent article published in *Aviation Week and Space Technology* magazine (18 March 1996) stated that the U.S. Air Defense Operations Center at Cheyenne Mountain, Colorado, acts as the interface with many other stations which serve as part of the North American Air Defense Command (NORAD). One of their missions is to track and identify the approximately eighty-five hundred "vehicles" in near and deep space. It also said, "These regions monitored about seven thousand aircraft tracts per day in 1994, and labeled approximately 880 of them as "unknowns" within the allotted two minutes required for identification....*Others (unidentified contacts) warranted intercepts by fighters scrambled from airfields around the continent's periphery."* This is quite an admission. What do these interceptor pilots see if they catch up with their targets? Which nation is now considered to be an enemy of the United States which also possesses a credible air force that continues to evoke jet-interceptor scrambles? The following current event indicates both that Air Force interceptors continue to pursue UFO over America and that the UFO seem to be playing a cat and mouse game.

Case E. 97-05-04 Three metallic UFO were seen through binoculars by Andrew Cavaseno and his girlfriend. The objects were traveling toward the southeast over Nassau, New York, at an estimated thrity to thirty-five thousand feet altitude and seemed to change shape as they glowed with an orange hue. They suddenly disappeared when two F-16 jets approached them from the southwest. Fifteen minutes later, two black helicopters arrived in the area and released "a large quantity of (white) balloons" in groups of three. (Filer, G. 1997. Filer's Files Week 19, 17 May; also cf. CSETI case 3 April 1997: http://www.cseti.org/freports/sreps/srep2.htm)

Foreign military and civilian activities

The fact that many foreign ministries of defense look to our country for direction is common knowledge (e.g., Fawcett and Greenwood 1984). One such reason for this could be that since the U.S. possesses the most powerful military capability of any nation, it should set policy in confronting the UFO phenomenon.

In a British report (Gross 1955, 9) British Air Marshal Lord Dowding, Chief of Fighter Command in the famous World War II Battle of Britain, is claimed to have remarked during a Flying Saucer Research Society meeting in London on 22 January 1955, "It's rude to fire AA guns and send fighters to shoot them down....Besides, you never know what they could do to you." He went on, "There is no material we know of that could travel nine thousand miles an hour—the recorded radar speed of one saucer—without becoming white hot." An *Australian Post* magazine article (12 January 1956) featured another article by Lord Dowding titled "Enemies from Space." In it he wrote, "Foes infest the skies above our earth. To these invincible, god-like invaders our massed planetary power is puny, our countries peopled by barbarians. Don't anger them. They could destroy all our works, enslave mankind." *(APRO Bulletin, 15 July 1956)*

There is good reason to believe that the armed forces of America's military allies use the same strategy toward UFOs as the U.S. High-level coordination meetings likely have been held to cope with the phenomena in a consistent, and hopefully more effective manner (Bowen 1977). An article published in the June 16, 1957, issue of London's *Reynolds News,* for example, stated that Britain's Air Ministry is studying UFO and that airplanes all over the country are to intercept and, "if necessary, engage, any unidentified flying object within combat range."

A letter signed by Roeshin Nurjadin, Air Marshall of the Indonesian Air Force dated May 5, 1967, states:

> In response to your letter about Unidentified Flying Objects I wish to state, that up to now the Indonesian Air Force has no official opinion regarding said subject. It does not imply however, that we disregard the intrusion of our skies by illegal air- or spacecraft. UFOs sighted in Indonesia are identical with those sighted in other countries. Sometimes they pose a problem for our air defense and once we were obliged to open fire on them. Contrary to foreign reports, however, the so called "contact stories" are never heard of in our country....We Indonesians have a world outlook based on the Panchasila philosophy, the first principle of which is the belief in God. Hence we believe that God is omnipotent enough to have created also other worlds and other human races, some of which may be more advanced than we are. That's why we have an open mind on the possibilities of UFOs from space also.

175

The September 19, 1976, two-jet interception with a UFO over Tehran, Iran, led to an official inquiry of all the facts. Lt. Gen. Abdulah Azarbarzin of the Iranian Imperial Air Force admitted to U.S. reporters that the UFO encounter had been carefully documented and passed on to the U.S. Air Force. "This was the request from the U.S.," he said. "They have this procedure, if we have some information on UFOs, we're just exchanging all this information, and we did it" (Pratt 1996, 144). Nevertheless, the U.S. Air Force's public position at that time was, as of the end of 1969, that they no longer studied UFO. Whom should we believe?

NATO commanders issued a shoot-down order to two Phantom jet pilots who were scrambled to identify a slow moving unidentified object seen on both

Greek and American radar. This account, received through military channels, made it clear that "the planes had gone up with orders to down the object, but these were canceled before visual contact with the object could be made" for unknown reasons. The UFO passed over the Turkish border moving inland. During their intercept, the pilots saw an "illuminated, multi-finned, rocket-shaped" vehicle. Radar showed that it came to a full stop above the largest radar station in Northern Greece near Thessalonika. It was sighted visually as a glowing sphere which dropped to within thirty meters of the ground, causing a total power failure and interrupting radar and communication capabilities. It changed to a rhomboidal shape and flew slowly in the direction of the Albanian border. Two more interceptors were scrambled and, when they caught up with the object, the pilots said that they were following a "normal looking airplane with conventional red, green, and white lights." The jets returned to base when the UFO crossed the border (*Flying Saucer Review* 28, no. 1, August 1982, p. iii).

While the following cases include many different kinds of human hostility, they all have one thing in common, the aggressive human behavior is overt. It is interesting to speculate on how the other party might interpret our behavior. Do "they" think humans are simply trying to communicate with them or attempting to evoke a knee-jerk type of reaction? Do they interpret bullets fired at them as some crude form of scientific or medical experiment? Perhaps they are so advanced that they are, if not oblivious to the projectiles, at least immune from their harmful effects. Many of the cases reviewed here suggest that the aerial objects are: (a) physical in the sense of having a hard outer surface, (b) energetic in the sense of being able to travel at very high rates within our atmosphere, and (c) highly advanced in terms of their flight control and navigation over large distances. If this assessment is accurate then it is likely that UFOs will not be harmed by our relatively inneffective acts of hostility. Would the situation be any different if we used nuclear weapons?

Another provocative event occurred in the 1970s. The U.S. Air Force Academy offered Physics 370 to its young officers-to-be. This class used a particularly revealing text in which section 33.4 of volume 2 is titled "Human Fear and Hostility." It states:

> The UFOs were luminous and moved very fast. We too have fired
> on UFOs. About ten o'clock one morning, a radar site near a

fighter base picked up a UFO doing 700 mph. The UFO then slowed to 100 mph, and two F-86's were scrambled to intercept. Eventually one F-86 closed on the UFO at about 3000 feet altitude. The UFO began to accelerate away but the pilot still managed to get to within 500 yards of the target for a short period of time. It was definitely saucer-shaped. As the pilot pushed the F-86 at top speed, the UFO began to pull away. When the range reached 1000 yards, the pilot armed his guns and fired in an attempt to down the saucer. He failed, and *"the UFO pulled away rapidly, vanishing in the distance* (Carpenter 1968, 462).

Here we have an official air force classroom text read by young men and women who will one day exercise leadership within the air force. This remarkably candid text was included deliberately, but for what reason?

American intelligence agencies

There is almost nothing in the open, i.e., unclassified literature, which presents any public position taken by America's intelligence agencies on UFO. This is supported by a recent review which claims to trace interest and involvement in the UFO controversy by the CIA. G.K. Haines (1997) claims that since the 1950s, the CIA has "paid only limited and peripheral attention to the phenomena." He states that the CIA had no official project in the 1980s and that their files on UFO subjects were purposefully kept to a minimum. But, the Central Intelligence Agency is not America's only secret intelligence gathering agency. D. Goudie and J. Klotz (1992), uncovered a letter dated January 20, 1953 from Dr. H.P. Robertson to Dr. H. Marshall Chadwell, Assistant Director of the Office of Scientific Intelligence, CIA, only several days after the famous Robertson Panel meetings had ended. In it, he writes, "Schedule: NSA group meeting on Thurs. 5 Feb." This little-known reference to the National Security Agency occurred only a few months after its creation.

Also consider the following statement by H.T. Wilkins (1955, 67). "February 13, 1954: American airline pilots report seeing five to ten flying saucers every night of their flights. American intelligence service asks the pilots not to discuss these sightings publicly. On March 24, 1954, an American Air Force spokesman admits that reports of saucer sightings reach the U.S.A.F., but the public is to be told nothing." No mention is made of which intelligence agency was referred to but it was very likely the CIA, probably given the responsibility from the air force to conduct technical analyses on UFO after 1970.

A U.S. Air Force "INOZA Alert Officer Log Extract" dated October 31, 1975, (1620L) indicated that the CIA was informed about unidentified object flight activity "over two SAC bases near Canadian border. CIA indicated appreciation and requested they be informed of any follow up activity." Notwithstanding the recent assessment by Haines (1997), CIA involvement in UFO phenomena appears to be continuing. Despite CIA silence, there is good reason to believe they were following UFO activities from their inception. And, if they weren't, they should have been!

Civilian government agencies

The following quote is from a civilian air route traffic controller in the airport tower at Washington, D.C.'s National Airport during July 1952, one of the most active UFO months in America's history: "We were too busy with other things and besides these objects aren't hurting anybody." The controllers working in the tower observed unusual radar contacts which disappeared and then "came back in behind the plane" (Gross 1986, 38). Many news reporters were present in the control tower trying to get more information about the Eastern Airlines plane which was involved.

Lewis Barton wrote an article titled "If it Moves—Shoot" (*Flying Saucer Review* 1, no. 3. July-August 1955, 3) in which he said, "We shall never establish proper contact with our mystery visitors from outer space until we can convince them that we have ceased to be trigger-happy—at least, as far as they are concerned." He speculates that these extraterrestrials have probably "progressed far beyond our state of jungle law. They must be alarmed, and possibly saddened, by what they have learned of our reaction to the appearance of strange objects— and they will, sensibly, keep their distance until they feel sure that a stream of bullets is not going to greet any attempts to contact us." Barton's view displays a common anthropological bias, i.e., that our bullets constitute a physical threat to UFOs. If E.T.s are as technologically advanced as accounts suggest, then our bullets, arrows, rocks, and missiles probably have little or no effect on them as Haines described (1990).

Frank Edwards (1966) recounts an incident in 1897 where a Kansas farmer ran toward a UFO "swinging an ax." Edwards asks, "how would you like to meet a race of people who said 'hello' with an ax?" On the whole, humans are truly a violent lot, not only in Kansas but everywhere else on the planet. As we shall see in most of the cases in this chapter, the UFO does not respond in-kind to aggressive human behavior. What does this mean? We will have occasion to discuss the implications of these findings later.

In a more recent article, Dennett *(MUFON UFO Journal* 299, March 1993, 3-6) has reviewed incidents where humans fired guns at UFO, five where they fired at humanoids, and five more where humans were prevented from firing their weapons or where the UFO responded with an overtly hostile act. All are reviewed here. Dennett concludes that UFOs would prefer to flee than fight. He states, "There is not a single case wherein gunfire resulted in the permanent damage of either a UFO or its reported occupants; nor is there any significant indication of the latter having ever opened fire first."

Provoked human aggression against UFOs

There are many hundreds of documented instances of UFOs which do something that provokes the human being to behave aggressively. Here are two examples.

Case F. 75-?-? Mr. Jose Morais Da Silva was with his brother-in-law Santiago riding horseback to the village of Bom Jesus, Brazil, at midnight. A huge intense beam of light suddenly appeared ahead of them. Then, the object which they though was giving off the beam approached them, hovering about thirty meters overhead. "I had a revolver and tried to shoot at it, *but the UFO disappeared*." Very relieved, the two men rode on to a friend's house and went in. But while they were there talking, the UFO returned and "shined a light on our horses. The horses stayed quiet, and the UFO disappeared" (Pratt 1996, 203). As Pratt states, "almost everyone knows someone who has been chased, hurt, or killed by these things" (xiii). Why did Da Silva draw his revolver? Had he heard the tales of other terrifying encounters in the area?

Case G. 91-03-09 Mr. and Mrs. Reinar Silva, were riding their motorcycle home about 1:30 A.M. and were traveling near Fortaleza, Brazil, where they owned and operated a sidewalk lunch stand. Upon reaching a point about 14 miles south of Fortaleza they both noticed a big light which came down out of the dark sky. *It stopped* ten to fifteen meters overhead and slightly ahead of them.

Then, their motor died. Both admitted being "paralyzed with fright" as the light remained above them. They noticed smaller lights revolving around the two-story-tall, "sombrero-shaped" object. From its "tail" was seen a "pretty rainbow" from which was emitted high-intensity colored lights. On its top was a transparent, yellowish, luminous dome. After several more minutes, it disappeared very fast. *Then, the object returned* quickly and silently, *stopping over them again*. It disappeared a second time and they were able to restart their motor. About two miles later, *it came back a third time* but did not stop. *It flew over them and dropped a "transparent ball*, like a big soap bubble, one and a half meters around," Reiner Silva said. The headlight of the motorcycle reflected off its surface giving it a fiery appearance. The bubble descended and exploded only about 1.5 meters ahead of the terrified couple. Reiner put his brakes on just in time to keep from driving into the object. The color of the ball was blue and it made no noise at all when it exploded (Pratt 1996, 216-19). Was this an act of open hostility by the UFO or perhaps of frustration in not getting the couple to "behave" as they were supposed to? Was it some kind of scientific experiment? What if Silva had a gun with him? It is likely that he would have tried to use it out of both fear and self-protection.

A research hypothesis

The following research hypothesis is presented to help us sort and analyze the available evidence.

Hypothesis: If UFOs are only natural phenomena, they will not respond in any consistent manner to overt human behavior.

This hypothesis leads to the following postulate that is related to some cases presented in this chapter: if UFOs respond with a relatively consistent pattern of actions to being fired at, there is more reason to accept the proposition that they are under some kind of intelligent control. And, if UFOs are not apparently damaged by acts of human aggression, it is reasonable to think that they were not provoked sufficiently to evoke a hostile response rather than that they could not respond.

Case Files

Case No. 92

Date: 17-04-14
Local Time: 0230
Duration: estimated 5+ minutes
Location: Portsmouth, New Hampshire
No. Witnesses: 3
DB: v   CB: a   PB-1: a   PB-2: x   UFOD: a

Abstract: A "mystery airplane" was seen in the early hours flying down the Piscataqua River toward the west. As it reached a point near the bridge connecting Portsmouth New Hampshire, with Kittery, Maine, it was heard by three guards from "L" Company of the Sixth Massachusetts National Guard who were stationed there on the lookout for German saboteurs.

The object was seen by the soldiers circling at a high altitude, and then appeared to be descending toward their location. A soldier, Corporal Davis, stationed at Post 1 near the bridge, "upon the suggestion of the corporal on duty, opened fire. *The aircraft immediately reversed course and moved up the river.*" He stated, "It was an actual thing. At considerable height, it swam in the upper air. I was astonished at its size. While I was looking, there came the well known slat, slat of a motor. As it started to move up the river, I let go a shot" (*Dover* (N.H.) *Democrat* 17 April 1917). The soldiers then immediately telephoned the Portsmouth Naval Shipyard with news of the aerial intruder. They were informed that "no government war machine was flying from the navy yard" (*Boston American* 13 April 1917).

Comments: *A Manchester Union* article claimed that a "strange airplane" was heard twenty miles to the northwest at Rochester, New Hampshire, between 1:00 A.M. and 1:15 A.M. by night officer Ferdinand Sylvain, a local policeman. He heard the "purr of a motor overhead. The noise was loud enough to cause dogs to bark and thus a number of citizens were awakened by the commotion." The object was never seen by Sylvain but he was certain that it was an aerial vehicle heading south towards Portsmouth.

Another eye witness was motorman James Walker of the Dover Rochester and Somerworth Street railway. He pointed the object, appearing like a toy balloon

181

surrounded by a circle of smoke, to his passengers. It was flying at a high altitude toward the north at that time according to Walker. He was convinced that the object was an airplane circling at great altitude and "belching dense smoke from either side and the rear. It disappeared into heavy clouds (*Foster's Daily Democrat* (Dover, N.H.) 14 April 1917).

Still, others witnessed the mystery airplane on April 12 in Dover, New Hampshire, hearing a "whirring noise," and on April 14 to April 16 in Milton, New Hampshire. Another witness, Mr. J. M. Bennett, said that the object was more "covered in" than the average flying machine. It isn't clear exactly what he meant by this.

Here, we find numerous eye witnesses to a single unusual aerial object which flys in circles, emits dense smoke and a whirring noise, descends, and immediately reverses course. Since the Portsmouth Naval Shipyard was near these aerial flights and was then one of the most important naval facilities on the Atlantic coast, the U.S. military command issued an alert to find out the "source of the aircraft."

References: Anon. 1917. "Just Cause." *Manchester (N.H.) Union* no. 45. 14 April, 2-3.

Cross references: 100 (4); 101 (4); 111 (4); 116 (4); 120 (4); 122 (4); 126 (4); 130 (4); 142 (4); Case C (4)

Case No. 93

Date: 42-02-24,25
Local Time: night
Duration: 3+ hours
Location: Los Angeles, California
No. Witnesses: thousands
DB: kk   CB: b   PB-1: a   PB-2: x   UFOD: O

Abstract: Several large glowing objects were seen hovering over western Los Angeles by thousands of Californians. Outlined in the light of several search-lights, the Army's anti-aircraft guns began firing at them. At least 1,430 rounds were fired but the objects did not move, nor were they affected. Unfortunately, a number of citizens were killed by the falling shrapnel.

Figure 25

Photograph of searchlights and unidentified aerial objects
over Los Angeles, California on February 24-25, 1942

Comments: Interestingly, the aerial objects did not depart when illuminated by
the searchlight beams or when fired at.

Reference: Story, R.D. ed. 1980. *The Encyclopedia of UFOs.* Garden City, New
York: Doubleday & Co.

Case No. 94

Date: 44-09-?
Local Time: night
Duration: estimated 5 minutes
Location: Epinal (Vosges Department), France
No. Witnesses: 500+
DB: i   CB: a   PB-1: a   PB-2: x   UFOD: a

Abstract: Private Erich Immel in the German army and five hundred other soldiers were near Epinal in the northeast part of France, waiting in trenches for the
Allied advance. Immel saw a spherical, silver-white object, like "dull aluminum"
hovering a few feet above the ground above nearby railroad tracks. He estimated
the object to be about six feet in diameter, but as it approached his position it
seemed to change into an oval shape. He later wrote, "I raised my rifle and fired.
As soon as the bullet struck it, it flared up into the brightest sunlight I'd ever seen.

"Every one of the five hundred men in my company opened fire on the ball of light. However, the next morning, when we searched the area, we found nothing."

Comments: Some readers will find reason to categorize this luminous phenomenon as ball lightning, however, its large size, relatively long duration, and change of shape might argue against this view.

Reference: *National Enquirer,* 17 September 1972.
Cross references: 98 (4); 115 (4); 116 (4); 124 (4); 46 (4); 150 (4)

Case No. 95

Date: 45-05-?
Local Time: night
Duration: estimated 10+ minutes
Location: Marianas area, South Pacific ocean
No. Witnesses: several
DB: kk   CB: a   PB-1: a   PB-2: x   UFOD: 0

184

Abstract: Reports of "fireballs" were received at headquarters "from every B-29 base in the Marianas. As time went on, the fireballs became more maneuverable and followed the Superforts farther out to sea." Excited gunners on board these airplanes would fire at the self-luminous globes. "Missiles would miss their targets and fall into the sea."[1] No mention is made by this eye witness of any particular evasive movement or other change in appearance of the UFO resulting from being fired at. However, in another letter from a William J. Roberts to the National Investigations Committee on Aerial Phenomena dated September 11, 1950, the writer stated that he was a tactical reconnaissance pilot during World War II. He remembered speaking with pilot Major Koser who flew in the South Pacific theater and who told of night bomber crews who spotted "lights" pacing their aircraft. When their gunners fired at these lights *"they went out."* These lights were seen so far out to sea from mainland Japan that it was not likely they were Japanese jet airplanes.

Comments: According to Mr. Dumphy, his first "fireball" report was made during a night bombing raid on Tokyo when he saw the round globe approach his airplane and then follow it out to sea as they were returning to base after dropping their bombs. He described them as "speedy balls of fire, fast as B-29, but not as maneuverable, burning warheads suspended from parachutes, or like molten chunks of steel."

> Note. 1. This comment suggests that tracer shells were used. Such shells leave a visible trajectory in the sky to more accurately "lead" and hit the moving target. These tracer shells could be seen at night falling downward toward the ocean's surface .

References: Dumphy, Gerry, dtd. 1963. Report from 13 October. CSI files, case 2187 .

(Madison, Wis.) *State Journal,* 8 July 1947.

Cross references: Case B (4); 126 (4); 141 (4)

Case No. 96

Date: 45-06-early
Local Time: 1400
Duration: 5+ hours
Location: Marshall-Carolines Islands, Pacific Ocean
No. Witnesses: estimated 25
DB: kk   CB: b   PB-1: a   PB-2: x   UFOD: 0

Abstract: The reporter of this important incident, Navy Security Officer W. Brown, was on the open bridge of the Victory ship "Calvin Victory" at about 2:00 P.M. The weather was clear and bright. Suddenly, a lookout yelled, "Silver object directly overhead." Brown used his 7 x 50 binoculars to view what looked like a brilliant, round, silver object hovering at a relatively high altitude over the ship.[1]

The Captain was called immediately. He looked at the object and said it was "one of those new-fangled Japanese magnetic balloons," and directed one of the gun crews to fire at it. A range burst was fired for 40,000 feet altitude and the puff of smoke appeared above the object. A second shot burst directly below the UFO and was pinpointed at 35,000 feet. Then, they opened up with eight or ten more rounds, "but the bursts close by the object *had no effect that we could observe. Since the object did not move off but stayed directly overhead, took no evasive action or attacked the ship,* the skipper called the firing off, since it wasn't doing any good anyway."

The ship was cruising at about 16 knots toward the west, yet the object stayed directly overhead for the rest of the afternoon. The UFO finally disappeared in the darkness of night.

Comments: The fact that the UFO followed the ship over at least five hours indicates it had some deliberate purpose, most likely remote surveillance; U.S. remotely piloted vehicles can do this today for many hours at a time. They send multiple channels of tactical information back to the ground. However, neither America nor Japan possessed this capability during World War II. During wartime, ships at sea typically change course irregularly in a zig-zag pattern to help avoid being tracked by submarines. If this ship was also doing this, finding a prosaic explanation for the UFO becomes even more difficult.

> Note: 1. The witness said the UFO subtended an angle equivalent to an aspirin tablet held at arm's length or about 20 minutes arc. The object would measure 204 feet across at 35,000 feet distance.

Reference: Brown, W. Jr., dtd. to J. Allen Hynek, 28 August 1974, Northwestern University, Evanston, Ill.

Cross references: 93 (4)

Case No. 97

Date: 48-10-01
Local Time: 2100
Duration: 27+ minutes
Location: Fargo, North Dakota
No. Witnesses: 1
DB: cc   CB: e   PB-1: a   PB-2: x   UFOD: c

Abstract: This famous air-intercept case involved Lt. George Gorman, pilot of an F-51 interceptor airplane of the North Dakota National Guard, who was flying a routine air patrol. On final approach to land at the Fargo air field he sighted a small white light about 1,000 yards ahead of him. Believing it was the tail light of a preceding aircraft, Gorman called the tower. He was told about a private airplane flying near the field which he easily located. Clearly, the light was not this airplane. It was perfectly round, white, and small (estimated 8 inches diameter). Gorman approached *the light which performed a sharp left-hand turn.* Even after adding full power he could not gain on the light. Then, *it started climbing in yet another left turn,* still somewhat below him. Gorman performed a sharp left banking turn to cut it off at about 7,000 feet altitude. *"Suddenly, the thing made a sharp turn right, and we headed straight for each other!* Just when we were about to collide, I guess I got scared. I went into a dive and the light passed over my canopy at about 500 feet. Then the thing made a left circle about 1,000 feet above, and I gave chase again.

"I cut sharply towards the light, which once more was coming at me! When collision seemed imminent, *the object shot straight up into the air*...the thing now turned in a north-northwest heading, and vanished."

Comments: The pilot's account is familiar to anyone who has studied the literature of aerial close encounters. UFOs very often fly precise maneuvers relative to the nearby airplane, indeed, giving the clear impression "they" are interested in the airplane. Gorman had flown previously in World War II as a flight instructor for French pilots. He said, "I am convinced there was *thought* behind the thing's maneuvers. I am also certain that it was governed by the laws of inertia, because its acceleration was rapid, but not immediate...(and also) followed a natural curve."

Reference: Wilkins, H.T. 1954. *Flying Saucers on the Attack*. New York: Citadel Press, 105-109.

Cross references: 107 (4); 108 (4); 110 (4); 141 (4)

Case No. 98

    Date: 51-early spring
    Local Time: evening
    Duration: estimated 25 minutes
    Location: near Chorwon, Korea
    No. Witnesses: 76
    DB: cc   CB: e   PB-1: a   PB-2: x   UFOD: 0

Abstract: "We were in the 25th Division, 27th Regiment, 2nd Battalion, 'Easy' Company," explained Francis P. Wall during an interview. "We were in what is known on the military maps as the Iron Triangle, near Chorwon. We were to the left of Chorwon, just across the mountain ridge from this city, town, whatever you want to call it. It is night. We are located upon the slopes of a mountain, between the fingers of a mountain as they run down toward the valley below where there is a Korean village. Previously, we have sent our men into this village to warn the populous that we are going to bombard it with artillery. Upon this night that I'm talkin' about, we were doin' just that. We had aerial artillery bursts comin' in. And we suddenly noticed down, with the mountains to our backs, we noticed on our right-hand side what appeared to be a jack-o-lantern come wafting down across the mountain. At first, no one thought anything about it. So, we noticed that this thing continued on down to the village to where, indeed, the artillery air bursts were exploding. And we further noted…that this object would get right into [an airburst]…it was that quick, that *it could get into the center of an airburst of artillery and yet remain unharmed*. And, subsequently, this time element on this…I would say anywhere from, oh, forty-five minutes to an hour all told.

"But then *this object approached us*. And it turned a blue-green brilliant light. It's hard to distinguish the size of it, there's no way to compare it. *It pulsated*. The light, that is, was pulsating. It wasn't regular.…I asked for, and received permission from Lt. Evans, our company commander at that time,[1] to fire upon this object, of which I did with an M-1 rifle with armor-piercing bullets, or rounds in it. And I did hit it. It must have been metallic because you could hear when the projectile slammed into it.

"*But the object went wild and the light was goin' on and off, and it went off completely once, briefly. It was moving erratically from side to side as though it might crash to the ground. Then, a sound, which we had heard no sound previous to this, the sound of…diesel locomotives revving up. That's the way this thing sounded.* And, then, we were attacked, I guess you would call it. In any event, *we were swept by some form of a ray that was emitted in pulses, in waves that you could visually see only when it was aiming directly at you*…like a searchlight sweeps around and the segments of light you would see it coming at you.

"You would feel a burning, tingling sensation all over your body, as though something were penetrating you. So, the company commander, Lt. Evans hauled us into our bunkers. We didn't know what was going to happen. We were scared. We did this. These are underground dugouts where you have peep holes to look out to fire at the enemy. I'm in my bunker with another man. We're peeping out at this thing. It hovered over us for a while, lit up the whole area with its light that I'm telling you about, and then *I saw it shoot off at a 45 degree angle*, it's that quick, just, *it was there and was gone*. That quick. And it was as though that was the end of it. But, three days later the entire company of men had to be evacuated by ambulance.[2] They had to cut roads in there and haul them out, they were too weak to walk. And they had dysentery and then subsequently, when the doc-

188

tors did see them, they had an extremely high white blood cell count which the doctors could not account for."

Comments: This classic case involves many valuable and important details. Consider the different responses of the UFO to being struck by a bullet. The large object moved erratically, emitted a throbbing noise, changed its luminance intermittently, and emitted a ray of light which allegedly caused serious physical injuries to a company of soldiers. The official Army records indicated (in hand writing) that over fifty men had been evacuated to a hospital as claimed by this witness. No other information is given.

Note: 1. This fact was verified from official U.S. Army records stored in St. Louis, Mo. 2. Ibid.

Reference: Haines, R.F. 1990. *Advanced Aerial Devices Reported During the Korean War*. Los Altos, Calif.: LDA Press.

Cross references: 94 (4); 100 (4); 115 (4); 116 (4); 120 (4); 121 (4); 124 (4); 126 (4); 46 (4); 147 (4)

Case No. 99

189

Date: 52-07-28,29
Local Time: unknown
Duration: estimated 10+ minutes
Location: region around Washington, D.C.
No. Witnesses: several
DB: kk   CB: O   PB-1: O   PB-2: x   UFOD: O

Abstract: The UFOs involved here were first picked up on air traffic control ground radar sets in the Washington, D.C., area. An airline pilot flying in the vicinity was radioed and asked to keep an eye out for the eight to twelve objects registered on radar. He did not see them but the authorities on the ground said that the UFO blips *disappeared from their screens when the airplane was in their area and then came in behind his plane.*

Comments: The rather common game of cat and mouse played by UFO with airplanes occurred here. Do temperature inversions, which can cause uncorrelated radar targets, disappear as airplanes approach and then fall in behind the airplanes? Proof of such dynamics clearly rests with the skeptically minded.

Reference: Hall, R.H. 1964. *The UFO Evidence*. Washington, D.C.: NICAP, 77.

Case No. 100

Date: 53-01-29
Local Time: 2318
Duration: 23+ minutes
Location: Conway, South Carolina
No. Witnesses: 1
DB: kk   CB: e   PB-1: a   PB-2: x   UFOD: O

Abstract: Mr. Lloyd C. Booth, 29, a farmer, had just returned home at about 11:15 P.M. from his many chores. He lived just north of Conway, South Carolina, with his parents. The sky was clear, the moon full, and a slight wind swept through the trees nearby. Inside his house he first heard the braying of some mules in a nearby barn and then a panic of squaks by chickens and ducks. Thinking that a prowler might be in the area, he got his .22-caliber pistol and went outside.[1] He found no one in the barn. Then, he moved quietly around the side of the barn to the back side where he provided the following account prepared by Kittler (1953). There are slight differences with other accounts, however they do not affect the overall interpretation of this event.

"Then I saw it. It was in front of me, not more than 10 feet above the tree tops, approaching slowly over the clearing beyond. I had never seen anything like it in my whole life. The eeriness of it made me feel terribly insignificant, and suddenly I felt as if I were the only person on earth. It was almost still, drifting towards me like a balloon, moving no faster than the slow walk of a man. I could see it plainly and could detect a slight humming noise. It was a light grayish color and lit up on the inside...."

"I stood there, stunned, watching as it approached from the eastern sky....I hollered to the house, about 200 yards away, hoping to arouse my family....I gripped my pistol firmly and looked back at the strange thing over my head. It looked like an egg, cut end to end. It was about 24 feet long, perhaps 14 feet across, and it seemed 10 feet high. The front sloped at approximately a 60-degree angle, and rear dropped away at about 40 degrees. The entire craft was highly streamlined."

The UFO flew smoothly over Booth's head and moved to a location above some nearby trees. He followed it and entered the woods, watching it for about twenty minutes. He claims to have run ahead of the object to a small clearing and then turned to wait for it to arrive near his location. "Again it passed over me. I raised my arm and aimed the pistol at it. When it was slightly to the west of me, I fired. I heard the bullet hit the craft. *It made a metallic noise and bounced off.*[2] I fired again, immediately, but this time I didn't hear the bullet strike because the moment the first bullet hit, *the craft suddenly made a much louder noise, like a stepped-up electric motor, and took off with terrific speed at a 65-degree angle. It moved much faster than any craft I've seen before, traveling out of sight without changing its course.* It virtually vanished into the sky." At this point, the witness waited several minutes and then returned to his house. His whole family had heard the gun shots and asked what had happened. He told them the details but they all agreed not to tell anyone else. A week later, Booth told Pastor Elwell Jones, minister of the local Baptist church, about the event. It was Jones who brought the story to the public's attention.

Comments: The witness said he was in the U.S. Army in an anti-aircraft unit and also trained to recognize "every type of aircraft" before this event. Following the event, personnel of the Civil Aeronautics Authority told the witness that a flight of U.S. Navy blimps had flown over the vicinity of the farm from Georgia to a base in North Carolina (i.e., from southwest to northeast) 191 at particularly low altitude due to high winds.[3] They suggested that he had fired at one of them. However, Booth is certain that the object he fired at traveled toward the west and continued doing so after being struck by his bullet.[4] In Fawcett's account, the UFO also became brighter when the first bullet struck it.

> Notes: 1. Booth had found one of his cows dead the previous evening. He could not find any obvious cause of its death; this event may have influenced his behavior on this night. 2. The account, written by Keyhoe and Lore, mentions a "sharp ping" sound at this point. 3. The witness remembered the winds as being light, while the Navy blimps had to fly at low altitude because of high winds. Also, blimps cannot accelerate like this object was seen to do. Finally, back in 1953 blimps were not internally illuminated as this object was described to have been. Other eye witnesses also reported seeing a similarly shaped object within a fifteen day period near Marion, South Carolina. 4. The witness later said, "I have seen many blimps and I've even been in one. I'd certainly know a blimp when I see one 80 feet over my head."

References: Fawcett, G.D. 1986. *Human Reactions to UFOs Worldwide* (1940-1983). Published privately, 11.

Keyhoe, D.E. and G.I.R. Lore, Jr., eds. 1969. *Strange Effects from UFOs.* Washington, D.C.: NICAP, 64-65.

State (Columbia, South Carolina), 7 February 1953.

Kittler, G.D. 1953. I shot a flying saucer. *Male.* (September): 16, 19, 86-87.

Cross references: 21 (3); 98 (4); 100 (4); 104 (4); 115 (4); 116 (4); 120 (4); 124 (4); 130 (4); 46 (4); 147 (4)

Case No. 101

Date: 54-12-early
Local Time: unknown
Duration: 20+ minutes
Location: near Guanare, Venezuela
No. Witnesses: 1+
DB: cc   CB: a   PB-1: a   PB-2: x   UFOD: b

Abstract: This incident involved the Director of Barquisimeto College in Guanare, Venezuela. The witness said that a bright, luminous globe approached him three times as he was driving in his car toward Guanare. Becoming alarmed, he "wildly" fired a handgun at the object. When he did so, *its lights went out.* It also disappeared and seemed to reappear in a different location. He claimed that the saucer-shaped UFO chased his car for more than twenty minutes. It then *flew away toward the south.* Both local police and an attorney also witnessed its departure.

Comments: Unfortunately, this is another abbreviated multi-witness account with little substantive information. We are not told why the man fired at the UFO or what else took place.

References: Anon. 1981. Telegram No. Dec 15-6-54; 708-103-075-10PM, RAULCOM, Barquismeto. *Saga Magazine.* 52.

Gross, L.E. 1955. *UFOs: A History, 1954 November-December.* Fremont, Calif.: Published privately, 44.

Lorenzen, C. 1962. *The Great Flying Saucer Hoax.* Tucson, Ariz.: William-Frederick Press, 43.

Cross references: 92 (4); 122 (4); 142 (4)

Case No. 102

Date: 56-01-11
Local Time: 1840
Duration: 25 minutes
Location: Wurtsmith Air Force Base, Michigan
No. Witnesses: 2+
DB: cc   CB: b   PB-1: a   PB-2: x   UFOD: c

Abstract: This Air Force Project Blue Book radar-visual interception report includes both the ground radar observer's account at the air base as well as the F-89 pilot's account. Both men agree about the UFO's actions.[1] This multiple witness case is included here because: (a) there were two independent radar locks on the object, (b) it was seen visually from the ground at the same time as the radar lock, (c) the motivation for the jet fighter's approach ranged from curiosity to potential hostility, and (d) it is representative of many scores of other cases. Only the pilot's account is given for brevity.

A luminous object was hovering between 6,500 and 7,000 feet altitude over the approach end of runway 06 at 6:40 P.M.[2] The jet was returning from a scramble (no further information given) when requested to check out the intruder. The F-89 pilot was radioed a ground-radar-based heading toward a UFO by Staff Sgt. Paul Porter at the Wurtsmith tower. The UFO was moving away from the air base on a 240 degree heading; the jet was on a 230 degree heading intercept and at a slightly higher altitude than the UFO at the time. A positive airplane radar lock was made on the unknown,[3] eleven miles ahead, with the airplane at 7,000 feet altitude. The airplane was overtaking the UFO by fifty knots.[4] The time was 6:50 P.M. At 6:52 P.M., *the "target" started to climb* as did the F-89 at 450 knots. *"The target was climbing much faster than we were"* and left the screen. The pilot took flight control manually but could not regain contact with the UFO. He returned to the field and landed at about 7:10 P.M.

193

Comment: The report does not contain any mention that the pilot had visual contact with the UFO.

Notes: 1. At one point the ground radar operator clearly discriminated the jet's radar return as separate from the unknown's. 2. To him, it looked like a reddish-orange ball, somewhat higher than wide. It subtended just over 2 degrees arc from the tower. 3. Is it possible that the UFO detected the positive radar lock and took evasive actions? 4. The jet would have reduced the distance to the UFO by over 10,100 feet during this two minute period.

This range reduction of 1.9 miles would not have increased the angular size of the UFO significantly if it could have been seen.

Reference: Gross, L.E. 1993. *The Fifth Horseman of the Apocalypse UFOs: A History 1956 January–April.* Privately published, 6-7.

Case No. 103

Date: 56-09-07
Local Time: day
Duration: estimated 10+ minutes
Location: near Loch Neagh, Ireland
No. Witnesses: 2
DB: cc   CB: d   PB-1: a   PB-2: x   UFOD: a

Abstract: This interesting account involved a husband and wife as the only witnesses and a small egg-shaped object which the man tried to grab and take to the local police station near their home in Ireland. Mr. and Mrs. Thomas J. Hutchinson watched in amazement as a flaming red, egg-shaped object descended out of a low cloud, landing on the "only dry piece of land in the middle of a bog about 200 yards from the Hutchinson front door." They left their house and trudged across the wet marshland, finding the silent and motionless object on the ground. It was about thirty-six inches high and eighteen inches in diameter. One end had two dark red marks and three dark red stripes. It rested on a saucer-shaped base. After watching the object for a while Hutchinson kicked it onto its side but was surprised to see it *right itself again.*[1]

As he got down on his knees to observe it more closely, it started to spin. "Then, he put a hammer lock on the object, but it was pretty powerful." He managed to pick it up, and both witnesses started toward the nearby police station because, as he said, "The police station was the only place for such a wicked looking thing as this." Along the way, he had to set the object down in order to get through a thick hedge. When he did, *it immediately started to spin again and rise into the air. Soon it was invisible* in the low cloud overhead.

Comments: The local police at Loup telephoned the Royal Air Force (RAF) base near Aldergrove and was informed that the object "did not belong to the RAF." A high-ranking RAF official said, "These weather balloons are almost identical

194

with the shape of the object that the Hutchinson's saw: it could have dropped to earth when it encountered some change in the air currents." Mrs. Hutchinson claimed that she saw her husband almost lifted up off the ground as he grasped the object although he did not comment on this detail. It is useless to speculate on how the object "interpreted" the behavior of the husband. Was it programmed to escape? Was spinning a part of its propulsion mechanism? Did either witness experience other physical symptoms later? Unfortunately, we have none of these answers.

Notes: 1. This response suggests a gyroscopic type effect.

Reference: *APRO Bulletin*, 15 September 1956, 2-3.

Omaha World Herald, 8 September 1956.

Register (Des Moines, Iowa), 9 September 1956.

Cross references: 106 (4)

Case No. 104

Date: 56-12-28
Local Time: early morning
Duration: 5+ minutes
Location: Wickford, England
No. Witnesses: 1
DB: v   CB: a   PB-1: a   PB-2: x   UFOD: c

Abstract: Mr. Maurice Waddope left his house early in the morning to shoot a troublesome sparrowhawk with his A40 shotgun. As he peered through the misty air, he saw a circular object at an estimated 120 feet altitude. Its diameter was "six times the size of a penny held at arms length."[1] He ran back to a clearing, loaded, and fired his shotgun at the object. According to the newspaper account of the incident, "The shot hit against the metal craft, rebounding and hitting him in the chest." *The object continued to hover over the nearby treetops for four to five minutes more and then accelerated away to the west* toward London.

Note: 1. An old English penny is 1.2 inches in diameter and, when held at arm's length from the eye (approximately 20 inches), subtends an angle of 20 degree arc. If the UFO was at 120 feet distance it would measure 43 feet across.

Reference: Gross, L. E. 1994. *UFO: A History 1956 Nov.–Dec.* Privately published, 64.

Case No. 105

Date: 56-12-31
Local Time: 0210
Duration: 10 minutes
Location: near Agana, Guam
No. Witnesses: 1
DB: cc   CB: e   PB-1: a   PB-2: x   UFOD: c

Abstract: This U.S. Air Force F-86-D jet-attempted interception of a UFO near Guam was unsuccessful because the object was able to out-fly the pilot. The aerial object not only out-accelerated the jet, but flew outboard of the airplane during turns.[1] The pilot knew there were no other fighters airborne. While at an altitude of 20,000 feet, the pilot saw the UFO at his altitude and proceeding south about ten miles. The UFO then rose to about 30,000 feet above the sea, turned around, and "came directly over interceptor. Pilot turned aircraft to keep it in sight and *it started circling him.* The interceptor was placed in afterburner and turned into the object trying to head it off. *It continued to turn around aircraft* and he was unable to gain on it or close with it."

196

The UFO then flew to the west at the same altitude as the jet, turned around and approached the airplane from 10,000 to 15,000 feet higher altitude. The pilot once again turned to keep it in sight and it "circled the fighter again." The pilot then returned to Anderson Air Force Base due to low fuel.

Comments: The object in this case was evaluated by Project Blue Book staff as unidentified!

Note: 1. The pilot acknowledges in his report that no other airplanes stationed in the vicinity were capable of turning outside his airplane's turning radius while maintaining the same velocity.

References: Gross, L.E. 1994. *The Fifth Horseman of the Apocalypse, UFOs: A History 1956 Nov./Dec.* November–December. Privately published, 65-66.

Empire News (London), 6 January 1957.

USAF Teletype from COMDR 7AF ADV GUAM MI to COMDR AIR DEFENSE COMD ENT AFB COLO, 31 December 1956.

Case No. 106

Date: 57-01-13
Local Time: 0310
Duration: estimated 4+ minutes
Location: Balfour, New Zealand
No. Witnesses: 2
DB: p   CB: d   PB-1: a   PB-2: x   UFOD: a

Abstract: This incident involved two men who saw and attempted to capture a small diameter globe of light which approached them. William West, 47, and Wallace Liddell, 34, were standing near West's garage in a fine drizzling rain. The air was calm and the sky dark. They noticed a point of light falling downward in the southwest sky from a high angle. It disappeared only to reappear several seconds later. It flew just above the tops of nearby trees (estimated to be about 60 to 70 feet tall), slowed down, leveled off at about 12 feet above the ground, and then seemed to avoid landing on a nearby house. It finally came down, hovering only about three feet off the ground near the astonished men. It was a "fluorescent blue-dull white light." As it hovered, it seemed to take on the form of a "bird," and was oblong in shape.

197

West went forward to catch it but *it suddenly jumped back about six feet from him, blinding him with its intense light.*[1] Its shape then transformed to a sphere 15 to 18 inches in diameter and it turned "the color of the moon." According to a signed statement by both witnesses, West tried to grab the sphere two more times "*but it ducked back from him each time.*"[2] Then, *the object jumped back a greater distance and "cleared the eight-foot iron fence."*[3] *The spherical light continued moving away, growing in intensity, "developed a red center and grew smaller in apparent size."* By the time the two men reached the opposite side of the fence, the object was gone. Meteorological officials in Invercargill's airport weather office had no explanation for the event.

Notes: 1. No heat was felt at any time. 2. The original report states that "the jumping back could be likened to the action of a balloon when one tries to catch it." 3. Such a fence would act as an electrical ground. It would be expected that an electrically charged plasma might well discharge itself when this near such a ground, but this did not happen.

References: Gross, L.E. 1995. *The Fifth Horseman of the Apocalypse, UFOs: A History 1957 January–March 22nd.* Privately published, 13.
The Southland (Invergargill) Times, 15 January 1957.

Cross references: 103 (4)

Case No. 107

Date: 57-05-24
Local Time: 1945
Duration: estimated 2 to 5 minutes
Location: north of Cincinnati, Ohio
No. Witnesses: 3
DB: p   CB: e   PB-1: O   PB-2: x   UFOD: O

Abstract: This U.S. Air Force interception incident involved six jet airplanes over Cincinnati about sundown on an otherwise clear evening. The jets were seen first in the northwest sky performing graceful loops, "S" curves, and other maneuvers. The primary witness wrote, "I saw clearly a silvery, spherical object in the center of the trails....The jets left trails of vapor, while the object left no visible exhaust. Two of the jets were heading toward the silvery object as though it were the target, but when they got near, *the object shot away to the northwest. Its speed was almost double that of the jets*, although I cannot say if the latter were at full throttle."

198

The third witness, Mr. George Wright, had watched the dog fight from another location and noted that: (a) the UFO was larger than the airplanes and was disc-shaped, (b) it appeared as would polished aluminum, (c) it had no protruding parts, (d) at one point the UFO did not move while two jets approached it but *then accelerated away to the west*, easily out-distancing the planes and, when other jets tried to approach the object from the west, it changed direction almost instantaneously again, out-running them. Then, it simply disappeared from sight. All witnesses estimated that the altitude of the airplanes and UFO was about 25,000 feet. **Reference:** *Civilian Saucer Intelligence (CSI) Bulletin* 20, 25 July 1957.

Cross references: 97 (4); 108 (4); 141 (4)

Case No. 108

Date: 57-07-24
Local Time: 1000
Duration: estimated 1 hour
Location: Just east of Hokkaido Island, Japan
No. Witnesses: several
DB: x   CB: e   PB-1: a   PB-2: x   UFOD: O

Abstract: Both the U.S. and Soviet Air Forces scrambled jet fighters to investigate UFOs sighted visually and on radar. UFOs were traveling at 700 mph according to radar returns. Two U.S. F-86 jets were scrambled to identify the objects, when one of them slowed to about 100 mph. The two jets flew in a different direction at first, then, one of the airplanes approached the doughnut-shaped object from above (then at an altitude of about 3,000 feet) over Nemuro Strait, Japan. The disk-shaped UFO *began to accelerate away* and the pilot opened fire with his machine guns. Nothing is known about what happened except that the UFO *"pulled away rapidly, vanishing into the distance"* in a steep, *impossible climb.*

According to one account, Soviet ground anti-aircraft guns fired at the large group of UFOs which were traveling at high speed. No hits were reported. (*Lumieres dans la Nuit* 334, 31-32).

Comments: This familiar story underscores two things, the great interest the American air force had in capturing or disabling a UFO, and their predilection to use lethal force to do so. The ability of UFO to out-fly U.S. airplanes of the time continues to be a distinguishing and potentially ominous trait. Clearly, the U.S. pilot was authorized to fire upon the UFO before he took off since he probably did not have time to radio for permission.

199

References: *Introductory Space Science II*, Dept. of Physics, U. S. Air Force Academy.
Ruppelt, E. J. 1956. *The Report on Unidentified Flying Objects.* Garden City, New York: Doubleday & Co.

Cross references: 107 (4); 110 (4); 141 (4)

Case No. 109

Date: 57-08-20
Local Time: unknown
Duration: unknown
Location: Quilino, Argentina
No. Witnesses: 1
DB: kk   CB: a   PB-1: a   PB-2: x   UFOD: 0

Abstract: An air force man drew his gun when he saw a saucer descending toward him. He heard a voice saying, "Don't be afraid," and, subsequently, "we have a base in the Salta Province" in Argentina. *A strange force stopped him from firing his weapon.*

Comment: Was the strange force externally applied or only an internal psychological reaction?
Reference: Bowen, C., ed. 1969. *Humanoids.* London: Futura Publications.

Case No. 110

Date: 59-Spring
Time: unknown
Duration: 24 hours
Location: Sverdlovsk, USSR
No. Witnesses: many
DB: dd   CB: e   PB-1: a   PB-2: x   UFOD: 0

200 **Abstract:** Personnel of the Soviet Union's Air Force and radar group at the headquarters of their Tactical Missile Command in Sverdlovsk sighted UFOs circling and hovering overhead over a twenty-four-hour period. Jet fighter aircraft were scrambled to identify the objects, but, as the airplanes approached and fired their machine guns, *the UFO reportedly zig-zagged position and readily out-maneuvered the airplanes.*

References: Good, T. 1988. *Above Top Secret: The Worldwide Coverup.* New York: William Morrow & Co., 227.
Hobana, I., and J. Weverbach. 1972. *UFOs Behind the Iron Curtain.* New York: Bantam Books.

Cross references: 141 (4)

Case No. 111

Date: 59-10-19
Local Time: 1815
Duration: 1.5 minutes

Location: near Poquoson, Virginia
No. Witnesses: 2
DB: v   CB: d   PB-1: a   PB-2: x   UFOD: b

Abstract: Two teenage boys[1] were hunting in a swampy area known as the Big Marsh[2] at about 6:15 P.M. in mid-October. The weather was clear. Mark Muza, 15, heard a sound like a "tornado" (another account uses the phrase "flock of wild birds") and looked upward in its direction. He sighted a small, dark object surrounded by a silvery ring which was hovering at about 80 feet altitude. It seemed to dip slightly from side to side and made a "whirring noise." Then, the four-foot-diameter object began to descend.

As the strange object was descending Muza became more alarmed. He raised his 12-gauge shotgun and fired a "Maximum 4" shell of #4 shot directly at it. He heard the "ring of metal striking metal" after each of his three shots. The object stopped at about 50 feet altitude. He quickly reloaded and fired a second time using a steel-bearing shell which *struck the UFO with a loud audible metallic clang.*" After his third shot, the object was "spinning like a top" and *shot straight up into the clear sky and disappeared from view.*"

201

Muza's friend, Harold Moore, Jr., 14, was about 100 yards away at the time. He said he heard the gunfire, saw the UFO, and heard the metallic sound of the ricochet and also a continuous "whirring noise" until the UFO departed. He also saw the object shoot straight up into the sky.

Comments: Here is another account of a solid aerial object whose surface produces a metallic reverberation sound when struck by a bullet. The boys said that the object had a dark center which was surrounded by a six-inch-wide silvery rim which glowed brightly "as if self-illuminated."

Notes: 1. A reporter from the *Newport News* met the boys and some neighbors and discovered that they were well-known and that there was no reason to consider them to be unreliable. 2. The marsh was used as a military bombing range and a warning sign was duly posted. The boys ignored the warning.

References: *Flying Saucer Review* 7, no. 6, (November-December 1961): 27.
MUFON UFO J. no. 258, 1989.
Newport News Daily Press, 21 October 1959.
NICAP Bulletin. November 1959, 2.

Cross references: Case C (4)

Case No. 112

Date: 61-Summer
Local Time: unknown
Duration: estimated 30 minutes
Location: Voronezh (Rybinsk), USSR
No. Witnesses: several
DB: kk   CB: e   PB-1: a   PB-2: x   UFOD: O

Abstract: There are at least two different versions of this alleged singular event; they differ in both large and small ways. The details common to both accounts are that the personnel of an anti-aircraft missile battery saw a huge UFO which "swooped down over the city." A number of smaller objects (referred to as "scout craft") came out of it. The commander ordered (apparently without authorization) missiles to be launched. "A salvo of anti-aircraft missiles roared up—only to explode harmlessly when still two kilometers from [the object]." It was as if the missiles had hit an invisible wall. He ordered a second salvo which suffered the same fate as the first. Then, the electrical system at the entire base malfunctioned for some unexplained reason so that a third firing could not occur. *The smaller objects seemed to return to the huge object in a "leisurely" fashion and then all flew away.* In one report *(Saga)* the event took place near Voronezh while in the other *(Flying Saucer Review)*, near Rybinsk, 150 kilometers north of Moscow. This article also claimed that the smaller objects descended toward the rocket batteries and somehow "stalled the electrical apparatus of the whole missile base."

Comments: While it is impossible to verify this account the events described are not unlike several others reported in the literature.

Reference: Creighton, G.W. 1962. *Amazing News from Russia. Flying Saucer Review* 8, no. 6 (November-December): 27-28.

Cross references: 93 (4); 127 (4)

Case No. 113

Date: 61-11-26
Local Time: approximately 0200
Duration: estimated 1 hour
Location: within about 150 miles from Minot, North Dakota

No. Witnesses: 4
DB: r   CB: a   PB-1: a   PB-2: x   UFOD: a

Abstract: Four hunters were driving home in freezing rain from a hunting expedition. Two were asleep in the back seat and a third in the front passenger seat. Only the driver was awake. All four were wearing hunting garb. Suddenly, through the falling rain and sleet, the driver saw a self-luminous object hovering in the sky about one-half mile distance. He yelled at the others to wake up and one man in the back did. The driver pulled off the road and stopped to see the thing better. Then, four "human-like" beings were seen standing on the ground near what looked like a violet-colored, vertical grain silo sitting in a nearby field. One of the witnesses plugged a powerful spotlight into the cigarette lighter socket and got out. He aimed the beam at the object which promptly "*exploded in a flash of light*" and everything disappeared. He thought that an airplane had just exploded. So they started the car and drove toward the strange object, thinking they might lend some assistance. By now the third hunter awoke. Upon arriving in the middle of the field they found nothing. They had to awaken the fourth man in the back who happened to be an Air Force medic stationed at Minot Air Force Base. When they explained what had just happened, he told them to go back to the original location on the road to regain their bearings more accurately. When they finally arrived at the original spot on the road, they all saw a lighted UFO and several beings standing near it.[1]

203

Then, the medic used the spotlight to illuminate the silo-shaped UFO up and down its vertical height. One of the two "beings" in the field yelled, "get out of here." The driver drove on about two more miles, thinking to himself that what they had stumbled upon was some new Air Force test device. But to their surprise, the UFO flew on ahead of them and settled down gently on the road about 150 yards away. Again, two human-like figures were seen. And again, the medic got out, only this time he had his rifle. He lay prone on the cold wet ground and took careful aim at the nearest of the two beings. He fired and the bullet struck the being in the shoulder. Then someone screamed, "Now what in the hell did you do that for?" The remainder of the events are hazy at best;[2] two of the witnesses denied that a gun had been fired at all. Nothing more is known about this strange encounter.[3]

Comments: This event contains several details that point to the possibility of a terrestrial activity (e.g., human-like beings; use of a clear and forceful English language). But how are we to explain the overflight of the automobile by the object, and why would a military test or training exercise be conducted on public property where civilians might be injured or otherwise compromise the operation? Why would the four hunters not be able to clearly recall what happened unless they were either drunk, guilty of some infraction of the law, or in an altered state of consciousness at the time? Indeed, the mind can be clouded upon emerging from a deep sleep. Why would a medic shoot at humanoids at all? Did he feel threatened? Was he intoxicated? Was he hallucinating? Interestingly, Dr. J. Allen Hynek investigated this incident and felt that the men were telling the truth.

> Notes: 1. The sudden disappearance and reappearance of seemingly solid objects is common in the UFO literature. This particular aspect has all the earmarks of a holographic image that only can be viewed from a limited range of angles. 2. The men's only clear memory was arriving home in the morning, their wives were clearly worried about them. 3. Later, the medic claimed that he had been visited by "strange men" at Minot Air Force Base where he was on duty. They asked him several questions; whether or not he got out of the car and what kind of clothing he had worn. He claimed that the men accompanied him back to his home to examine his clothes and shoes. But, we are not told why he was so accommodating—unless he thought they were government personnel on official business. They never mentioned anything about the other three men in the car. When they left in their car the medic was stranded, i.e., they didn't give him a ride back. They told him not to talk about anything.

204

References: Berlitz, C. 1988. *World of Strange Phenomena*. New York: Fawcett Crest Books; Ballantine Books, 243.

Hynek, J.A., and J. Vallee. 1975. *The Edge of Reality: A progress report on unidentified flying objects*. Chicago: Henry Regnery Co., 129.

Case No. 114

Date: 62-01-29
Local Time: unknown
Duration: estimated 50+ minutes
Location: eastern Netherlands
No. Witnesses: 6+
DB: kk   CB: b   PB-1: a   PB-2: x   UFOD: O

Abstract: A Netherlands Royal Air Force jet was scrambled to investigate a UFO which had been sighted. The pilot fired a Sidewinder missile at it, however, the instant the missile exploded the *UFO vanished from sight*. No debris fell to earth.

Comments: The report that the pilot fired a missile at the UFO indicates that his superiors had ordered it, either well in advance or during the engagement.

References: Hall, R., ed. 1964. *The UFO Evidence.* Washington, D.C.: NICAP, 122.

Keyhoe, D. 1973. *Aliens from Space.* Garden City, New York: Doubleday & Co., 51.

Cross references: 112 (4); 131 (4)

Case No. 115

Date: 65-09-22
Local Time: night
Duration: estimated 10+ minutes
Location: Rio Vista, California
No. Witnesses: 300+
DB: q   CB: b   PB-1: a   PB-2: x   UFOD: b

Abstract: This event took place in a small village of about two thousand people northeast of San Francisco in the Sacramento Valley. This series of events started in May 1964 when a strange aerial craft began appearing "rather frequently" near Rio Vista. Shaped like a dirigible or torpedo, the glowing red object was estimated to be from 3 to 5 feet in diameter and from 12 to 15 feet long.

The UFO reappeared so often that large crowds of curious people would wait for hours at the base of a water tower on a hill near the town to try to see it. Among the witnesses who did see the object on September 22, 1965, was Deputy Sheriff John Cruz of Fairfield, California. The object approached the hilltop in complete silence, glowing a dull reddish hue. It passed above the crowd at an altitude of only a few hundred feet.

One of the reports received by the Solano County Sheriff's Office concerning this event was that witnesses had seen some boys with .22 rifles shooting at it one night "and the bullets made a metallic *'twang'—and caused the object to flare up bright red for a second.*"

Comments: Here is yet another account both of a ricocheting bullet sound from the surface of a UFO and an increase in its brightness. These are small and rather unusual details that young boys probably would not think to include if they were making the story up.

References: Edwards, F. 1966. *Flying Saucers-Serious Business*. New York: Lyle Stuart, 32-33.

Cross references: 98 (4); 115 (4); 120 (4); 126 (4); 130 (4)

Case No. 116

Date: 66-03-23
Local Time: 2345
Duration: estimated 3+ minutes
Location: near Bangor, Maine
No. Witnesses: 1
DB: v   CB: d   PB-1: a   PB-2: x   UFOD: c

Abstract: John King, 22, was driving his automobile when he sighted a huge disc-like saucer flying toward him at a low altitude. He estimated its diameter at between 20 and 25 feet with a "clear bubble-type canopy on top." Becoming scared, King pulled off to the side of the road, reached over and took out a .22-magnum pistol from the glove box of his car then got out. Following is his own account as published in the magazine article.

"Then it began to hum or to whine...that was when I fired at it the first time, accidentally. But as it *continued to come toward me*....I fired again. *It turned away from me and...hovered* over a big puddle of water. That was when I shot at it the third time and I think I must have hit it, *for there was a 'spring' (sound) and its light flashed up real bright and it took off.*"

Comments: In spite of the young man's statement that his first shot was accidental, this seems to be another instance of a human act of self-protection using a gun.

References: Edwards, F. 1968. *Flying Saucers—Here and Now*. New York: Bantam Books, 26.

Keyhoe, D. 1973. *Aliens from Space*. New York: Doubleday & Co., 119.

Cross references: 98 (4); 115 (4); 120 (4); 126 (4); 130 (4)

Case No. 117

Date: 66-04-28
Local Time: unknown
Duration: estimated 4 minutes
Location: between Edgewood and Moriarty, New Mexico
No. Witnesses: 1
DB: kk   CB: a   PB-1: a   PB-2: x   UFOD: 0

Abstract: New Mexico businessman Don Adams, 20, was driving his car on U.S. Route 66 when he saw a UFO hovering ahead of him over the road. He continued on and eventually drove underneath the green-colored object. Suddenly, his engine stalled and he got out with a gun. Afraid, Adams fired six shots directly at the object without any obvious effect. He quickly reloaded and fired six more rounds; he said that he hit the object at least six separate times. It did not crash but *spun away at high speed into the distance.*

Comments: Why Adams continued driving forward beneath the hovering object isn't clear. It certainly is not the rational thing to do. Nevertheless, his overt aggression toward the object is both familiar and common.

207

Reference: *MUFON UFO Journal*, no. 252.

Rodeghier, M. 1981. *UFO Reports Involving Vehicle Interference: A Catalogue and Data Analysis.* Chicago: Center for UFO Studies.

Cross references: Case B (4); 126 (4); 141 (4)

Case No. 118

Date: 67-02-14
Local Time: 0700
Duration: estimated 5 minutes
Location: Miller County, Missouri
No. Witnesses: 1
DB: v   CB: e   PB-1: a   PB-2: x   UFOD: c

Abstract: At 7:00 A.M. on a clear and bright morning, the witness was walking toward a large barn in Miller County, Missouri. He noticed a large object through some nearby trees sitting in a field about 330 feet away. At first he thought it was a parachute. When he moved to the northeast corner of his barn, he could see several small objects moving rapidly beneath the object. He picked up two large rocks and got to within about thirty feet from the object. "It was sitting there kind of rocking slightly and I thought, 'Boy here goes, I'm going to knock a hole in that thing and see what the hell it is,'" Miller later told the investigator. "I cut down on it and the rock stopped along about 15 feet from it and just hit the ground. The next rock I thought I would throw on top of it and it just hit something and bounced," he said.

The witness moved closer to the object and saw that the smaller things "started moving around behind the shaft and into it." The UFO was a big greyish-green shell. Oblong holes on its underside emitted very bright light. He said later, "I thought I was going right up to it, I got to [about 15 feet from the object] and there it was, I just walked up against a wall, I couldn't see it at all, there was just a pressure." *The UFO then began rocking slightly off the vertical about six times and retracted the tube before it took off silently into the air, disappearing from sight in seconds.*

Comments: The witness's motive here seemed to be simple curiosity rather than fear or anger. The presence of an invisible force field has been reported before as people try to approach a UFO on the ground (cf. preliminary cases in chapter 3).

Reference: Lorenzen, C., & J. 1976. *Encounters with UFO Occupants.* New York: APRO/Berkley Medallion Books, 191.

Cross references: 131 (4)

208

Case No. 119

Date: 67-03-?
Local Time: unknown
Duration: estimated 10+ minutes
Location: northeast of Cuba over Caribbean
No. Witnesses: several
DB: kk   CB: 0   PB-1: 0   PB-2: x   UFOD: 0

Abstract: This interesting event became public knowledge in October 1979 through a Freedom of Information Act request. Cuban air defense radar control personnel reported a "bogey" approaching the Cuban mainland at 650 mph and about 30,000 feet altitude from the northeast. Two MIG-21 jet interceptors were vectored by ground radar toward the UFO. One of the jet pilots allegedly radioed back that he could see a "bright metallic craft." Then all UHF radio communication failed between the jet and the ground.[1]

Several seconds later, the second jet's pilot screamed to flight control that his flight leader's jet had just exploded! No smoke or flame was seen, however. The *UFO then rose to about 90,000 feet altitude and headed to the south-southeast* toward the South America continent.

Comments: American monitoring posts were not unaware of this event. The U.S. Air Force Security Service personnel who listened in on this terrifying event were directed to transmit all of their data tapes, log entries, and reports to the National Security Agency headquarters immediately. Fifteen members of Detachment A at the Florida monitoring post were fully informed about this unusual event which was not made public until twelve years later. It may be significant that several months later, the National Security Agency prepared a report entitled "*UFO—Hypothesis and Survival Questions*" (released October 1979). It said that "the leisurely scientific approach has too often taken precedence in dealing with the UFO question...(no matter what hypothesis is considered)...all of them have serious survival implications." This rather enigmatic statement can be interpreted in many ways, most of them frightening.

Note: 1. According to Gersten (1981), "when the pilot attempted to fire at the object, the MIG and its pilot were destroyed by the UFO."

References: Gersten, P. 1981. What the U.S. Government Knows about Unidentified Flying Objects. *Frontiers of Science* (May-June).

U.S. Air Force Security Service, 6947. Security Squadron, Detachment A, Key West, Florida. Original report from American Security Specialist of Detachment A Group, National Security Agency, Boca Chica Key Island, Florida. FOIA request by Peter Gersten.

Cross references: 144 (4)

Case No. 120

Date: 67-05-06
Local Time: 0143
Duration: estimated 3 minutes
Location: (U.S. Highway 91) 17 miles west of
        St. George, Utah
No. Witnesses: 1
DB: v   CB: b   PB-1: a   PB-2: x   UFOD: b

Abstract: This terrifying incident involved Mr. Michael Campeadore, 24, who was driving from National City, California, to Salt Lake City, Utah, by himself to attend the funeral of his grandmother. It was after midnight; he was on U.S. Route 91 when he heard a loud and unexpected humming noise. His initial thought was that a truck was trying to pass him, so he checked his rear view mirror. Nothing was there. Then, he saw directly ahead of him a circular object with a dome on top which glowed an amber color. He thought it was from 40 to 50 feet across and hovered above him at from 25 to 35 feet altitude. He slowed down and pulled off the roadway. What follows is his own account:

"I got back in the car, grabbed my .25-caliber pistol and put a clip in. I fired a full clip at the object and I heard the bullets ricochet, so I know it was solid." *The UFO accelerated away at high speed.*

Comments: This account is somewhat fragmented and leaves out many important details. We are not told, for instance, about his getting out of the car or what happened after he shot at the UFO. Without more facts such reports raise more questions than they settle.
References: *APRO Bulletin*, May-June 1967.

Lorenzen, J. & C. 1968. *UFOs Over the Americas.* New York: Signet Library, 46.

Cross references: 98 (4); 115 (4); 120 (4); 126 (4); 130 (4)

Case No. 121

Date: 67-05-20
Local Time: daytime
Duration: 45+ minutes

Location: Falcon Lake, Canada
No. Witnesses: 1
DB: x   CB: d   PB-1: a   PB-2: x   UFOD: c

Abstract: This now classic story of Mr. Steven Michalak's alleged encounter with a landed UFO near Falcon Lake, about 80 miles east of Winnipeg, Canada, is recounted here only in abbreviated form. This case has generated far more heat than light and continues to be debated as to its reliability. An amateur prospector, Michalak, 51, went hunting minerals and gems by himself in rocky terrain near Falcon Lake. At one point his attention was drawn to something behind him and, turning around, he saw two "scarlet-red" oval-shaped objects descending rapidly toward the ground. One landed about 160 feet away on the rocky surface of the ground but the other hovered nearby for several minutes before leaving at high speed.

For at least thirty minutes, the UFO simply changed colors (red to gray-red to light gray to "hot stainless steel."[1] Then, he saw a square-shaped opening appear in the side of the craft from which a "brilliant purple light" was emitted along with "wafts of warm air" in waves and the odor of sulphur.[2] Upon walking carefully toward the object, he reached a distance of about sixty feet where he thought he heard human voices coming from it. Michalak, Polish born, shouted out at the object using English, German, French, Russian, Polish, Ukrainian, and Italian to try to obtain a response! He heard no reply at all.

211

Figure 26

Eyewitness sketch of UFO seen on May 20, 1967, at Falcon Lake, Canada

Now standing within a foot of the UFO, the witness decided to take a look inside it; donning green protective lenses, which he used during rock prospecting, Michalak craned his neck into the opening. He saw a maze of lights moving horizontally and diagonally, and several series of flashing lights. *The UFO suddenly and without warning tilted slightly* and he felt a "scorching pain" in his chest. He reflexively touched the surface of the object and his rubber-coated glove melted and his shirt caught fire. He ripped his shirt off and watched as *the UFO rose upward* "with a sudden rush of air." *It was soon out of sight.*

Finally managing to reach the highway and get bus transport to a local hospital, he presented a number of severe physiological symptoms including chest burns in the pattern of a grill (see Figure 27). He said that both his outer shirt and undershirt caught fire, so he tore them off and stamped them out. Later, upon walking back to where the UFO had been, he felt nauseated and got a strong headache. "He broke out in a cold sweat, and began vomiting. Red marks began to appear on his chest and abdomen, burning and irritating" (Rutkowski 1967, 17).

The witness, who normally weighed 180 pounds, claimed to have lost twenty-two pounds over the next few days,[3] and his blood lymphocyte count dropped from 25 percent to 16 percent, only returning to normal after four weeks. During this time he felt nauseated, vomited, and had diarrhea. Three weeks later, he found a large V-shaped rash and skin blisters extending from his chest's mid-line to both ears. Five months later, he felt a burning feeling on his chest and neck and large red spots where the grill-shaped pattern had been.

Michalak and a friend, G. A. Hart, returned to the landing site six weeks after the encounter and found the "outline of the ship on the ground" along with the remains of his shirt, among other items. Soil and rock samples were collected and analyzed but nothing of importance was found, according to Story (1980).

Figure 27

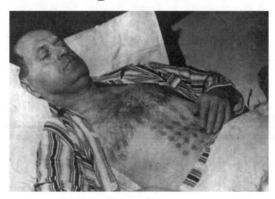

Photograph of Michalak's chest burns after his encounter on May 20, 1967

Comments: Michalak was examined on May 20 by a physician at the Misericordia Hospital in Winnipeg, by his own physician on May 22, who diagnosed first degree (thermal origin) burns on his abdomen, and by a radiologist on May 23. Interestingly, no symptoms of ionizing radiation were found.

As Rutkowski points out, the UFO's landing site was within sight of a forest ranger's tower. This ranger was later interviewed and said that he did not see the objects or the smoke from the alleged grass fire which the UFOs were supposed to have started. Unless Michalak was deliberately attempting to gain notoriety—there is no evidence to support this contention—it is patently absurd for him to make up such a story to explain his burned clothing and physical injuries when simpler explanations would suffice. The close similarity of the various physiological effects reported by Michalak and Wall [see 98(4)] are striking.

> Notes: 1. The exact meaning of "hot stainless steel" is not known, however, the description that there was a "golden glow" around it suggests it was incandescent. 2. Since it is unlikely that Michalak could have felt both the waves of warm air and smelled sulfur at such a large distance, it is more likely that he confused the order of events here. 3. Clearly an error was made here since, even with lethal doses of ionizing radiation and accompanying fulminating diarrhea, it is literally impossible for someone to lose this much body weight this rapidly. Since each pound of body weight corresponds to about 3,500 calories of energy, Michalak's body would have had to convert 77,000 calories of energy within three days, an impossible occurrence. If he had been exposed to a very large dose of ionizing radiation he would have died soon thereafter. Another type of penetrating energy may have been involved, but even it could not reasonably account for this weight loss or the skin pigmentation changes which occurred. Some amount of ultraviolet radiation would be required to alter the skin's pigment like this.

213

References: *APRO Bulletin*, March-April 1968, 4.

Clark, J. 1996. *High Strangeness: UFOs from 1960 through 1979, The UFO Encyclopedia* 3. Detroit, Mich.: Omnigraphics, 191-200.

Condon, E. U., ed. 1968. *Final Report of the Scientific Study of Unidentified Flying Objects*. New York: Bantam Books, New York.

Michalak, S. 1967. *My Encounter with the UFO*. Canada: Osnova Publications.

Rutkowski, C. 1967. The Falcon Lake Incident—Part I. *Flying Saucer Review* 27, no. 1 (Prologue): 14-16; Part 2, Vol. 27, no. 2, 15-18; Part 3, Vol. 27, no. 3, 21-25.

Story, R. 1980. Michalak encounter. *The Encyclopedia of UFOs*. Garden City, New York: Dolphin Books, Doubleday & Co., 230-231.

Case No. 122

Date: 67-07-19
Local Time: night
Duration: estimated 3 minutes
Location: Wilmington, California
No. Witnesses: 1
DB: p   CB: a   PB-1: a   PB-2: x   UFOD: O

Abstract: A night watchman at the Lumber Consolidated Co. in Wilmington, California, Mr. Jack Hill, 60, was patrolling the yard when he sighted a strange aerial object approaching him. He thought that it was about 80 feet long. He became alarmed as it seemed to approach him directly—within about 50 feet! He quickly took out his service revolver and fired six shots at the object. "*Then the lights of the craft went out and it flew away.*"
References: *Herald-Examiner* (Los Angeles, Calif.), 19 July 1967.

Saga Magazine. 1981, 52.

Cross references: 125 (4)

Case No. 123

Date: 68-11-24
Local Time: evening
Duration: estimated 5+ minutes
Location: near Ripley, West Virginia
No. Witnesses: 5+
DB: v   CB: d   PB-1: a   PB-2: x   UFOD: b

Abstract: This event began when two cousins were in their front yard killing a turkey for Thanksgiving. Eathel Southall, 16, and his 14-year-old cousin, noticed the family dog becoming agitated and looking up into the sky. They then saw a huge disc-shaped object (estimated 50 yards diameter) moving directly over their house and beyond into an "old, overgrown field."[1] The object's surface had many different colored lights (amber, green, red, and blue) on it (see Figure 28). Almost immediately, they saw things start to shoot skyward from the object. Mrs. Helen Southall and others in the family only saw these quickly rising "pods" from their vantage in the backyard. Intervening trees kept them from seeing the landed UFO. She recalled, "We started noticing these things popping up and down in midair...round and lighted and popping up to about the height an airplane would be allowed to fly at low altitude."[2]

Figure 28

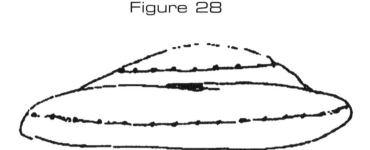

Sketch of UFO by Eathel Southall seen on November 24, 1968, West Virginia near Ripley
(Used with permission)

215

About then, the second boy grabbed his .22 rifle and started firing at the landed object. "When the first bullet hit the craft, *it went whomp*," Eathel recalls. His other shots probably missed the object because *it began darting one way or the other* "to avoid the shells." Then they watched it rise up and fly off out of sight toward Ripley.

Comments: This case is almost unique, in my estimation, with regard to the many aerial displays which shot out of the object. Were they electrostatic discharges as the craft suddenly grounded itself to earth? Were they part of an attention-getting mechanism or to warn the humans to stay away? Mrs. Southall said later, "We call them UFOs today, but they're in the Bible. Things can appear, can happen for a reason. It can be a warning, or it can bring good news....People are going to be saying, 'What an awesome God we've got!'"

Notes: 1. The silence of the UFO while overhead is suggested by the fact that the dog noticed it first. 2. No less than 1,000 feet above ground level.

Reference: Teets, B. 1995. *West Virginia UFOs: Close Encounters in the Mountain State.* Terra Alta, West Virginia: Headline Books, 87-88.

Case No. 124

Date: 70-08-30
Local Time: 2145
Duration: estimated 4 minutes
Location: Itatiaia (Rio state), Brazil
No. Witnesses: 1
DB: v   CB: d   PB-1: a   PB-2: x   UFOD: a

Abstract: Guard Almiro Martins De Freitas, 31, was patrolling a hydroelectric powerhouse at a dam when he saw a flying wing-shaped object on a Sunday night. He estimated the object to be about 50 feet away from him and about one meter in diameter. He was scared. He took out his .38-caliber pistol and fired at it point blank.

Its lights suddenly became brighter and the sound of a turbine engine on a jet plane filled the air. It seemed to dim and brighten alternately: yellow, near white, scintillating blue, yellow orange, yellow, blue, etc. He fired a second shot at it and the object emitted a blue ray of light. He felt a great heat and a prickling sensation like "pins and needles." He closed his eyes and when he opened them he could not see or move. For the sake of brevity, other details are omitted.

Comments: The witness claimed that he heard the bullet strike the surface of the object. Except for the witness's blindness, this sequence of events is remarkably similar to those reported by Francis Wall nineteen years earlier in Spring 1951 in Korea. If Wall did not know of this case, or vice versa, the two narratives provide strong evidence for a remarkably common pattern of UFO responses. We know nothing about the nature of the blindness experienced by the witness. Doctors at the hospital in Guanabara said that his blindness was "caused by shock." He regained his sight only after three days had elapsed according to Brookesmith. Did his symptoms include pain or hemorrhage effects? Was his vision affected in any other ways? What kind of small diameter process or phenomenon will produce a turbine engine-like noise when struck by a tiny piece metal?

References: Brookesmith, P., ed. 1984. *The UFO Casebook*. London: Orbis Publishing, 78.

APRO Bulletin, September-October 1970; 1, 5.

Cross references: 46 (4); 94 (4); 98 (4); 100 (4); 115 (4); 126 (4)

Case No. 125

Date: 72-03-24
Local Time: unknown
Duration: unknown
Location: Tukuran, Philippines
No. Witnesses: 1
DB: kk   CB: O   PB-1: O   PB-2: x   UFOD: O

Abstract: Mr. Tong Sanda Balinghingan said that he fired his .45-caliber pistol at a nearby UFO. *The object vanished* after the weapon fired.

Comments: While uncomfirmed, the weapon is said to have cracked in some unexplained manner as a result of this discharge. The event took place 480 miles south of Manila.

References: *APRO Bulletin.* May-June 1972, 1.

Philippine News Service (PNS).

Cross references: 236 (8)

217

Case No. 126

Date: 72-06-26
Local Time: 0900
Duration: 4 hours
Location: 9 miles from Ft. Beaufort, South Africa
No. Witnesses: 7
DB: dd   CB: e   PB-1: a   PB-2: x   UFOD: O

Abstract: This event took place on a farm near Fort Beaufort, the account is by the primary eye witness, Mr. Bennie Smit, 40. A Boer farm laborer arrived in Smit's office claiming he had just sighted some unusual smoke and then a large fireball around 9:00 A.M. The witness was frightened at the "ugly thing" he had seen. The two men returned quickly to the spot and saw a "fiery ball hovering at tree top height." It looked about 2.5 feet across with "flames shooting out."[1] The worker shouted something "and *it moved sideways for about 300 yards and disappeared behind a big bush.* All that was left was a greyish-white smoke trail. Later, it reappeared from behind a tree and kept changing color. When I (Smit) first looked it was a big red ball but now it was green and it suddenly changed to a yellowish-white. [2]

"I hurried back to the house to get my 0.303 rifle and phoned the police. I went back to the spot and fired several shots. I am sure my eighth shot hit it, as *I heard a thud. It moved up and down, and disappeared behind the trees again.*" [3]

Two police officers from Grahamstown arrived at the scene at about 10:00 A.M. and at least one policeman [4] and Smit fired more shots in the direction of the UFO. They noticed, according to Smit, a "star-like protuberance at the right of the oval-shaped thing—something that seemed to grow in size as the white light it emitted grew in intensity....I fired at the star. I'm a pretty good shot and the thing no longer changed color once the star had been hit. It stayed a darkish, gun-metal color." Officer van Rensburg said that he saw a round, black shiny object about 2.5 feet in diameter emerge from behind a tree. The Boer worker said that he heard "dull thuds" when the bullets did connect with the object. It slowly disappeared from sight and then reappeared. *Shots had no effect on it and when anybody approached it shied away behind the bushes.*" Smit then began to walk toward the object and, when he was within twenty yards of it, he saw it was a greyish-white. He said, "I fired two quick shots, but with *a loud whirring noise it veered off over the tree tops, cutting a pathway through the foliage. It disappeared quickly from sight.*" [5] As it crashed its way through the dense Fordyce Bush, it could be heard breaking the branches. Later, nine definite impressions were found in the ground near the location of the object during the event. Each set of three imprints was in a triangular arrangement; each individual mark was circular and about 5 inches across.

Comments: This case has all the aspects of most good UFO reports: color shift; reactivity to being shot at, yelled at, and approached; audible ricochet of the rifle round; hide and seek behavior; horizontal departure through dense brush rather than vertical; and disappearance/reappearance while in full view of multiple witnesses. An article in the *Pretoria News* indicated that investigators later found imprints in the very hard clay soil which were "definitely not animal or human." Very fine parallel streaks or lines were found which suggested to the investigating geologist from Fort Hare University that the imprints were probably made by metal. Each impression was round and about 0.4 inch deep at its center becoming shallower at the perimeter. Smit also recalled that he saw the trees and vegetation part to permit passage of the object, as if it possessed some kind of invisible force field.

> Notes: 1. Other newspaper accounts state that the UFO was estimated to be about 0.75 meters across and of gun-metal color. These minor size and appearance discrepancies may be the result of different interviewers obtaining data from different witnesses and at different times during the event. 2. A newspaper source indicated that Smit said the UFO changed from "bright red to dark green and then to whitish yellow." 3. According to newspaper reports, at least fifteen rounds were fired at the object, some by Smit, a "crack shot." Nevertheless, none of them seemed to affect the direction of flight of the object. 4. Either Warrant Officer P. R. van Rensburg, police station commander at Fort Beaufort, and/or Sgt. Kitching also saw and fired at the UFO with their weapons. 5. Smit said that he was so frightened, "I just thought of shooting at it."

References: *APRO Bulletin*, September-October 1972; 1, 3-4,

Daily News (So. Africa), 28 June 1972.

Flying Saucer Review 18, no. 5.

Natal Mercury (So. Africa), 29 June 1972.

Pretoria News, 5 July 1972.

Cross references: 10 (3); 115 (4); 120 (4); 130 (4); 147 (4)

Case No. 127

Date: 72-09-29
Local Time: day
Duration: 80+ minutes
Location: Hanoi, North Vietnam
No. Witnesses: several
DB: kk   CB: b   PB-1: a   PB-2: x   UFOD: 0

Abstract: This French Press agency article by eye witness Jean Thoraval stated that an orange ball appeared in the clear blue sky over Hanoi hovering in one spot for more than an hour, despite the presence of a slight breeze. At first, the object looked like a parachute in shape. Air raid sirens sounded, and three surface-to-air missiles were launched. Many citizens and military personnel watched as their vapor trails all converged on the same spot. "It was clear too that the missiles were unable to reach it, suggesting that the object had been released at a very high altitude."

Comments: If it is true that the missiles could not reach the altitude of the object, then it would have had to be very large indeed. If the term "ball" refers to an angle of even 20 minutes arc and the UFO was at an altitude of 200,000 feet, then the object would have been about 1,165 feet across.
Reference: *Flying Saucer Review* 18, no. 6.

The State Journal (Lansing, Mich.), 30 September 1972.

Case No. 128

Date: 73-10-17
Local Time: 2300
Duration: estimated 5+ minutes
Location: near Athens, Georgia
No. Witnesses: 1
DB: kk   CB: d   PB-1: a   PB-2: x   UFOD: c

Abstract: This close encounter took place as Mr. Paul Brown was on his way home from work in his automobile on U.S. Route 29 near Danielsville, Georgia. He noticed that the interior of his car suddenly lit up with an intensely bright white light; his radio also became a jumble of static. Very shortly thereafter, he saw a strange bright light come down in front of him and land on the roadway about three hundred feet ahead of him. It was about fifteen feet in diameter and blocked his path, forcing him to slam on his brakes. Then, he saw three (other accounts state "two") creatures emerge from the object. Each wore a silver suit of some kind.

Becoming afraid, he grabbed a pistol and stepped half-way out of his car to confront the beings who had begun to approach him. As he raised the gun in their direction, *the creatures got back inside the object and its lights went out. Then, the craft rose up from the ground making a "whooshing sound."* He fired several shots at the cone as it took off with no special effect.

Comments: What, if anything, did the creatures do other than approach the witness? Did they somehow threaten him? Was his response unprovoked? How did the creatures know to retreat when he aimed a pistol at them? This brief account contains telltale earmarks of an episode of amnesia or missing time. Of course there is no way to prove this.

References: Dennett, P.E. 1993. UFO: Don't Shoot. *MUFON UFO Journal,* no. 299 (March): 306.

Fowler, R. E. 1974. *UFO's: Interplanetary Visitors.* New York: Bantam Books, 311.

Machlin, M. 1979. *The Total UFO Story.* New York: Dale Books, 247.

Randle, K. 1988. *The October Scenario: UFO Abductions, Theories About Them and a Prediction of When They will Return.* Iowa City: Middle Coast Publishing, 10.

Cross references: 101 (4); 111 (4); 136 (4); Case C (4)

Case No. 129

```
Date: 74-Fall
Local Time: 1000
Duration: estimated 5+ minutes
Location: Binn, South Korea
No. Witnesses: estimated 10+
DB: v     CB: a     PB-1: a     PB-2: x     UFOD: O
```

221

Abstract: Soldiers of South Korea's Air Defense Artillery near the town of Binn were on ready alert should the enemy fire at them. Armed with U.S. built Hawk surface-to-air missiles, radar operators of Battery "D" noticed an unidentified return approaching their position at high speed from the ocean. The first visual contact occurred at a range of 700 yards through a lifting fog over the coastal waters. A huge oval form that glowed like a self-luminous metal was approaching them. It seemed to be about 100 yards in length and 10 yards thick and had red and green pulsating lights appearing to travel around its rim in a counter-clock-wise direction. Suddenly, it stopped and hovered at a range of about 700 yards from the missile battery. The object had no markings and was not a missile or any known type of airplane.

The battery commander decided that the object was hostile and ordered that one Hawk missile be fired. The rocket ignited normally and accelerated off the pad toward the UFO. The soldiers and officers watched from their control bunker. Researcher Stringfield stated, "according to my informant at the scene, *the missile never made it. It was hit by a beam of intense white light and destroyed! So was the launcher! Both were melted down like lead toys."* Then, *the UFO made a sound like a "swarm of bees" and flew away at a very high speed,* disappearing from the radar scope.

Reference: Stringfield, L. 1977. *Situation Red.—The UFO Siege.* New York: Fawcett Crest.

Cross references: 132 (4); 149 (4)

Case No. 130

Date: 74-03-20
Local Time: 2200
Duration: estimated 5 minutes
Location: Boshkung Lake, Canada
No. Witnesses: 50+
DB: o   CB: a   PB-1: a   PB-2: x   UFOD: b

Abstract: A group of over fifty people gathered to watch from the shore of a lake in Canada as an oval-shaped UFO sat on the ice and snow of its frozen surface. Six men on snowmobiles rode out onto the ice toward the object. "One of the pursuers fired shots at the UFO and the witnesses heard distinct 'clunks' as the bullets struck its metal surface. *The UFO then lifted straight up and quickly moved away from the area."*

Comments: Depending upon wind conditions, distance from the witnesses to the UFO, and the loudness of the gun's explosion, it may or may not be possible to hear such a ricochet noise.

References: Fawcett, G. 1975. The "unreported" UFO wave of 1974. *UFO Report* 2, no. 3 (Summer): 50-52.

Flying Saucer Review 20, no. 4.

Cross references: 94 (4); 100 (4); 115 (4); 116 (4); 120 (4); 121 (4); 124 (4); 126 (4); 46 (4); 147 (4)

Case No. 131

Date: 74-09-06
Local Time: 1800
Duration: 11.7 hours
Location: Khaishi, (Georgian SSR), USSR
No. Witnesses: 4
DB: jj   CB: f   PB-1: b   PB-2: x   UFOD: O

Abstract: While walking in the mountainous area beside the Inguri River,[1] four men came across a very strange, four-meter-diameter, sphere sitting in a "small glade among the bushes." The following sketch is based upon their verbal description of the silent object.

Walking up to the inert object, Drs. A.M. Gurgenidze (from Tbilisi; candidate in physics and mathematics), A.I. Nikolayev (candidate of historical sciences), I.I. Gershenzon (Moscow; candidate of technical sciences), and Mr. S.P. Lezhava (Moscow; engineer) attempted to "establish contact with the supposed crew of the UFO; they shouted loudly and waved their hands. There was no reaction, however."

223

Figure 29

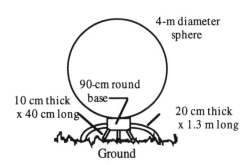

Reconstruction of spherical UFO seen on September 6, 1974, near Khaishi, USSR

Coming closer to the strange object, everyone began to *feel a fear* building inside themselves at about 5 meters distance from the object. Walking even nearer, this fear transformed itself into "*intolerable terror*" at about 4 meters distance.

Nikolayev continued to approach the sphere to a distance of about three and one-half meters when he lost consciousness and fell to the ground.[2] The other three men carried their unconscious colleague some distance (75 meters) away to set up camp and watch the UFO. Lezhava returned to the village to report the ongoing event. After lighting a campfire, those remaining at the site decided to carry out some more experiments. They obtained nineteen photos of the object while it was still twilight. Clearly, they were trying to evoke some overt response from the UFO.

"First, the witnesses threw stones at the object. *The stones hit the UFO and, as it were, disappeared inside it.* After the UFO departure, these stones were not found at the place." Then, they made a long stick out of several shorter ones and tried to "thrust the stick into the object. The stick's end plunged into the object's substance at a length of 3.5 centimeters. When the stick was carefully drawn out from the object, its end proved to have been considerably altered."[3] When it became dark they stopped their "experiments" as it was hard to see what was happening.

At 5:23 A.M., September 7, the two men saw five small—one meter diameter, 35 centimeters thick—saucer-shaped objects hovering about 6 meters away from them. All were at 1.5 meters above the ground. Gurgenidze and Gershenzon began to re-experience fear and stayed where they were. Four minutes later they heard some footsteps and saw a 2-meter-tall human-like figure approaching the UFO. The being was dressed in a "sort of overalls or space-suit. After a few minutes, the entity vanished inside the UFO."

At 5:39 A.M., the sphere emitted an orange light; the witnesses felt that it was hard for them to breathe. Three minutes later, they saw a reddish-golden fire appear under the object[4] which then emitted a whistling sound and "quickly vanished from sight to the northeast." It was followed by two of the small saucers. The other three saucers flew away to the west.

Comments: Soil samples were obtained and analyzed where the UFO had been. Signs of high temperature heating were found. An increase in ionic concentrations of lighter elements was also found along with slightly increased ionizing radiation. If substantiated, this case may provide valuable clues to the energy domain (power levels, frequencies) of UFO.

Notes: 1. The event took place 32 kilometers northeast of Khaishi in the Georgian SSR. 2. He remained unconscious until the UFO took off at 5:42 A.M. It isn't clear how the other men were able to retrieve his body without being similarly affected. 3. Later laboratory analyses were carried out on the stick; the wood's molecular structure had "collapsed." The exact meaning of this is not known. 4. The fire was not like a rocket exhaust but was "very flabby." This most likely refers to slow, undulating tongues of fire.

Reference: Private research files of Vladimir Rubtsov (used by permission). Original data received from Dr. Vadim Vilinbakhov (now deceased) of Leningrad.

Cross references: 118 (4); 133 (4); 208 (7)

Case No. 132

Date: 75-?-?
Local Time: unknown
Duration: estimated 10+ minutes
Location: Sardinia Island, Mediterranean Sea
No. Witnesses: several
DB: cc   CB: e   PB-1: a   PB-2: x   UFOD: c

Abstract: The soldiers of a missile battery of the North Atlantic Treaty Organization (NATO) were on duty on the island of Sardinia when they sighted an UFO hovering out over the sea at an unspecified distance from land. Subsequent motion pictures obtained at the missile launch site showed the object to be very large, disc-shaped with a dome on top, and emitting light of some unspecified type and color. Orders were given to fire missiles at the object. When the missiles reached a certain distance they suddenly exploded as if they hit "an invisible bulwark." The report seems to imply that more than two missiles were fired, all destroyed in the same manner. The movie allegedly showed the *UFO emitting a "laser-like ray, which disintegrated the missile."* The controlled missile launch was said to be "flawless" implying that the disintegration of the missiles was not due to some malfunction from the ground.

225

Comments: Retired Major Colman VonKeviczky received this remarkable report, dated November 3, 1976, from a captain in the NATO offices in Brussels, Belgium. When Maj. VonKeviczky requested a copy of the film from Supreme Headquarters, Allied Powers Europe as suggested by the Italian Ministry of Defense, the Air Attache informed him that, "the film is not available on account of the 'national security policy of Italy.'" The letter was signed by Brig. Gen. Fernando Buttelli.

Reference: Anonymous to Maj. Colman VonKeviczky, 3 November 1976. NATO Headquarters, Bruxelles, Belgium.

Cross references: 129 (4); 149 (4)

Case No. 133

Date: 75-06-14 to 16
Local Time: 0150[1]
Duration: 1 hour
Location: Gribanovka Region (Voronezh Oblast), USSR
No. Witnesses: 1
DB: jj   CB: e   PB-1: a   PB-2: x   UFOD: 0

Abstract: Capt. V. G. Paltsev of the Soviet Air Force was waiting for a ride to the town of Borisoglebsk at 1:50 A.M. The weather was cool and clear. A full moon illuminated the otherwise dark countryside. He noticed a "dark form in the field" across from him, between two stands of trees. His initial mental interpretation was that a tractor's or car's interior lights were on. Thinking that it might be broken down, he began to walk toward it.[2]

Crossing a field of green sprouts of wheat, he was about 80 meters from the object when he realized that the object was disk-shaped and had a dome on top that was internally illuminated. "Inside it there was seen a silhouette of a human (or human-like) figure." Its lower half was not visible, being obscured by the body of the craft. Upon reaching 30 meters distance, Paltsev could see "outlines of another human-like figure inside the dome." At about 25 meters distance, he "saw its thin edge sway wave-like. At the same time his body (not his face) felt an invisible barrier, something like a net or a hammock. Trying to move further, Paltsev pressed the barrier…and lost consciousness."

Awakening some (unknown) time later, the witness got up and "rushed to the attack at the enemies (the barrier and the object) with bare hands, but with raised fists. His only thought was: "To thrash the scoundrel!" But then he noticed that as he ran toward the object *the distance between them did not change. It was flying at the same speed away from him, the separation distance remaining 25 meters. Then, the object rose vertically and without any noise up to 3 meters altitude and hovered momentarily. Then, gaining more altitude (to about 15 meters), it flew across the highway*. He noticed that the bottom of the UFO was "very dark." It landed again on the opposite side of the roadway and, as Paltsev approached it again, *it took off toward Voronezh at a high acceleration*.

Comments: This CE-5 encounter involved physical contact with some type of barrier about 25 meters from the surface of the UFO. This boundary effect: (1) was invisible, (2) possessed some type of tactual qualities (net, hammock) (see preliminary cases 64-06-14; 68-07-? in chapter 3), and (3) produced unconsciousness (see preliminary cases 54-12-29; 69-02-06; 75-11-05; 76-11-14 in chapter 3). The thin edge of the object which swayed in a wave-like fashion is also highly interesting. It suggests either optical refraction, possibly due to heat, or a dynamic, non-solid surface.

> Notes: 1. The witness's judgment of the total duration of this event was "no more than fifteen minutes" although one hour went by. The exact date is not known. 2. He thought he could get a ride to his destination if he helped them fix the trouble.

Reference: Private research files of Vladimir Rubtsov, Kharkov, Ukraine. (Used by permission.)

Cross references: 131 (4)

Case No. 134

Date: 75-11-08
Local Time: 0420
Duration: 2+ hours
Location: 5 U.S. Air Force SAC nuclear missile sites,
          Montana
No. Witnesses: several
DB: kk   CB: b   PB-1: a   PB-2: x   UFOD: 0

Abstract: Height finder radar units at Malmstrom Air Force Base, Montana, received seven radar contacts at 2:53 A.M. at altitudes ranging from 9,500 and 15,500 feet. Air Force Sabotage Alert Team witnesses on the ground "simultaneously...observed lights in the sky and the sounds of jet engines" at missile sites K1, K3, L1, L3, and L6.[1] The unidentified objects moved at 7 knots speed over Lewistown, Montana.[2] At 2:57 A.M., two F-106 jet interceptors were scrambled from the 24th NORAD Region. The *"objects could not be intercepted. Fighters had to maintain a minimum of 12,000' because of mountainous terrain[3] and* broke off at 3:25 A.M. Sightings had turned west, increased speed up to 150 knots....SAC site K3 reported sightings between 300' and 1,000' while site L-4 reported sightings 5NM NW of their position."

At 4:20 A.M., the personnel at four SAC missile sites "reported observing intercepting F-106's arrive in area; *sighted objects turned off their lights upon arrival of interceptors, and back on upon their departure.*"[4] SAC site C-1 continued to see lights in the sky as late as 4:40 A.M. Visual sightings of both the UFO and the F-106 jets were made from sites L1 and L6 at 4:04 A.M. From site L5 the UFO "increased in speed—high velocity, raised in altitude and now cannot tell the object from the stars."

Comments: This interesting CE-5 event involved many highly credible military witnesses to a potentially serious game of cat and mouse. Were the "intruders" merely friendly forces testing the defensive response capabilities of SAC? If so, why would the air force even release these reports at all since they would later be explained? Clearly, SAC failed the test miserably. While NORAD raised the possibility that Northern Lights may "sometimes cause phenomena such as this on height-finder radars," a check with weather services revealed "no possibility of Northern Lights." Upon checking with the FAA in Washington, D.C., the National Military Command Center discovered that no commercial or other pilot reports had been received for this location or time.

Notes: 1. The FAA indicated no known air traffic within 100 miles. 2. Although the unidentified object sounded like a jet engine, it traveled at only 7 knots. 3. One wonders what good such interceptors are if they cannot fly low enough to the ground to engage low flying objects after dark. 4. Another document released under provisions of the FOIA stated that, "This same type of activity has been reported in the Malmstrom area for several days." Interestingly, the UFOs disappeared as the sun rose at about 7:00 A.M.

Reference: Memo for the Record, dtd. 8 November 1975, 0600 EST, *The National Military Command Center*, Washington, D.C. I am grateful to Jon Griffith who obtained this report for me and to Dale Goudie and James Klotz (1992) for sharing their Freedom of Information Act-obtained documents with me.

Case No. 135

Date: 76-09-19
Local Time: 0130

Duration: 4 hours
Location: Tehran, Iran
No. Witnesses: several
DB: cc   CB: h   PB-1: a   PB-2: x   UFOD: c

Abstract: This now classic air interception case involved two F-4 Phantom jet interceptors of the Imperial Iranian Air Force and at least one brightly illuminated UFO over the capital of Iran. The object was first seen flying over the southern section of Tehran at about 6,000 feet altitude by residents of the Shemiran district. They telephoned the Mehrabad Airport control tower.[1] The initially round object alternately flashed red, green, and blue lights. Its presence was unexpected; there were no scheduled airplanes at that location and time. So, Hossein Pirouzzi, chief air traffic controller and another controller present, looked for the object with binoculars from their position in the tower. They sighted something with blue lights on its two sides and a flashing red light in the middle between them.

Figure 30 is a drawing of this object seen by Pirouzzi showing the bright blue-white pulsating lights at each end and a small red light which appeared to rotate around its body, stopping momentarily every 90 degrees arc. This drawing was checked for accuracy by Pirouzzi; it was obtained by reporter John Cathcart. Also note the hazy yellow-blue arc extending above the body of the strange object. The two controllers instantly realized it could not be an airplane since blue lights are not permitted on any type of airplanes.

229

Perouzzi called his supervisor who called General Youssefi, then sound asleep in his home some distance from the airport. Youssefi got up, went out onto his balcony, and saw the UFO himself. He then contacted two separate military radar stations (Shaharoki, 135 nautical miles west-southwest; Babolsar, 88 nautical miles to the northeast). The relatively low altitude of the UFO precluded it from being picked up by either radar station.

Figure 30

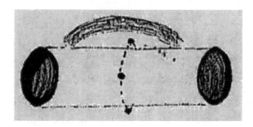

Sketch of cylindrical UFO seen by tower observer, Hossein Pirouzzi on September 19, 1976
(Used by permission)

At 1:30 a.m., an F-4 jet fighter was scrambled from Shaharoki and flew at high speed to the east-northeast and, when about 25 miles from the aerial phenomenon, *its radio and some instrumentation stopped working*. The panel flashed intermittently. For the following thirty minutes or more, the airplane tried to chase the object, gradually moving to the east near the border with Afghanistan.[2] Not wanting to cross the border, he turned back toward Tehran and, to his great surprise, he saw a brightly colored UFO ahead of him to the south! At a range of about 14 miles from the UFO, the pilot estimated its altitude to be 15,000 feet; the pilot decided to break off his chase because his fuel was running low.

A second F-4 airplane took off from Shaharoki at 1:40 A.M. and picked up the UFO on its on-board radar at a range of 27 miles. The return on the screen was equivalent in size to that of B-707 passenger plane. The second pilot described its lights as "that of a flashing strobe...arranged in a rectangular pattern and alternating blue, green, and orange in color." Both the jet and UFO were now south of Tehran when the pilot saw a second, smaller globe of light suddenly come out of the main object. *The smaller light flew directly toward the pilot at a very high rate of speed.* He armed his missile system, intending to fire an American-built AIM-9 missile at it but *his entire weapons control panel stopped working*. Also, his interphone and UHF radio communications systems failed to operate. He executed a high speed turn *and the smaller UFO quickly maneuvered behind him at a range of about 3 to 4 nautical miles.* Now flying away from the main UFO *the smaller object left him and made a perfect rendezvous with it.*

Immediately after the smaller UFO had rejoined the multi-colored globe, another smaller light came out of its opposite side and fell almost vertically down to Earth. Both crewmen watched as the light illuminated a large area of ground before coming to a gentle decelerating stop. The very high brightness of the UFO caused a dazzling and long-lasting loss of sensitivity to light within the pilot's visual system, causing him to decide to land at Mehrabad airport. But in order to allow his vision to regain its sensitivity to the dark, he decided to fly several large radius circles above the city to gain time. Significantly, on each circle *he noticed a transient gyro-compass deviation*[3] and *UHF interference when passing through a magnetic bearing of 150 degrees from Mehrabad airport.*

When the jet was on its final, straight-in approach to Mehrabad, the crew sighted another cylinder shaped object estimated to be about 8 meters long and 10 miles away. The many other interesting aspects of this important case will not be presented here.

Comments: The Central Intelligence Agency, U.S. State Department, and U.S. Air Force all showed a great deal of interest in this case for obvious reasons. Something had disabled the weapons and other electronic systems on an American-built fighter airplane. This fact took this particular UFO out of the realm of the theoretical very quickly. Of course, their subsequent findings have never been made public. Assuming that UFO have encountered jet fighter interceptors before, which this review and others has clearly documented, it is reasonable to assume that the UFO expected to be fired upon. Were they prepared to react in some particular manner? Were the small round globes of light some type of defense system, some kind of remotely guided sensors, a symbol of technological superiority, or something else entirely different?

231

> Notes. 1. I interviewed one of these ground eye witnesses, Mrs. S. Cashani, who said that she and her mother saw several grey, round objects in the hazy, early morning sky (5:00 A.M.) for about 10 to 15 seconds. They moved in circles and also moved up and down. They disappeared by descending behind a nearby building. 2. The distance from Tehran to this border is about 550 miles. Traveling at 550 miles per hour the jet would have traveled this distance in about sixty minutes, flying in a straight line. 3. It is the gyro compass deviation of from 30 to 50 degrees arc which is so extraordinary since this sensor system is only influenced by inertial and not magnetic field forces. A similar gyro-compass deviation case occurred on March 12, 1977, over New York by a DC-10 piloted by Capt. Neil Daniels.

References: Joint Chiefs of Staff (Department of Defense). 17337. Report 6 846 0380 78. 19 July 1978.

Fawcett, L., and B. J. Greenwood. 1984. *Clear Intent.* New Jersey: Prentice-Hall, 81-93.

Tehran Journal 23, no. 6692 (20 September 1976.)

UFO Investigator. Washington, D.C.: NICAP, September 1977.

Cross references: 153 (5)

Case No. 136

Date: 76-?-?
Local Time: approximately 2200
Duration: 5+ minutes
Location: Miltown Molbay (County Clare), Ireland
No. Witnesses: approximately 4
DB: p   CB: a   PB-1: a   PB-2: x   UFOD: c

Abstract: It was about 10:00 P.M. when Mrs. Ramsier saw a "great big ball of fire" land in her garden. She ran back inside the house and told her guests what she had just seen. They hurriedly ran to a nearby shed and got pitchforks, shovels, and sticks and then approached the thing. As they got closer *it simply rose up silently from the ground and began moving away from them*. They followed it for about a mile across the fields before returning home. They stopped at the cemetery fence and the object went over it. An inspection of the ground the next morning showed no marks or other evidence that the event had ever happened.
Reference: Paget, P. 1979. *The Welsh Triangle*. Granada, London: Panther Books, 167.

Cross references: 111 (4); Case C (4)

Case No. 137

Date: 77-?-?
Local Time: night
Duration: estimated 20+ minutes
Location: Fukien Province, People's Republic of China
No. Witnesses: (military company) estimated 150+
DB: w   CB: e   PB-1: a   PB-2: x   UFOD: 0

Abstract: This alleged incident involved the silent approach of a luminous globe or sphere over the Formosa Strait from the direction of Taiwan. It flew without making any noise and landed on the top of a hill on mainland China in Fukien Province. Witnesses quickly alerted officers at a nearby army garrison who ordered a company of men to investigate. They rushed to surround the entire hill on which the strange object had landed. A captain in charge of the operation supposedly said, "Do not fire unless I order it." A lieutenant replied, "Are we going to capture it?" At this the captain confidently replied, "I presume, it may be some kind of secret weapon that the United States is providing Taiwan."

The men began to approach the UFO and, as they did so, it seemed brighter than ever.[1] Those who were nearest to the object could not look directly at it due to its extreme brightness; many felt dizzy. The Lieutenant asked, "Captain, should we open fire? We cannot get close to it." The captain commanded, "Yes, but first our troops must retreat in security." It was at that moment that a *"fearful noise came out of the UFO—a sound that resembled the howling of the wind and the rattling of a running engines....the light also brightened as well."*

Then, the captain ordered his men to fire at the object. *The object was unaffected by the bullets fired at it.* At some point, *it began to rise gently off the ground,* "lighting up the whole vicinity. Twenty seconds later, it could no longer be seen."

233

Note: 1. It is not known whether the object became brighter because of a distance-related effect or because it emitted more luminous energy.

Reference: Dong, P. to Major Colman VonKeviczky. *"Barrage 'Farwell' to a Departing UFO."* Undated.

Cross references: 94 (4); 98 (4); 100 (4); 115 (4); 124 (4); 126 (4); 46 (4)

Case No. 138

Date: 77-10-29
Local Time: 1800
Duration: estimated 45+ minutes
Location: Tapiapaneema (on Mosqueiro Island), near
             Colares, Brazil
No. Witnesses: 4+
DB: dd   CB: a   PB-1: a   PB-2: x   UFOD: 0

Abstract: This terrifying event involved a young couple who where resting in their home. Silvia, 17, was well into her pregnancy; her husband, Benedito

Trinidade, 24, was asleep. She noticed something luminous moving in the dark evening sky outside and then a beam of light which came down to the ground. The light beam entered the house, and fell on her arm. She screamed in terror and awoke Benedito who took her into another room. He heard a neighbor's dog barking as he did this. But while helping Silvia, he was also hit by the light. This caused him to feel "strange...I couldn't talk, and Silvia fainted," he explained later. Hearing the young woman scream, a neighbor grabbed his rifle, ran outside, and shot at the UFO, "*but it disappeared*" (Pratt 1996, 190). Benedito found that he could talk again several minutes after the light left. He decided to take his wife to the hospital in Mosqueiro about an hour's journey down the river. After helping her into a small boat, the shaken but brave couple set out through the darkness. The UFO returned and "flew over the river several times....*It followed us and shined a light down on the river for ten or fifteen minutes. It didn't make any noise, and it didn't shine the light on us, just the river." The UFO traveled on over the jungle and disappeared from sight.[1] Hospital personnel examined the couple and admitted Silvia for a week.[2] Sadly, two months later, she lost the baby. Some neighbors had also seen "a light from the UFO passed over" another neighbor's dog that night; the dog stopped barking at that moment and never barked again. It died several weeks later.

Comments: Once again, visible radiation (light) seems to play a pivotal role in causing traumatic, physiological symptoms whether physical and/or psychological. The exact mechanism by which it effects humans so quickly is not yet known. Indeed, most electromagnetic radiation in and near the visual spectrum requires hours or even days to produce a biological effect (Haines 1976). As do others, this case includes many problematic events. For instance, how did the light enter the house? Where was the UFO while the light fell on the young couple inside the house? Why didn't the UFO shine its light on the couple later in the boat when it would have been so easy to do so? What caused the dog to die? What caused the baby's death? We are faced here with both fascinating, yet fearful, facts.

Notes: 1. Pratt documents many other frightening close encounters around the town of Colares near the mouth of the Amazon River (see chapter 20). 2. Silvia was found to have a bruise on the inside of her left elbow but no skin burns immediately after this event.

Reference: Haines, R. F. 1976. Psychophysical and biological aspects of viewing very bright objects. Proceedings of the 1976 CUFOS Conference, 30 April/ 2 May, at Lincolnwood, Ill.

Pratt, B. 1996. *UFO Danger Zone*. Madison, Wis.: Horus House Press, 189-191.

Case No. 139

Date: 77-11-?
Local Time: 1800
Duration: estimated 10 minutes
Location: Colares, Brazil
No. Witnesses: approximately 20
DB: kk   CB: 0   PB-1: 0   PB-2: x   UFOD: 0

Abstract: The witnesses to this close encounter included many inhabitants of Colares, one of numerous small villages on islands at the mouth of the Amazon river. In many instances field workers were struck from above by a hot, white beam of light while resting in their hammocks.[1] Dr. Wellaide Cecim Carvalho de Oliveira, 24, was director of a small public hospital in Colares and also an eye witness to the following encounter. She and her secretary (who fainted during the sighting) saw a cylindrical, metallic-silver object. It flew within about 40 meters of the two women and remained for several minutes. "It was very beautiful," she said later. People nearby shouted at her to run and hide but she didn't. "I was too fascinated," she explained. *It continued to fly in circles* and made "rings in the sky.

235

"One time a UFO flew very low.[2] It was going to land. People shot at it and threw stones to drive it away, but [military personnel][3] arrived at that moment and shouted: "No! No! No! Don't do that!" (Pratt 1996, 182).

Comments: Nothing is known about the response of the UFO to being shot at. It is likely that the soldiers present were trying to keep from provoking the UFO into more destructive actions. Carvalho treated approximately forty people for injuries allegedly related to encounters or sightings of UFO in 1977 in and around Colares. Most of the trauma presented consisted of superficial burns on the chest, face, and throat, typically from 10 to 20 centimeters across. The skin became very red and swollen and sensitive to touch. Healing was faster than normal; skin peeled off readily. Significantly, she noticed that almost all of these burn areas had two small puncture wounds near their center. These patients told her that this was where the ray of light had struck them. It would be valuable to know whether she treated other patients with the same burn symptoms, yet did not see a ray of light. Could the two puncture marks have been a poisonous bite? Perhaps the light ray was hallucinated after they were bitten while in a high fever state. During interviews in the field, Vallee also discovered that every patient seen by Carvalho presented a reduced red blood cell count. Many also evidenced headaches, dizziness, weakness, involuntary tremor, reduced arterial blood pressure, anemia, and other symptoms.

Notes: 1. Vallee (1990) discovered that the beam struck each victim suddenly and without searching around first. 2. It isn't clear whether this event took place at the same time. 3. The officer in charge of investigating UFO sightings was Capt. Uyrange Hollanda of the Brazilian Air Force.

References: Pratt, B. 1996. *UFO Danger Zone.* Madison, Wis.: Horus House Press, 182-183.

Vallee, J. 1990. *Confrontations.* New York: Ballantine Books, 198-203.

Cross references: 135 (4)

Case No. 140

236

Date: 78-summer
Time: evening
Duration: estimated 2 minutes
Location: Eagle Creek, Oregon
No. Witnesses: 1+
DB: v   CB: a   PB-1: a   PB-2: x   UFOD: a

Abstract: The witness, a well-known and highly respected citizen of Eagle Creek, saw a self-luminous orange ball hovering above his lawn. He ran inside and got his shotgun, came back outside, took aim, and fired a shot at it. *It immediately disappeared.* When a reporter asked him why he shot at the UFO he replied, "I am kind of like my dog, I am territorial." The object returned the following night. Sightings continued in the area for at least another twelve months.

Comments: The witness was claimed to be "a professional man trained to think in scientific and practical term [and thought to be an] astute observer of his surroundings."

Reference: *Clackamas County News* (Estacada, Oregon), 14 February 1979.

Case No. 141

Date: 80-05-09
Local Time: approximately 0700

Duration: 12 minutes
Location: La Joya, Peru
No. Witnesses: several
DB: cc   CB: b   PB-1: a   PB-2: x   UFOD: 0

Abstract: A UFO was spotted on two separate occasions, the first during the morning of May 9 and the second in the early evening hours of May 10[1] in southern Peru. A group of Peruvian Air Force officers at the Mariano Melgar Air Base were outside standing in formation when they sighted a round object "hovering [in the sky] near the airfield." The commander quickly authorized a Soviet-built Sukoi 22 jet fighter to take off to try to intercept and positively identify the object.[2] The jet pilot "fired upon it at a very close range *without causing any apparent damage. The pilot tried to make a second pass on the vehicle, but the UFO out-ran the SU-22." The UFO accelerated out of sight.*

A UFO reappeared on May 10 after dark and another SU-22 jet was scrambled. But the unknown object again *"out-ran the aircraft."*[3]

Comments: This case is interesting in that a UFO returned a second time to the same area at which it had been fired at before. Was this the same UFO returning? Is this a deliberate attempt to show their technological superiority or to give evidence of non-belligerent friendship? One wonders what the Air Base Commander felt like following this second unsuccessful intercept and also how the American Joint Chiefs analyzed and reacted to these events.

237

Notes: 1. It is more likely that the second event occurred in the early morning hours. The report later states that the second event occurred during hours of darkness, when the UFO was "lighted." 2. These facts clearly indicate that: (a) the witnesses did not know what it was and (b) considered it to be a potential threat. 3. The terse telegram sent to Washington, D.C., does not indicate whether the witnesses thought it was the same object on both days. The reporting officer did comment that, "Apparently, some vehicle was spotted, but its origin remains unknown."

Reference: U.S. Joint Chiefs of Staff memo from USDAO, Lima Peru, No. 18134, dtd. June 3, 1980

Good, T. 1988. *Above Top Secret: The Worldwide UFO Coverup.* New York: William Morrow & Co., 324.

Cross references: 97 (4); 107 (4); 110 (4)

Case No. 142

Date: Before 81
Local Time: daytime
Duration: estimated 5 minutes
Location: Venezuela
No. Witnesses: 2
DB: v   CB: a   PB-1: a   PB-2: x   UFOD: 0

Abstract: Mr. Almiro A. Finol and another man were duck hunting in a marshy area. They noticed two saucer-shaped objects approaching them. One came within about 250 feet. Finol shouted, "Go away or I fire!" The object continued to approach them so Finol aimed his double-barreled shotgun at it and fired. He then reloaded a second shell and fired again at the second object somewhat farther away. *Both UFO then flew away.*

Comments: It is likely that the objects' departure was a result of being fired upon.

Reference: *Saga Magazine*, 1981. p. 52, 54.

Cross references: 100 (4); 101 (4); 111 (4); 116 (4); 120 (4); 122 (4); 126 (4); 130 (4); 142 (4); Case C (4)

Case No. 143

Date: 81-10-17
Local Time: evening or night
Duration: estimated 5+ minutes
Location: Parnarama, Brazil
No. Witnesses: 2
DB: kk   CB: 0   PB-1: a   PB-2: x   UFOD: 0

Abstract: This event occurred shortly after a series of highly unusual and suspect deaths of three hunters[1] allegedly caused by an intense ray of light from a hovering, box-like or rectangular UFO estimated to be from 4 to 6 feet long. "It was like a giant truck tire, all lit up and spinning around....It emitted a light which surrounded its victims causing their bodies to become "glittering," said Mr. Ribamar Ferreira who was present when his friend Abel Boro was killed. The witness to another killing was Jose dos Santos who was being chased by the UFO. He fired five shots at it and managed to escape. He saw a friend working at the top of a nearby hill when the object approached him and "*shone a beam of light at him.*" Santos claimed, "*He received a shock* and came rolling down the hill. He went crazy for three days. He was terrorized. Then he died."

Comments: This story was published in the U.S. *National Enquirer* on December 29, 1981, and should be accepted accordingly. It is presented here only because of its general similarity with other cases of human hostility shown toward UFO. If the three prior deaths of the hunters are true, it would readily explain why dos Santos fired at the object. Unfortunately, nothing is said about the response of the UFO to being fired at.

Note: 1. Vallee (1990) states that at least five deer hunters were killed, including Raimundo Souze (on October 19, 1981), Dionizio General, Jose Vitorio, and Ramon (from Parnarama).

References: Vallee, J. 1990. *Confrontations. A Scientist's Search for Alien Contact.* New York: Ballantine Books, 118.
World round-up. *Flying Saucer Review* 27, no. 5 (March 1982): 25-26.

Cross references: 98 (4); 129 (4); 145(4)

Case No. 144

Date: 83-08-26
Local Time: unknown
Duration: 20 minutes
Location: Ventspils, Latvia
No. Witnesses: several
DB: kk   CB: e   PB-1: a   PB-2: x   UFOD: 0

Abstract: This newspaper account (of questionable reliability) describes an event of both military and political significance if true. Four Russian jet interceptors allegedly fired rockets at a UFO. Allegedly, *the missiles exploded so fast that three of the airplanes were destroyed.*

Comments: If this account is accurate, why would Soviet officials permit any report of such an event to mention UFO? To do so would not only cause greater doubt about the whole event, but also bring discredit upon the commander who ordered the jet scramble. In short, UFOs may have been used as a "cover" for some military testing in the Soviet Union, as James Oberg and others maintain from time to time, but certainly not for every reported UFO event.

Reference: *National Enquirer UFO Report.* 1984.

Case No. 145

Date: 85-08-?
Local Time: 2000
Duration: approximately 4 minutes
Location: Parnarama, Brazil
No. Witnesses: 2
DB: v   CB: d   PB-1: a   PB-2: x   UFOD: b

Abstract: Two men were hunting near Parnarama in the relatively cool evening hours. One of them, Luis Silva Da Silveira, 32, provided investigator Pratt with the following narrative of what happened. "We saw two green lights coming toward us....We were scared, [so] we jumped out of our hammocks and ran and hid in the bushes," he said. "I tried to shoot at it, but when I did, *I got a shock. I felt paralyzed and couldn't fire my gun.*" The paralysis lasted under a minute but the UFO had left the area by then. The men ran for over two hours to get back to the safety of town.

240

Comments: It isn't known whether an electrical shock was experienced or something else. No mention was made of seeing a bolt of lightning or flash of light which might point to an electrostatic discharge effect. The rapid paralysis duration is also very interesting, suggesting a stun-gun type of neurological stimulation which leaves one unable to function for some minutes, depending upon the amount of voltage applied.

Reference: Pratt, B. 1996. *UFO Danger Zone.* Madison, Wis.: Horus House Press, 198-201

Vallee, J. 1990. *Confrontations.* New York: Ballantine Books, 191.

Cross references: 143 (4)

Case No. 146

Date: 85-08-05,06
Local Time: 2015
Duration: estimated 10+ minutes
Location: Tehran, Iran
No. Witnesses: several
DB: kk   CB: 0   PB-1: a   PB-2: x   UFOD: 0

Abstract: According to an Islamic Republic News Agency article, a UFO flew from west to east above the northeastern sector of Tehran. It appeared to be self-luminous. Anti-aircraft batteries opened fire on it, believing it was launched from Iraq. The article stated, "There was no immediate comment from Iraq and there were no reports that the flying object had been hit or shot down."

Comments: Iran and Iraq had been at war with other for many years and this type of reaction is "normal."

Reference: United Press International (combined dispatch); Islamic Republic News Agency release.

241

Cross references: 93 (4); 127 (4)

Case No. 147

Date: 87-07-15
Local Time: unknown
Duration: 5 minutes
Location: Santana do Acarau, Brazil
No. Witnesses: 1
DB: kk   CB: a   PB-1: a   PB-2: x   UFOD: 0

Abstract: A UFO with two lights on it approached a fisherman who shot at it. He said he heard the bullet hit the object with a "ping." Unfortunately, little more is known about this incident.

Reference: Vallee, J. 1990. *Confrontations. A Scientist's Search for Alien Contact.* New York: Ballantine Books.

Cross references: 98 (4); 115 (4); 120 (4); 126 (4); 130 (4)

Case No. 148

Date: 88-12-28
Local Time: 1950
Duration: estimated 3+ minutes
Location: east of Betances, southwest Puerto Rico
No. Witnesses: 2+
DB: v CB: f PB-1: a PB-2: x UFOD: c

Abstract: Wilson Soza and Carlos Mercado watched in amazement as a very large triangular UFO flew across the sky. They then noticed two military jet interceptors approaching it. One of the jets was nearer to the UFO and seemed to "disappear into the huge UFO and then, shortly after, the second also disappeared, with the sound of the jet engines ceasing the moment the jets vanished" (Pratt 1996, 223).[1]

The witnesses saw the UFO descend to just above the surface of Laguna Cartagena, a lake about one mile away. There, *it blew up in a "massive but totally silent explosion."* Small pieces of flaming debris were seen falling to the ground. Moments later, both witnesses again saw something above the lake's surface; there were now two smaller objects present which accelerated out of sight in opposite directions.

Comments: Investigator Jorge Martin discovered that none of the military bases in Puerto Rico acknowledge losing any airplanes that day. Soza tried to find pieces of the debris he had seen falling the day before but couldn't. This aerial "show" is typical of many others in which the sensory evidence violates the laws of physics as we now know them. Did the two jets merely collide with each other and fall into the lake? Whose airplanes were they and what kind were they? If air crew died, what were the families told about their deaths? How can something blow up visually without producing a sound shock wave?

Note: 1. If the jet engine noises ceased simultaneously with the visual disappearance of the airplanes how does one account for the much slower speed of sound that is involved? If the event took place even two or three miles away from the listeners there should be some delay between the two events. Otherwise, the UFO and jets would have had to be very near the witnesses. This discrepancy must be cleared up.

Reference: Pratt, B. 1996. *UFO Danger Zone.* Madison, Wis.: Horus House Press, 222-224.

Case No. 149

Date: 89-04-21
Local Time: 2115
Duration: 2 minutes
Location: Crestview, Florida
No. Witnesses: 1
DB: p   CB: e   PB-1: a   PB-2: x   UFOD: b

Abstract: A man saw a large unidentified disc-shaped object in the sky approaching him. It made a humming noise and stopped several hundred feet away. It was about ninety feet across, had "window-like sections around the perimeter," and a bright white light centered on its bottom. He had a .22-caliber rifle with him. He took aim at the hovering object when *"a beam of light engulfed him and the weapon misfired."* The UFO then flew away.

References: Good, T., ed. 1991. *The UFO Report.* London: Sidgwick & Jackson, 223-24.
MUFON UFO Journal, no. 259.

Cross references: 70 (3); 78 (3); 83 (3); 129 (4); 132 (4)

243

Case No. 150

Date: 91-11-23
Local Time: day
Duration: estimated 5 minutes
Location: approximately 90 km from Riga, Latvia
No. Witnesses: 1
DB: v   CB: d   PB-1: a   PB-2: x   UFOD: b

Abstract: A lone hunter watched in amazement as a gray disc about 8 meters diameter and 4 meters thick with parallel top and bottom surfaces flew by only one meter above the ground. He noticed flame coming from it. Then, he approached to within about 15 meters. There were no surface markings or detail of any kind. The UFO investigator wrote, "Instead of greeting [the object], that stupid hunter shot at the UFO, firing two shots which came back with a loud noise. *The object began to become white very quickly and, in a few seconds, it* *dematerialized*. The hunter walked over to the place where the UFO had been. He had no strange feeling. [Shooting at the object] was obviously a stupid action, and it is really strange that no harm occurred to him."

Comments: It is likely that the disc's color change to white was related to its disappearance. Since gray is but a mixture of dark and light, this shift could represent nothing more than an increase in luminous energy; its emitted wavelengths did not take on any perceptible colors.

Reference: Anon. 1992. What do these entities from other worlds want from us? Joint USA-CIS Aerial Anomaly Fed. General Translation 92-12.

Some concluding thoughts and observations

Why does mankind have to "shoot first and then ask questions?" Anthropologists and psychologists have pondered this question for a long time. There seems to be some kind of instinct buried in the heart of most men to hurt others rather than help them. The Christian concept of original sin fits quite nicely as an explanation. But no one really knows. What we have learned from this review of reports of human aggression against UFOs and their alleged "occupants" is that they are remarkably peaceful, almost unprovokable. Herein lies a great dilemma for humankind.

244

Figure 31

═══*Уфологи улыбаются*═══

— БЕЗ СЛОВ.

Рис. В. Быстрова и Ю Ткаченко.

Ufologists' Humor (From: Russian Newspaper *M-Скии Treygolnik*, No. 5, 1990)
Words of hunter beneath saucer: "I have no excuse."

What if UFOs really exist? What if alien beings really exist? And what if they are highly advanced in comparison with our technology? Are they going to continue to remain passive and congenial toward our aggressive behavior, or are they going to be provoked into open retaliation someday? Are we treading a dangerous path through our fear and ignorance. A recent interesting chapter by C.A. Rutkowski (1997) touches upon this matter.

In a thought-provoking article by the highly respected ufologist Aime Michel (*APRO Bulletin*, January–February 1970), he offers an analogue between a lamb which he used to raise and how humans and aliens might respond to each other. He said that he loved this animal. He protected it from wolves, and it loved him in return. After all of this he sheared it, killed it, and ate it "before the eyes of his brothers who have seen all that for thousands of years and continue to love me (Michel) and to lick my hand, never having understood.... *There is no way in the world to make the lamb understand that the real wolf is me.*"

He wrote, "If the UFOs dominate us by an inequality comparable to that by which we dominate the lamb, to claim to make a decision concerning their hostility or their goodwill toward us is equal to claiming to translate Spinoza's Ethics into bleatings. We can then legitimately speculate on their goodwill or their hostility only on the condition of previously admitting that there exists between them and us no inequality at all" (7).

If our review of cases involving overt human aggression against UFO has shown us anything it is that we are the wolves!

Overview

UFO Responses to Being Fired Upon from the Air or Ground
Of the thirty-three cases (55.9% of total) where these characteristics were reported:

Speed and Motion Features:
   The UFO reacted immediately in some way in fourteen cases (42.4%)
   The UFO never moved or reacted at all in six cases (18.2%)
   The UFO reacted in an unspecified manner in six cases (18.2%)
   The UFO reacted but only after a significant delay in four cases (12.1%)

Color-Luminance-Visibility Features:

The UFO reacted by brightening significantly in five cases (15.2%)

The UFO reacted by emitting a visible beam or ray in three cases (9.1%)

The UFO reacted by not changing at all in three cases (9.1%)

The UFO reacted by becoming invisible in two cases (6.1%)

The UFO reacted by changing color in two cases (6.1%)

Auditory Features:

The witnesses heard audible ricochet noise in seven cases (21.2%)

The UFO reacted by emitting some audible noise in two cases (6.1%)

UFO Responses to Military Aircraft Approaching and/or Firing

Of the seventeen cases presented here:

The UFO reacted by out-maneuvering airplane in seven cases (41.2%)

The UFO reacted by destroying the airplane in four cases (23.5%)

The UFO reacted in other ways in three cases (17.6%)

The UFO reacted by accelerating away in two cases (11.8%)

The UFO reacted by becoming invisible in one case (5.9%)

246   Some features appear in more than one case so that the above numbers do not always add up to the total cited. We are faced once again with the basic question, can a natural phenomenon explain all of these reactions? The immediate reaction to being struck by one or more bullets suggests that the UFO did not want to be struck again and took rapid action to avoid this. Is this an intelligent response, a sign of invincibility, total disinterest, or something else entirely?

It seems reasonable to conclude from these cases that UFO continue to approach earth, sometimes landing on its surface, to fly at slow speeds and approach highly populated areas, even military installations, all in spite of their presumed knowledge that if we had an effective weapon against them we would not hesitate to use it!

References

Ash, B., ed. 1977. *The Visual Encyclopedia of Science Fiction*. London: Pan Books.

Bowen, C. 1977. One Day in Mendoza. In *Encounter Cases from Flying Saucer Review*, ed. C. Bowen. New York: Signet Book, New American Library, 132.

Carpenter, Donald G. and Edward R. Therkelson. 1968. *Introductory Space Science 2*, section 33.4.

Census of Population and Housing, U.S. Dept. Commerce, Washington, D.C., November 1992.

Edwards, F. 1966. *Flying Saucers—Serious Business*. New York: Bantam Books.

Fawcett, L., and B. J. Greenwood. 1984. *Clear Intent*. New Jersey: Prentice-Hall.

Goudie, D., and J. Klotz. 1992. *The Confirmation Paper*. The UFO Reporting and Information Service. Mercer Island, Wash: Privately published, 19 February.

Gross, L.E. 1955. *UFOs: A History, 1954 November–December*. Fremont, Calif.: Published privately.

Gross, L.E. 1968. *UFO's: A History 1952 July 21st–July 31st*. Fremont, Calif.: Published privately.

Gross, L.E. 1991. *The Fifth Horseman of the Apocalypse UFOs: A History 1954 November–December*. Published privately, 16.

Haines, G.K. 1997. CIA's role in the study of UFOs, 1947-90. (July): http://www.odci.gov/csi/studies/97unclas/UFO.htmlgerald.html

Haines, R.F. 1990. *Advanced Aerial Devices Reported During the Korean War*. Los Altos, Calif.: LDA Press.

Harrison, H. 1977. Warfare and Weaponry, Section 02.04 In *The Visual Encyclopedia of Science Fiction*. ed. B. Ash. London: Pan Books.

Moravec, M. 1980. *Summary of psychological reactions to UFO events from Australia* (1980). Paper presented at UFOCON 5, November, at Canberra, Australia.

Paget, P. 1979. *The Welsh Triangle*. Granada, London: A Panther Book.

Pratt, B. 1996. *UFO Danger Zone*. Madison, Wis.:Horus House Press, 1996.

Rutkowski, C. A. 1997. Take me to your leader: The ups and downs of contact. In *Making Contact*. ed. B. Fawcett. New York: Wm. Morrow and Co., 91-122.

Rux, B. 1997. *Hollywood vs. the Aliens: The Motion Picture Industry's Participation in UFO Disinformation*. Berkeley, Calif.: Frog, Ltd.

Wilkins, H.T. 1955. *Flying Saucers Uncensored*. New York: Pyramid Books.

To think is to engage in one of life's most mysterious events.

Chapter 5

UFO Responses to Human Thought

(13 cases)

What happens when someone sees an unidentified light in the sky and then merely thinks that it should do something specific like performing a loop or flashing a light on and off in reply? This is the general subject addressed in this chapter. Methodologically, there are no scientific means to verify any of the alleged UFO responses. In most cases, neither the human's nor the UFO's reported behavior can be measured. We are left with the witness's own testimony and little else. Therefore, narratives presented in this chapter are offered only as interesting stories and not as verified fact.

What are some of the methodological and analytical problems associated with alleged thought projection? Even if we accept that something highly anomalous took place (as witnesses claim), the details still may not be accurate. All kinds of perceptual errors can creep into an ambiguous sighting situation (Haines 1980; Hendry 1979). And, if people claim to have projected a thought of some kind to the UFO, how do they know that the subsequent response of the phenomenon was in direct reply to their thoughts? Again, the subject of "coincidence" becomes relevant. It seems as if it is easier for most people, particularly scientists, to accept coincidence as a "causal factor" than it is to accept that one's thoughts can be projected and received in another location.

Focused thought

It may be useful to apply an analogy from optics to human thought. Regardless of whether light is viewed as consisting of waves or particles, it can be manipulated in many different and useful ways. Passing a parallel (collimated) beam of light through a lens or reflecting it off the surface of a mirror, for example, will change its direction of travel and, ultimately, the location and/or shape of the beam. Use of an optical filter will subtract certain wavelengths to yield a different color. Use of other light-diffusing materials can cause the light to be scattered upon passing through it. And, if multiple beams of light are optically combined, the result can be even greater luminance. Perhaps human thought, then, is somewhat akin to light.

Subjectively speaking, human thought may be focused through concentration. Are the coordinated thoughts of several people more "intense" or effectual than the thought of one person? Indeed, organizations have been founded on such a possibility in order to try to effect a change in the conscience and behavior of society. The so-called "Give Peace a Chance" movement of the 1970s in America is an example. People can affect social change if they think and then act alike on a subject. But affecting social change is one thing. Controlling the behavior of alien life forms is likely to be quite another!

The very private nature of one's thoughts make them impossible to monitor accurately so that we must accept the word of the thinker as to their content and temporal course. If two people agree to think about the same thing at the same time there is absolutely no guarantee that: (1) they will think about the identical thing, (2) they will think about it during precisely the same period of time, (3) they will not be distracted by other thoughts, and (4) they won't allow their concentration to wander back and forth over the agreed-upon thought. All of these problems may be referred to as the "anisotropy of thought," analogous to incoherent light. As the beautiful hologram derives from two precisely reconstructed coherent wave fronts, so too a truly isotropic thought field might well result in a powerful tool for communication if it could be achieved.

The Center for the Study of Extraterrestrial Intelligence (CSETI) uses coherent-thought sequencing (CTS) to attempt to communicate with alien life forms through projected thought. Based upon the idea of multiple emitters (i.e., brains) all working in concert to send the same imagery to some other place—e.g., where Earth is in the Solar System, where America is on the Earth, and where the CSETI group happens to be at that moment—the CSETI belief is that highly advanced beings will be able to sense, decode, understand, and hopefully respond to what is being transmitted to them. Participants attempting to use CTS try to follow instructions like, "Visualize the United States and the outline of its coastlines," "See in your mind's eye the place you are in right now with its mountains and rivers, towns and highways as they might be seen from outer space," and, "See yourself right now in the desert sitting next to this particular cactus."

Of course the practical difficulties of carrying out each of these visualizations are great and subject to many sources of variability including personal motivations, distractions, limitations in one's knowledge of geography, deficiencies in spatial perception, perspective errors, and others. Each of these sources of variability adds to the anisotropy (incoherence) of thought from the assembled group. The mixture of all thoughts, even if they do radiate outward from the skull, are not only extremely weak in transmissive "power" (in the normal sense of communications technology) but also are likely to be mutually canceling, at least if time plays a direct role. Consider the following examples of two people attempting to attract a UFO with CTS:

Person A:

1)	0 seconds	"...starry heavens with beautiful galaxies...
2)	1.5 seconds	(notices a tiny point of white light suddenly appearing...)
3)	1.8 seconds	"the point of light is an approaching spacecraft...
4)	4.4 seconds	"it is getting bigger so it must be coming to Earth...
5)	17.7 seconds	"since it is coming to Earth and I am here waiting for it, it must be coming to me...
6)	40.3 seconds	"I don't need to do anything more." (and the person merely sits and waits patiently while also thinking random, unrelated thoughts)

Person B:

1)	0 seconds	"...starry heavens with beautiful galaxies...
2)	1.9 seconds	(notices a tiny point of white light suddenly appearing...)
3)	2.6 seconds	"it is getting bigger so it must be coming to Earth...
4)	4.3 seconds	"what is it?
5)	5.2 seconds	"It will come to where I am if I think about guiding it down...

6) 12.7 seconds	"I am now sitting in the middle of a flat desert area that is bounded on the North by a mountain range, on East by a winding river, on the West by ...
7) 14.9 seconds	"I must not stop sending my thoughts out to the light...
8) 21.6 seconds	"Oh, why aren't you coming faster? Don't you know how important it is to me to personally experience you?
9) 29.1 seconds	"Come to the right... no over this way... I don't want to get frustrated... I must remain calm..."

We face almost insurmountable practical problems in accepting the efficacy of CTS. For instance, while the initial motivation and the end goals of both people may be the same, the particular thought sequence of each person after event two are clearly different. Indeed, the times of occurrence are different after event one! And, before whole concepts are considered, we also must be certain that individual words mean the same thing to everyone involved. How can we be sure that everyone's concept of "starry" or "heavens" or "beautiful" are the same? Truly coherent or isotropic thought requires an identical sequence of ideas, concepts, visual images, and even emotional responses at identical instants. In short, everyone in the group must think identical thoughts at the same time in order for there to be a theoretical amplification effect. To behave otherwise is merely to add noise to the signal.

It is also unlikely that one's higher motives (e.g., to achieve some "greater good") are sufficient to boost one's thought out into space. Using the now familiar analogy of the hologram once more, the two coherent optical wavefronts must meet at precisely the correct angle in order to yield the new virtual image known as a hologram. Even the slightest variation in either two-dimensional image plane will reduce the intensity, color, and/or form of the resulting hologram. And so distractions and other variability of thought of one kind or another are the enemies of CTS.

In the Hollywood movie *Close Encounters of the Third Kind*, the scientists employed colored lights which flashed in temporal sequence with auditory tones. Hollywood writers seemed to have understood the inherent power of time-synchronized emissions of sound and light energy. Of course, sound travels only as far as air molecules will permit, falling to near zero energy at stratospheric altitudes and certainly to zero in the vacuum of space. Let us now consider thirteen cases which seem to involve UFO response to human thought.

Case Files

Case No. 151

Date: 65-09-20
Local Time: 1600
Duration: 30 seconds
Location: Selsey, Sussex, England
No. Witnesses: 1
DB: u   CB: O   PB-1: a   PB-2: x   UFOD: O

Abstract: The witness, an artist, had gone to the beach to relax for a while. The weather was clear, the air very still, and the sunshine was hot. As he lay on the sand, he shut his eyes for a moment to eliminate the sun's glare. "Suddenly, the thought came to me: 'What a day to see UFOs.' I opened my eyes, looking up at an angle of 45 degrees. It was then…that I saw it," he said. The completely silent object was elliptical and reflected sunlight like polished metal. Its edges were sharply defined but he thought there was a "dusky haze" beneath it, a faint "wispy cloud" came from its rear end, and some "sparkling spots" were seen "close to its stern." It traveled at an estimated 100 mph or more[1] and finally disappeared southward.

253

Comments: This witness was an artist and demonstrated exceptional powers of observation and memory. The very unusual nature of this particular flying object also must have contributed to his keen memory. Whether or not the aerial object somehow led him to have this unusual thought or vice versa cannot be determined. The boldness of the witness even to report such a bizarre event is also notable and tends to lend support the validity of his sighting.

Note: 1. The UFO took 30 seconds to travel 1.5 miles or more off Selsey Bill.

Reference: Anon. 1966. World Round-up, England: Ovalloid object off Selsey. *Flying Saucer Review* 12, no. 1 (January-February): 30.

Case No. 152

Date: 73-09-mid
Local Time: 2130
Duration: estimated 60 minutes
Location: Groveton, New Hampshire
No. Witnesses: 2
DB: kk   CB: i   PB-1: a   PB-2: x   UFOD: O

Abstract: Mr. Joseph C.[1] was in his kitchen studying for his private pilots exam while his wife and 19-year-old son were watching television in the living room. Mr. C. got up to step outside for "a breath of fresh air" and approached the back door which had a window in it. He was astonished to see, about 1,500 feet away, something that resembled a "huge ball of fireflies" in the air.

Becoming curious, Mr. C. left his house and walked toward the luminous display. After walking a few more yards, the lights appeared to burst into seven "much smaller six-foot to nine-foot diameter (balls)" which quietly and rapidly approached him. He saw that all of the lights were green—similar to a firefly's light—but changed alternately back and forth from white to warm red. The balls of light approached to within about 100 feet from him at an altitude of about 50 feet. They were arranged in a "4-3 set of positions." At this point we turn to Mr. C's own narrative as reported in the 1982 *New England UFO Newsletter*.

"I could not believe it was happening! I realized that there were alien beings inside the glowing balls. As I thought to myself or was saying out loud, 'Why me, Why me, Why me?' A telepathic thought in my brain said, 'Why not you?' in an amusing, friendly thought. I viewed their craft, wondering at its composition and energy source, hearing no noise. I marveled, wondered, was amazed and scared, all at the same time. I never did see their persons.[2] We continued looking each other over, they, undoubtedly receiving incredibly more than I, (my two eyes to their fourteen). This went on for ten to fifteen minutes."

He thought of taking a picture of the things and ran to the front door 30 to 40 feet away and yelled to his wife and son, "Come on, come on, look" He loaded his camera and his son Wayne grabbed his .22-caliber rifle. Mr. C. grabbed a shotgun propped in the corner. His wife continued watching television, strangely showing no interest in even looking out the window. "Wayne and I bolted out the door and there they were, waiting. But (the things) had moved, the four units (were now) 150 to 200 feet away and 35 feet above the ground. The three units had moved northerly 40 to 50 feet away.…Wayne and I got back-to-back, he watching three units and I watching four.[3] Suddenly, they asked, 'Are you going to shoot us?' I was standing with a camera and gun and, strangely, never did use either.

"I felt ridiculous, and said, 'No, that is not how we humans are, silly!' They then started a beam of energy at my feet. They asked if they could examine me. I said, 'Will there be any pain?' as I didn't like the idea, and they were already starting with my feet and toes and continuing rapidly, it seemed, to examine." Mr. C. noticed that as the beam moved over his body it seemed to spend more time in areas of old injuries he had had. The beam moved up to his head. "I could feel the beam as it was adjusted to varied intensities, widths, 'speeds,' and so on," he said. "It was apparent to me that they were gathering a Xerox copy of me and everything about me, down to the cell structure, a spectrograph (sic.) analysis on the highest order.

"Continuing to probe and copy all areas of my brain, they seemed to be acting with greed and glee, so to speak.[4] I was not overly apprehensive, but the beam began to increase its energy to the point of becoming very uncomfortable," Mr. C. recalled. "'Hey, you said it wouldn't hurt!' Immediately, I felt the beam energy lessen at least 40 percent. Continuing on all the while 'talking,' I was thrilled and amazed that it was possible to telepathically communicate. They were friendly! Of that I was happy."

After the examination and "talking" were over, the strange things came nearer to the two men. It was evident that they were going to want the men to go with them. "They actually wanted us to go with them," he said. "Leave to what? Be back when? [These were] thoughts in my head as they were very close. I started to get apprehensive and downright scared.[5] I then held my hand out, [as a signal for them to stop] as [it seemed] they would be right where we were in only a few seconds. The 'signal' worked. They immediately returned to the previous position for a few more minutes and…out went their glow. They were gone."

Comments: The aftermath of this encounter for Mr. C. was typical of many other people including highly strange psychic events. Since the event took place, he has had an ongoing desire to "be in continual communication and become friends with more of them." He began investigating UFOs at that time. An article about Mr. C., published in the *Lancaster (New Hampshire) Coos County Democrat*, claimed that he had spent all his savings and much of his pension on travel around America researching UFO observers and that this had led he and his wife to divorce after twenty-three years of marriage.

Notes: 1. At this point Mr. C. had worked for eight years after his federal government retirement as a steam generator inspector. 2. This puzzling statement raises the question how he perceived the entities at all and how he knew there were exactly seven beings each with two eyes? Was his only communication mental? 3. This strongly implies that Wayne also witnessed these luminous phenomena, but where is his testimony? 4. It isn't clear how Mr. C. knew exactly what they were doing to him. 5. Mr. C's emotions seemed to shift back and forth several times regarding the beings. He is curious, amazed, questioning, marveled, scared "all at the same time." It is interesting to speculate on the possible cause of such a broad and almost simultaneous array of emotions. Could it have resulted from stimulation of the central nervous system by carefully selected, multi-frequency radiation, or something else?

Reference: *New England UFO Newsletter,* April-July 1982, 16-17.

Case No. 59 (from chapter 3)

Date: 77-early Autumn
Local Time: 1720
Duration: estimated 8 minutes
Location: Isfield, near Lewes, Sussex. U.K.
No. Witnesses: 1

Note: This case is inserted here for cross reference purposes only. See the full account in chapter 3. The witness waved at and claimed to have projected specific thoughts toward the UFO.

Case No. 153

Date: 78-07-22
Local Time: 2315, 2322
Duration: approximately 10
Location: Chalandri District, near Athens, Greece
No. Witnesses: 19
DB: q   CB: e   PB-1: a   PB-2: x   UFOD: a

Abstract: This interesting case involved a family of three, two of whom worked at the British embassy in Athens. Their daughter, Catherine Dawson, 27, heard and saw the object first while she was sitting at a dining table in a restaurant near

Athens. It emitted a humming noise and looked to her like a large, dark, round-shaped object hovering at a very low altitude (almost touching the TV antennae on the nearby buildings). There was a single red light on top and "lots of square windows around it." She brought the object to her parents' attention. To them it appeared to have flashing lights that seemed to float from left to right. Then it suddenly vanished from sight.[1]

The Dawson family hurriedly paid their bill and returned to their parked car to try to follow the object. Mr. Raymond Dawson caught sight of the object in several minutes when they reached a stop light. "Everything was still and quiet" and there was no traffic at that place, which is very unusual.[2] He pulled over and set the brake but left the engine running.[3] The three witnesses got out and walked to the side of the road to look at the strange thing. Mr. Dawson recalls seeing it "gently circling" then stopping to hover. It had circumferential lights which rotated counter clockwise, its top was shaped like a bell (see Figure 32) and had a red light on top of a small dome or protrusion. A beam of light suddenly appeared from the bottom center of the object, diverging slightly as it shown downward. "As the object rocked from side to side, the beam seemed to be scanning the area. He recalls wishing he could see the underside—and at that precise   257
moment *[he] felt quite eerie as the dome tilted away to reveal the flat underside at an angle."* He then noticed that its lights were "spinning in a really complex way."[4]

Figure 32

Sketch of UFO Seen on July 22, 1978, near Athens, Greece
(Based on Drawing by Mr. Dawson; Courtesy of
Flying Saucer Review. Used by permission)

Mrs. Dawson and Catherine generally agreed that the UFO was a round disc with a red light on its top and a row of windows along its side which seemed to rotate. It rocked as it hovered, "as if floating on a breeze." The object was estimated to have been at an altitude of only 150 feet during the second observation and was about 360 feet across (i.e., "three times longer than a medium-sized jet airplane"). A group of sixteen other eye witnesses also saw this same object.

Comments: Of course the tilt response of the UFO could have been coincidental with the reported thoughts of Mr. Dawson. Differences in perceived visual features of the same UFO by different eye witnesses is common. While the skeptically minded point to these differences as proof that none of the evidence is worthy of acceptance or further study, just the opposite is true. Perceptual psychologists can find much useful information here concerning the complex processes that underlie perception and later memory extraction. On the other hand, they may also discover, as I have, that the core phenomenon has a great deal of consistency and stability.

Note: 1. As far as is known, no one else in the restaurant saw the object. 2. Periods of extreme quiet and calm are commonly reported during such close encounters. Is it possible that time is distorted in the immediate vicinity of the UFO so that sensory perceptions are also distorted? This might explain the apparent absence of other people, traffic, and sensory distractions which are reported so often. 3. He said he wanted to find out if the UFO would affect the engine. It did not! 4. The spinning lights are reminiscent of the cylindrical UFO seen in the September 19, 1976, Tehran case presented in chapter 4, case 135.

Reference: Randles, J. 1979. UFO over a Greek taverna. *Flying Saucer Review* 25, no. 1 (January-February): 31-32.

Cross references: 155 (5); 156 (5); 157 (5); 158 (5); 160 (5); 162 (5)

Case No. 154

Date: 78-12-12
Local Time: 0800
Duration: estimated 5 minutes
Location: Burghausen, Bavaria, Germany
No. Witnesses: 1
DB: h   CB: h   PB-1: a   PB-2: O   UFOD: a

Abstract: Ms. Adele Holzer was in her car traveling to work when she first sighted a hemispherically shaped white disc. More puzzled than afraid, she pulled off the road, got out, and saw a "row of shining green ports along its periphery." *Then, she allegedly thought to herself, "If I could only see its bottom...when suddenly the object turned."*[1] She reported that there were three spheres on its bottom surface, the middle one being larger. It possessed a bright orange spot at its center. As the UFO hovered over some nearby trees, six beams of green light shot from it; she felt "slightly paralyzed" by one of them. She lost track of time as the object hovered above her. When the light beams went out, the object rose up silently into the sky. Her watch had stopped at 8:01 A.M. and the car's ignition key in her hand the entire time was found to be bent. Later, she told investigator Hesemann (1994) that she felt "somebody was trying to get into telepathic contact with me." She claimed to have been given a message saying that they had friendly intentions...and wanted to stop us from destroying the earth."

Comments: The total encounter duration is not known. Here is yet another instance of psychological disturbance as a result of being in close proximity to a UFO. Nothing was said about any physical injury produced.

> Note: 1. The meaning of the word "turned" is unclear here but is thought to refer to a change in the direction of flight of the object.

Reference: Hesemann, M. 1994. *Geheimsache UFO*. Neuwied, 463.

259

Case No. 155

Date: 82-12-31
Local Time: 2350
Duration: estimated 4 minutes
Location: Kent, New York
No. Witnesses: 4
DB: v   CB: d   PB-1: a   PB-2: x   UFOD: b

Abstract: A Mr. Edwin Hansen, 55, was on his way home by himself on Interstate Highway 84 in southern New York on New Year's Eve. He observed a very large boomerang-shaped aerial object hovering over the highway just ahead of him. Other drivers had already pulled off the highway to watch it. It had a bright searchlight that projected a white beam of light down to the ground. He claimed to have thought to himself, "I wish it would come closer so I can get a better look at it." *Almost immediately, he said, the object started to descend and move directly toward him.* As it passed by very close to his position, he became scared. He said, "I felt thoughts that weren't my own, but a kind of voice telling me not to be afraid." *The aerial object flew in "slow, tight circles" before turning away and disappearing into the night sky.*

Comments: Like many other such cases, we have to accept the word of the witness as being accurate. These reported UFO responses are not uncommon.

Reference: Hynek, J.A., and P.J. Imbrogno 1987. *Night Siege: The Hudson Valley UFO Sightings.* New York: Ballantine Books, 7-8.

Cross references: 153 (5); 156 (5); 157 (5); 158 (5); 160 (5); 162 (5)

Case No. 156

Date: 83-02-26
Local Time: 2000
Duration: estimated 3+ minutes
Location: near Lake Carmel, New York
No. Witnesses: 2
DB: p   CB: d   PB-1: a   PB-2: x   UFOD: a

Abstract: Mrs. Monique O'Driscoll and her 17-year-old daughter, Monique, were returning from Kent in the family car; the weather was clear and calm. Their radio began to make a hissing sound. That was when the younger Monique first sighted a cluster of about fifty bright, colored lights "going like crazy, flashing." After watching it for several minutes, the objects seemed to be moving away from the two women. The mother is reported to have said to herself, "Oh, please don't go, I want to look at you some more. *At that split second, it stopped, made a complete turn, and then it was facing toward me. Then it started moving toward me, very slowly."*

Comments: During the field investigation, conducted by Phil Imbrogno and Allen Hynek, it was discovered that this seeming mental communication with the UFO was not a unique event but occurred a number of times during this particular UFO flap. The object was estimated by the women to be from 200 to 300 feet in size tip to tip. Its underside had many repeating geometric patterns of raised detail, a fact that is not at all unusual in UFO lore.

Reference: Hynek, J.A., and P.J. Imbrogno 1987. *Night Siege: The Hudson Valley UFO Sightings.* New York: Ballantine Books, 8-11.

Cross references: 153 (5); 155 (5); 157 (5); 158 (5); 160 (5); 162 (5)

Case No. 157

Date: 83-03-17
Local Time: 2040
Duration: 3+ minutes
Location: Brewster, New York
No. Witnesses: several
DB: p   CB: d   PB-1: a   PB-2: x   UFOD: a

Abstract: Scores of car and truck drivers stopped their vehicles along Interstate 84 in southern New York state to watch a huge triangular-shaped UFO as it moved only about twenty feet above the top of a truck traveling on the highway! One of the witnesses, Dennis Sant, saw the object from his home. He remembered saying to himself at that moment, "I wish I could get a better look at it." Sant recalled, "As I was thinking that, *it made a 360 degree turn, as if rotating on a wheel, stopped, and started to float in my direction.* It continued to approach me, and I just stood there transfixed. It stopped 40 feet from me and was hovering 20 feet above a telephone pole in front of my house."

The object was V-shaped and about 130 feet long. Interestingly, its center point was facing downward toward the ground. Numerous colored—red, green, and white—lights covered its surface and flashed in a regular sequence. All of the lights increased greatly in brightness at one point and then returned to their original intensity. For comparison, the following photograph is of the Air Force's F-117A stealth fighter as seen from the front. Its wing span is 43' 4", height : 12' 5", and length : 65' 11." Its maximum speed is still classified. If this particular military airplane was the cause of this sighting it would have had to: (a) hover silently (which it cannot do), (b) turn on a dime and stand on its nose vertically (which it cannot do), and (c) possess multiple external colored lights (which is possible).

261

Comments: As with virtually every case of this kind, the response of the aerial object may have been only coincidental with the thought of the witness. Nevertheless, how many instances are there where a thought occurs and the UFO does not "react" in any obvious way? There are no statistics available on this important experimental control condition.

Reference: Hynek, J.A., and P.J. Imbrogno 1987. *Night Siege: The Hudson Valley UFO Sightings.* New York: Ballantine Books, 19.

Cross references: 153 (5); 155 (5); 156 (5); 158 (5); 160 (5); 162 (5)

Figure 33

Photograph of F-117A Stealth Fighter (Courtesy of Lockheed Systems)

Case No. 158

Date: 83-03-24
Local Time: night
Duration: estimated 25 minutes
Location: near Stormville, New York
No. Witnesses: 1
DB: v   CB: a   PB-1: a   PB-2: x   UFOD: a

Abstract: Mr. Mark Galli, a bus driver, was driving south on Interstate 84 in southern New York state one night when he noticed something unusual in the sky. Then, he lost sight of it. Some 15 miles north of the first location he saw the lights again (near Mount Storm ski area near Stormville).[1]

"It appeared to be hovering right over the ski slope," he said. "I pulled off to the shoulder of the highway and got out. One of the red lights had a beam on it, a red light that came on. There were quite a few clouds in the sky, and the red light was bouncing off one of the clouds.

"I said to myself, 'I wish this thing would come my way so I could get a better look,' and as I did, it started drifting toward me as if it was responding to my thoughts. I then got back in my car and left as the UFO approached because I had the feeling it was watching me. It was as big as a football field. I hit the gas pedal and sped away. I was upset and frightened."

Notes: 1. Stormville, New York, is about 15 miles southeast of Poughkeepsie, New York, in Dutchess County.

Reference: Hynek, J.A., and P.J. Imbrogno 1987. *Night Siege: The Hudson Valley UFO Sightings.* New York: Ballantine Books, 43-44.

Cross references: 153 (5); 155 (5); 156 (5); 157 (5); 160 (5); 162 (5)

Case No. 159

Date: 84 - Fall
Local Time: late afternoon
Duration: estimated 2+ minutes
Location: Cabell County, West Virginia
No. Witnesses: 2
DB: p   CB: a   PB-1: a   PB-2: x   UFOD: 0

Abstract: A fifty-year-old man and his young nephew were hunting squirrels. The older man was looking for the moon when he spotted a UFO flying quickly toward them. He used his 10 x 50 binoculars to see it better. Then, a second object came into sight. "Instantly, I knew what was about to happen," he said later. "I had the distinct impression it was going to come over and land, so I tried to tell it telepathically that my nephew would die [of fright] if it did.[1] Apparently, it worked.... *It took off.*"

263

Comments: Like other accounts of this nature, the connection between the reported departure of the object following the man's thoughts is problematic.

Note: 1. Later he claimed that his mother used extrasensory perception all the time and he was familiar with this ability. He remarked that, "UFOs have expanded our horizons."

Reference: Teets, B. 1995. West Virginia *UFOs: Close Encounters in the Mountain State.* Terra Alta, West Virginia: Headline Books, 123-124.

Case No. 160

Date: 89-09-17
Local Time: 2305
Duration: estimated 5+ minutes
Location: between Brinklow and Highway A5, England

No. Witnesses: 2
DB: v   CB: d   PB-1: a   PB-2: x   UFOD: 0

Abstract: This frightening event took place while Rita Goold and a male passenger were driving back to Leicester just after 11:00 P.M. from Brinklow. The event is presented in abbreviated form here. They both noticed an odd light in the sky about 600 feet ahead of them as they drove along. There was almost no other traffic at the time nor were there street lights in the area. The very bright white light came within about 325 feet from them and then shifted over to the center of the roadway. "And it started, very slowly, to come towards us," she said.

The passenger had a camera but seemed frozen, immobilized, and never used it. Both listened intently after Goold rolled her window down and turned the engine off. They knew it was a UFO; it seemed to approach with a "feathery, bouncing sort of effect....I knew that I simply had to get away from it at all costs, as it was about to engulf or envelop the car!

"As it drew closer and closer, but still not quite down to road-level, I mentally pleaded with it to leave us alone!" she cried. *"Suddenly it dropped right down on to the road,[1] and seemingly was wide enough to block the entire road. Then, it moved to the right of us.....Then, it took off again towards us, at a breakneck speed, with no perceptible engine noise, apart from a "whooshing" sound as it went past us."* They both saw a solid object enveloped in light. They said the object had a "broken edge" with a "starry effect," however the meaning of these terms is not clear.

The UFO vanished from sight after passing the automobile. Goold started the engine and accelerated down the road "for both of us felt that it had been sinister," she explained. As they were driving on the A5, lights from an unknown vehicle appeared behind them[2] and quickly seemed to pull to within feet of the rear bumper. The interior of their car was flooded with bright white light. The object did not change its position left or right. Then, as both occupants were trembling and shaking with fear, the object left them, suddenly turning right. The rest of their journey was relatively uneventful, for which they were thankful.

Comments: This close encounter seems to have taken place despite the (intense) thought by Goold that the UFO should leave them alone. It is not clear how much time elapsed between her pleading and the UFO's subsequent response.

Notes: 1. The passenger commented that, "It was as though it then changed shape and course, and materialized into something else...like a car!" 2. They were traveling at about 60 mph at the time toward Narborough.

Reference: Whitehead, P. 1991. An alarming close encounter on an English highway. *Flying Saucer Review* 36, no. 1 (March Quarter): 22-24.

Cross references: 153 (5); 155 (5); 156 (5); 157 (5); 158 (5); 162 (5)

Case No. 161

Date: 90-11-11
Local Time: estimated 0200
Duration: 1 hour 30 minutes
Location: Pensacola, Florida
No. Witnesses: 3
DB: kk   CB: a   PB-1: a   PB-2: x   UFOD: O

Abstract: This interesting incident occurred above Big Lagoon, near the coast-side city of Pensacola and involved two fishermen. The nocturnal light first appeared as a single white star to the south of them over the gulf of Mexico. But as it came closer, a yellow, red, and blue light could be seen in a regular, repeating sequence. It seemed to move vertically at times and continued to approach the men. They saw a dull orange light from the bottom of the object reflecting from the water's surface and a bright red light on its top. Near its middle was one row of four windows and a rotating ring near its base. Its surface reflected moonlight as polished metal would. On several occasions, it emitted "blue beams to the water....*One fisherman tried to mentally 'talk' to the object, and it appeared to react to his thoughts.*" The other man went to find a telephone to report the event.

265

The UFO flew away toward the southwest and, while doing so, an airplane circled it. "The object shot out a 'red ball of light,' which the airplane followed."

Comments: The alleged CE-5 component of this event is as problematic as most of the others are. This area of the U.S. has witnessed a surprisingly large number of sightings over the past ten years. The presence of an unspecified airplane is also interesting since it too seemed to evoke a response from the strange light. Why the pilot would follow the red light and not the main object isn't clear, nor is it reasonable.

Reference: Current Case Log. 1992. *MUFON UFO Journal* 294 (October) 18.

Case No. 162

Date: 94-07-20
Local Time: 2300 & 2330
Duration: 2+ minutes
Location: Milk Hill, Alton Barnes, England
No. Witnesses: 20
DB: v   CB: a   PB-1: a   PB-2: x   UFOD: c

Abstract: Participating as a field investigator with a group of about eighteen others interested in British crop formations and UFO,[1] Reuben Uriarte had climbed to the top of Milk Hill, located near the famous Monument to Marconi in southwest England, by himself at about 11:00 P.M. in order to meditate and be alone for a time. It was during this period that he thought,[2] "If you read my thoughts, please appear and, if you should appear, I thank you." Then he walked back down the hill to where his group was waiting; this took only about one to two minutes. Remaining with the group another two or three minutes Reuben and a friend Terry climbed back up the same hill to begin their sky watch. At about *11:08 P.M., both of them sighted a bright golden-amber partial circle* (see Figure 34) *rise vertically upward from beneath a distant hill. It remained in sight about a minute and then slowly descended out of sight.* Looking through his 7 x 40 binoculars, Uriarte said the light source almost filled its field of view. It subtended an angle to the naked eye equivalent to ½ inch at arm's length, about 1.5 degrees arc. The top ⅕ of the object seemed to be a brighter gold than the rest of the object.

Both men shouted to the rest of the group to come up to their location right away. "You guys are missing something," Reuben yelled. He also blew his whistle to attract their attention. Soon the entire group of spectators had arrived. But it wasn't until 11:30 P.M. that the same, or a similar, UFO rose up into sight again. It was followed in less than thirty seconds by a second identical light located to the right which ascended slowly and deliberately to a slightly higher altitude than the first. Then, within another thirty seconds, a third identical yellow-white light rose into view slightly to the right of the second object and ascended to an altitude only slightly less than the first. Then, the third light faded completely from sight and other two moved toward each other, seeming to merge into one. The third light reappeared in its original location, and all three bobbed slowly up and down to different altitudes over a period of about two minutes.

They finally disappeared by descending vertically behind the hill

Comments: Interestingly, Uriarte thought that a good deal of disbelief was evidenced in the group during and after the three lights had appeared. Several men present, each of whom had had prior military experience, said that they were only flares.[3] Later, a member of the group who was a physicist made a video animation of the behavior of the lights. It shows an almost continuous rising and falling of the three lights to different heights during each cycle and some limited lateral motion of one or two of the lights. All angular accelerations were extremely slow and smooth. Mr. Michael Lloyd viewed the UFO through an infrared spotting scope and said they were not only domed on top but had a geometric protrusion on the bottom approximately as shown in Figure 35. He raised his camera to get a picture of them and Uriarte thought to himself, "They're not going to allow themselves to be photographed!" Sure enough, all three suddenly disappeared from sight before he was able to take a shot.

Figure 34

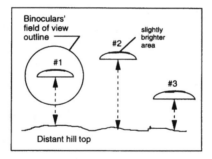

Reconstruction of three UFOs sighted on July 20, 1994, at Alton Barnes, England

Figure 35

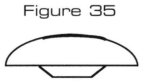

Drawing of one UFO as seen through an infrared "night" scope.

Whether these rather prolonged UFO responses were in direct relation to Uriarte's thoughts is impossible to verify. A stronger relationship is the attempted photography which failed, a not uncommon event [see 78 (3)]. The relatively great distance involved here is also very interesting since most of the close encounter events presented in this book are much closer.

Notes: 1. Others in the group included Pamela Jones, Dr. Jean Noel and his wife Mary, George Knapp, Michael Lloyd, Collin Andrews, Terry D., Pike Murphy, and Mary Aubren. 2. He said that he tried to project these thoughts outward into the sky. 3. Just before the lights appeared to Uriarte the second time, he said he was thinking that he really wanted to see them again. He felt very excited and also curious. He explained, "I wanted to validate what I had seen the first time by myself, I wanted the others to see them also so that they would believe me."

Reference: Uriarte, Reuben. 1996. Interview by author. 15 August.

Cross references: 153 (5); 155 (5); 156 (5); 157 (5); 158 (5); 160 (5)

Case No. 163

Date: 95-06-19
Local Time: 2200
Duration: 3+ hours
Location: Iowa
No. Witnesses: 2
DB: x   CB: b   PB-1: a   PB-2: x   UFOD: O

Abstract: At about 10:00 P.M., two men saw a light oscillating laterally above a field about two and one-half miles away. After driving to the area out of curiosity, they were able to see "four, chartreuse green glows" shining upward from the ground. Each source of light was about 100 feet apart and about 20 feet in diameter. The two men then drove to a different side of the wooded area to get a better look. From there they saw two "greenish lights and one intensely bright white light floating about twenty feet above the ground not more than 150 feet from them." A beam of green light came out of it and swept back and forth in an arc. Then, the white light "swung around from in front of them to their side, stopped, and appeared to focus on the car." They drove back home at high speed in fright.

Then at about 1:00 A.M., both men, having returned home, saw a ball of light hovering across the road less than 600 feet away. They both walked toward it but it disappeared several minutes later.[1]

Remaining outside, the two men continued to look for the strange lights until the early morning hours. One of them sighted similar appearing lights both to the northwest and the southwest of their location. "Bill began talking to the light, making statements such as, 'Please don't scare me too much,' and 'Can you flash once for "yes" and twice for "no?"' *He was sure a light west-southwest of him responded by flashing once. Another question resulted in another flash.*[2] Finally, he asked, 'Can't you leave me a sign or some kind of evidence?' Bill claims that *the light 'zipped' up to him then backed off and performed a bouncing motion about* $\frac{1}{4}$ *mile away.*"

Comments: Unfortunately, we don't know whether the second man saw these events. The fact that the two men admitted being frightened is not only a realistic response but suggests some degree of reporting honesty. Their description of the encounter seems reasonable and has adequate detail when considered in the larger context of the world wide UFO literature.

Notes: 1. Both witnesses claimed that they stood at the road watching the lights for no more than fifteen minutes, however, more than an hour had passed before they returned to the house. 2. The fact that the light flashed only once each time suggests that its first occurrence may not necessarily have been in reply to his request.

Reference: Current Cases. *MUFON UFO Journal* 332, December 1995, 15-16.

While they are not repeated above, there are another six cases from chapter 3 which qualify as pleasant, welcoming, human thought-provoked UFO response. They are: 74 (3); 75 (3); 76 (3); 77 (3); 78 (3); and 90 (3).

Overview

Pleasant, welcoming thoughts

In seven cases the witnesses transmitted such thoughts: the UFO reacted by approaching the witness in all seven case (100%)!

Fearful, frightened thoughts

Of the two cases in which the witness was afraid and pleaded to be left alone, the UFO reacted by:

• flashing a light in one case (50%)

• departing from the area in one case (50%)

Thoughts requesting a specific UFO reaction

Of the nine cases in which the witness asked it to perform a particular response: the UFO performed the requested response in all nine cases (100%)!

Certainly, we usually hear about the successes and not about the disappointments of people trying to communicate with aerial phenomena. Nevertheless, these relatively few cases are revealing; they provide more evidence to support the assertion that UFO are intelligently controlled and possess sophisticated sensing and response systems. Far more data will be needed to test the statistical "null" hypothesis that none of these reported events are caused by intelligent causal agents. If the UFO involved in these thirteen cases are all misidentified natural phenomena, then one still must explain what these phenomena are and why they reacted in the particular manner they did. Further insights about these data are presented in chapter 10.

References

Haines, R.F. 1980. *Observing UFOs*. Chicago: Nelson-Hall.

Hendry, A. 1979. *The UFO Handbook: A Guide to Investigating, Evaluating, and Reporting UFO Sightings*. Garden City, New York: Doubleday & Co.

"Everything should be made as simple as possible, but not simpler."

—Albert Einstein

Chapter 6

UFO Responses to Miscellaneous Kinds of Human Behavior

(43 cases).

When I reviewed much of the available literature on the subject of CE-5, I found many narratives where the human behavior could not be placed in a particular category such as friendy or hostile. Nonetheless, such diverse behavior still seemed to evoke some definite response by the UFO. It is this broad collection of reports that is presented here.

One of the more common types of close encounters is that of vehicle pacing, that is, when a UFO flies along with a moving vehicle. Some of these reports qualify as CE-5 events, particularly when it can be shown that the UFO changed its speed when the vehicle changed speed, or the UFO changed its position relative to the vehicle while the vehicle was traveling in a straight line—this eliminates the well-known "pacing" illusion that is produced when a light source is seen at a very great distance at night. When the vehicle changes direction—as when turning a corner to get away from the UFO—and the object subsequently turns the same corner angle, then there is more reason to consider this response as a CE-5 event. Of course, the longer the pacing continues, the more eye witnesses there are, and the more corners turned, the more important the case becomes. A number of such pacing incidents involving surface vehicles and airplanes are presented here.

The following incident illustrates how easy it is to identify a natural phenomenon after the fact, in this case, ball lightning.

Case A. 1890-summer This fascinating event took place in May or June in East Prussia when a Mr. Reich was driving his open carriage along a country road; a wire fence lay on each side. Both the wheel rims and axles of the carriage were made of iron. The sky was heavily overcast but rain had not yet started to fall. At about 8:00 P.M., he suddenly noticed two balls of intense light (each about 8 inches across) appear to travel along the fence wire on his left and right sides. They moved at the same speed as his carriage and sparks bounced back and forth over to his axles, frightening the two horses. As they ran faster, the luminous balls kept pace. Then, upon reaching the end of the fences, "both balls collapsed into nothing...with a noise like crumbling of a sheet of paper" (Silberg 1965). This interesting account has little in common with the CE-5 stories recounted here since the behavior of these luminous balls conforms so well with known laws of physics. Almost the opposite can be said of the following UFO narratives.

The following cases illustrate the methodological difficulties faced in this field, not only in trying to reconstruct what occurred during a complex experience, but also to decipher the actual temporal order of the events.

Case B. 54-09-07 This incident took place in the Aisne Department of France when Mr. and Mrs. Robert Chovel and Robert's father-in-law were returning home at about 12:30 A.M. in their automobile. Driving from the town of Hirson, they all saw a red-orange, luminous disk traveling over nearby railroad tracks. It stopped at an altitude of from 300 to 400 meters. Upon reaching the bridge at Buire, "they saw what they believed to be a flying saucer" increasing its altitude....As soon as the car lights beamed on it, *the object started flying at great speed towards LaHerie 3 to 4 kilometers away and disappeared."* (CIA files 1954; Fawcett and Greenwood 1984). It is not clear how the headlights could have illuminated the UFO if it was this high in the air. Was the translation from the French language faulty? Was the field investigation incomplete?

Case C. 72-01-late This boat-pacing incident took place near Blount Island, Jacksonville, Florida. Norman Chastain was in his boat fishing with bright lights to attract fish at 3:00 A.M. when he noticed a light in the sky moving toward him. Soon he made out very intense lights around the circumference of a domed disk with a body estimated to be seventy-five feet diameter and eight feet thick with a five-foot-tall dome on top. *Becoming afraid, he turned all of the lights off on his boat and, as he did, the object left.* Now moving through darkened waters, he ran aground. While he was working to get his boat free, he saw a being nearby about five feet tall wearing a tight-fitting silver suit of some kind. Its head was large and its ears looked pointed. They stood staring at each other for a moment before the creature "fired a light which caused the witness to feel a tingling sensation and fall to the ground." He couldn't move until dawn and "full mobility did not return for another day" (Slate 1974). This event includes many details that are also found elsewhere in this book.

Case D. 76-06-20 This close encounter involved an alleged abduction of a husband and wife and their four-month-old child from their automobile near Goodland, Kansas, by strange beings. This traumatic event began while the family was driving on Interstate 70 across western Kansas on their way to Colorado just west of Colby, Kansas. It was after midnight when nineteen-year-old Joe, and his wife Carol, 18, first sighted "unusual lights" in the clear, star-filled sky. Their baby was asleep in the back seat. Both adults recall the following identical events.

Three orange-blue lights—like bright stars—were seen coming together and then breaking apart again. One of the lights traveled south, one went north, and the third rose vertically upward. According to a review of the case in the *International UFO Reporter* (Vol. 2, no. 10, October 1977), Joe said, "No one can tell me there's no such thing as UFOs." Sometime later they saw "four large oblong 'stars' moving through the sky." Two were above and the other two below, in a kind of diamond formation. These lights looked different than the earlier ones, larger and whitish. Joe expressed out loud, "how great it would be to go up in one of the UFOs." His wife answered that she would like to go as well, "if the baby wasn't in the car."

The Miller's car had experienced mechanical problems and Joe said that he thought to himself, "if they can make flying saucers like that, I can imagine what they could do to my car. If they'd pick us up and use us, if we could be guinea pigs for them, maybe they'll fix our car for us...because it's broke." They continued to drive toward their destination.

Some distance farther they saw a yellow rod appear in the sky. It had two intense, glowing balls on either end of it which suddenly shot upward and over the top of their car. Carol watched it through a rear window as Joe pulled over and stopped the car. They watched the strange lighted object climb into the dark sky until it was nothing more than a point of light. At this point neither of them could recall

273

further details consciously. It was only through the careful hypnosis conducted by Dr. Richard Sigismund that these details were recovered.

This rather typical case exemplifies the night approach of several UFOs to an automobile traveling on a lonely road. The strange UFO shapes, lighting patterns, and colors, now so familiar, are present here. These characteristics almost appear like attention-getting behavior displayed by young children. The statements allegedly made out loud by Joe: (a) may be accurate, (b) may be inaccurate, (c) may be totally fabricated, (d) may or may not have been sensed by the UFO, and (e) may or may not have been a direct contributor to subsequent appearances and dynamics of the UFOs. The same may be said for the alleged thoughts that Joe shared later with the investigator. Interestingly, both adults provided almost the same pre-amnesia details and temporal order of events which suggests either that an actual event took place as claimed, that one person influenced the testimony of the other, or that both people conspired together in an intricate hoax. Sketches of the large disc-shaped UFO and one of the humanoids, made separately by Joe and Carol, are remarkably similar. Those who try to explain the cause of reports such as this must not only contend with their own personal uncertainty and doubts about the validity of the narratives, but also with the vagaries of human memory.

274

Case E. 93-09-04 This narrative suggests that missing or perceptually distorted time may have played a part, particularly at the points marked (*) below. This also illustrates the difficulty in establishing whether the humans did something to cause the UFO to approach them. About 3:00 A.M. near Newstead Abbey, Mansfield, England, two young couples were on a double date in a car parked beneath some oak trees. The moon was full and the sky clear. All four noticed three aerial lights oscillating in the distance, suggesting that the lights were not airplanes. Everyone got out of the car to see the bright lights better. *Within only a few seconds the lights traveled directly to their position, passing over them * with a slight whooshing sound.* Mr. Darren Dove recalled that the lights came to a full stop over the oak trees where the car was parked *. The girls had walked back and gotten into the car by this time *. Although somewhat obscured by the trees, the object looked to Dove like a large triangle with different colored lights along its sides and one large red light on its bottom. Both young men said they stood watching the UFO from some distance away for about five minutes and then returned to the car and drove away. The time was then about 4:00 A.M.

As they drove home, they looked back in the direction of the trees but the UFO had disappeared.

Readers familiar with the so-called alien abduction literature will find many familiar threads here. But, according to this brief account, neither the humans nor the UFO signaled each other in any obvious way. Therefore, this case does not qualify as a CE-5 event. Yet, how did the aerial vehicle "know" to travel directly to the grove of trees where the witnesses were and why did it stop there? And, if the UFO took only a few seconds to move from its initial distant location to the position above the trees, it is difficult to understand how the young women had time to get back to the car. There are several areas where time may have been missing or distorted, approximately one hour cannot be tracked (Dodd 1993). This and many other accounts raise the pivotal issue of the reliability of human memory.

Case F. The exact date of this interesting close encounter isn't known (it probably occurred in the 1970s) and took place near Ismeraldas in a region known as Vale Das Velhas (Valley of the Old Women) in Minas Gerais, north of Belo Horizonte, Brazil. Prominent UFO investigator Dr. Hulvio Aleixo Brant interviewed the witness, Mr. Joao de Jesus, who told him he was on his way home on a very dark night when he sighted a light in the sky. The light seemed to be flying in his direction along a broad arc-like flight path. Then the light went out and several seconds later he was aware of a "dark silhouette in front of him six meters away. It was a spherical object with two landing legs" (Pratt 1996, 228). When he tried to turn his flashlight on to better see the object, the light wouldn't work. Then, he got a match out and started to light it. *Immediately, "he was pushed backwards about ten meters.* He was not hurt. It was fast but not violent." No tactual contact was ever felt with anything. He fell to the ground and he saw nothing more. He couldn't see and had to crawl home. *He was found to be extremely light sensitive (photophobic) for three days.* From that day on, he suffered partial blindness of an unspecified nature.

This encounter has many earmarks of a loss of time, i.e., a state where one's experiences cannot be retrieved from one's memory. I suspect that the instant he felt thrown backward was the instant his perception of time began to function again after earlier unaccounted events had taken place. Of course there is no way to verify this assertion.

Memory is said to be reconstructive far more than duplicative. In the reconstruction process, memory may exchange the temporal order of some events, leave events out altogether, and/or distort others to some unknown degree. Yet, assuming that the entire event has been recovered, memory will not normally add totally new details thereafter. And so how are we to approach the following cases? I suggest that the best approach is to read them carefully and then let them lie unresolved in your consciousness. If their truth is to emerge, it will.

Case Files

Case No. 164

Date: 51-06-24
Local Time: 0330
Duration: 15 minutes
Location: Richmond-Henrico Turnpike, Virginia
No. Witnesses: 1
DB: cc   CB: e   PB-1: a   PB-2: x   UFOD: c

276

Abstract: Police officer William L. Stevens, Jr. was on duty in his cruiser well after midnight when he sighted lights low in the sky. Upon driving closer, he made out a dirigible-shaped object with alternating white and yellow lights around its circumference. When he drove nearer, it began to move away from him. *Then, it stopped and allowed Stevens to catch up with it, and then flew off again.* He accelerated to over 100 mph at one point but couldn't keep up with the object. *He watched in amazement as the UFO followed each turn in the roadway rather than flying in a straight line.* He stated, "Out in Hanover it kind of stopped for a little bit, then went into a high speed vertical climb [and] it climbed out of sight."

Comments: This cat-and-mouse game of "tag" qualifies as a CE-5 since the UFO responded to Seven's approach. Clearly, both the UFO and the witness were physically involved in this pacing incident.

Reference: Gribble, R. 1991. Looking Back. *MUFON UFO Journal*, no. 278 (June): 12.

Cross references: 169 (6); 184 (6)

Case No. 165

Date: 54-04-23
Local Time: night
Duration: estimated 20+ minutes
Location: near Pearcy, Arkansas
No. Witnesses: 6
DB: v   CB: a   PB-1: a   PB-2: x   UFOD: a

Abstract: Six workers at a Reynolds Metal Co. plant gathered at the home of one of the men to ride to work. Suddenly, everyone saw a glowing ball of white light circle slowly and stop at irregular intervals above the house over a twenty-minute period. Les Reatherford said that the ten-foot-diameter ball dived toward him as he walked out into the yard. "If I hadn't ducked it would have hit me," he said. One witness said the UFO followed their stationwagon as they drove toward Pearcy. *"We tried to put a spotlight on it,...but every time, it would dodge the beam," the witness said.* The UFO disappeared as they reached the town. Other eye witnesses included Fred and Harley Skeets, John Vaughn, and Tom and Dayton Henderson.

277

Comments: This light avoidance response is very common.

Reference: Associated Press. 1954. *Six in car pool report ball of light circling house, followed their auto.* 23 April.

Cross references: 6 (3); 7 (3); 8 (3); 14 (3); 18 (3); 27 (3); 46 (3); 47 (3); 48 (3); 51 (3); 62 (3); 79 (3)

Case No. 166

Date: 54-06-23
Local Time: 2045
Duration: approximately 1 hour
Location: NW of Dayton, Ohio
No. Witnesses: 1
DB: jj   CB: e   PB-1: a   PB-2: x   UFOD: b

Abstract: This F-51 propeller airplane encounter took place at between 7,000 and 9,000 feet altitude and began when First Lt. Harry Roe, an Ohio Air

National Guard pilot, sighted a white light at his eight o-clock position. It looked similar to an airplane tail light; it did not change in intensity and remained slightly above his own altitude. *He executed several 360-degree circles and 90-degree turns and the UFO remained at the same relative position on his plane's canopy, "on wing tip" almost all of the time. "As I started to change my position, he would go right with me."* [1] He remarked, "At one time, I chopped the throttle, put gear and flaps down in hopes that it would pass me. I was down to 100 to 110 mph, but the damn thing acted like there was a pilot out there. When I slowed down all of a sudden, he gave a little surge at first, came practically parallel with me, and then maintained the same speed as I did off my right wing. [2] The UFO "chased" the airplane from Columbus to Dayton and Roe "chased it back to Columbus," [3] a total of about sixty-three miles. The UFO was last seen near Columbus with the airplane at about 12,000 feet altitude. It left the airplane in linear flight.

Comments: Lt. Roe said that although he tried many different maneuvers to get the object silhouetted against a lighter sky in order to discriminate its size and shape, he could not do so. However, this admission seems to contradict another statement he made later that the UFO followed his airplane around several circles at the nine o-clock relative position which would eventually bring the UFO into a position in front of the lighter sky. No explanation is given for this discrepancy. Perhaps by the time he did this the sky had darkened too much. This case further illustrates the high degree of coordinated flight that UFOs exhibit.

Reference: USAF Blue Book Files, Investigator's Narrative, Tech. Info. Sheet, Pp. 1-7, June 25, 1954.

Gross, L.E. 1990. *UFOs: A History 1954 June–August.* Fremont, Calif.: Privately published, 29-32.

Notes: 1. It isn't clear whether this UFO response was simultaneous with the airplane's motion. At one point, the pilot flew directly over Patterson tower and asked the controllers to see if they could see the light. They couldn't! 2. The fact that the UFO was on the right side of the airplane is consistent with Lt. Roe's statement that "sometimes he would be on my left and sometimes he was on my right." 3. Actually, the airplane's turn around point was about 40 miles southwest of Columbus.

Case No. 167

Date: 55-07-10
Local Time: 1430

Duration: 6+ minutes

Location: 7-9 mi. west of Newport, California
(Pacific Ocean)

No. Witnesses: 3+

DB: cc   CB: a   PB-1: a   PB-2: x   UFOD: O

Abstract: This pleasure boat-following incident took place off the Southern California coast on a clear sunny day. Mr. and Mrs. George Washington and their daughter Maria, 11, were on their way to the port of Avalon on Catalina Island thirteen miles from Newport Beach. Mrs. Washington first sighted the strange object hovering above them at an estimated altitude of 2,500 feet. Her husband immediately throttled back before going out on deck to see the UFO for himself. Both adults viewed the "perfectly round, gray-white" object for about six minutes through field glasses. It seemed to spin and was "surrounded by a 'haze of fumes' apparently blowing out from the object."

As Mr. Washington continued his journey he radioed the Coast Guard about the object. They asked him to keep it in sight until they could launch an airplane. *The object "maintained its position above him as he moved through the water."* Then, it suddenly rose upward in a zig-zag motion, disappearing into a cloud bank to the southeast. It isn't clear how soon thereafter the airplane arrived. The object was gone by the time the Coast Guard plane arrived.

279

Comments: A number of fishermen in the same area also reported seeing a cigar-shaped object at about 11:00 A.M. which was pale blue on top and aluminum color on its bottom. It traveled at a "moderate speed and medium altitude" and flew above the Catalina Island Channel the same day (*Santa Ana Register*, 11 July 1955).

Reference: Gross, L.E. 1992. *The Fifth Horseman of the Apocalypse, UFOs: A History 1955 July–September 15th*. Fremont, Calif.: Published privately, 13-14.

Case No. 168

Date: 55-08-01

Local Time: 2105

Duration: estimated 4 minutes

CE-5

Location: Willoughby, Ohio
No. Witnesses: 2+
DB: v   CB: d   PB-1: a   PB-2: x   UFOD: b

Abstract: The witness drove home from his radio and television store after dark. Upon getting out of his car he saw a "large circular object, with a red light on the front rim, come down rapidly over a nearby field." It stopped its descent at about 800 feet altitude and two light beams came on and illuminated the ground.

Brown became badly frightened and ran toward his house. "I was so scared I ran past my car. I parked with the motor running. I felt the occupants of the object were following me. *As I ran up the drive it was right overhead, moving in the same direction I was running,"* he said.

His wife met him at their back door and said, "turn the outside lights on before it hits the house." He replied, "Hell no, turn them off and maybe it will miss it." They turned the lights off and ran through the house to the back porch area where they looked out at the UFO; *but by then it had extinguished all of its lights, including "several windows around the edge…of the disc."* Their children were very afraid and hid beneath the dining room table. The huge, almost silent[1] dark craft was hovering less than 200 feet above the ground and seemed to be from 80 to a 100 feet in diameter (see Figure 36).

280

Figure 36

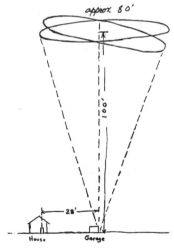

Sketch of size of UFO relative to Mr. Brown's house and garage seen on January 8, 1955

Then, the object moved over a nearby wooded area which permitted Mr. Brown to see a dome on its top with a white glow. The object then accelerated away into the night sky.

Comments: It is impossible to tell whether the UFO extinguished its lights in response to the Brown's turning their outside lights off.

Note: 1. Mrs. Brown said that she could hear a faint, soft humming sound coming from the object.

References: Keyhoe, D.E. 1960. *Flying Saucers Top Secret.* New York: Putnam, 237.

Gross, L.E. 1992. *The Fifth Horseman of the Apocalypse UFOs: A History 1955: July-September 15th.* Fremont, Calif.: Privately published, 34-35.

Cross references: 205 (6)

Case No. 169

Date: 56-11-25
Local Time: 1215
Duration: 10 minutes
Location: E. of Pierre, South Dakota
No. Witnesses: 2
DB: cc   CB: a   PB-1: a   PB-2: x   UFOD: 0

281

Abstract: The eye witnesses of this bizarre incident were a South Dakota State Policeman, officer Donald Kelm and Jack Peters, a radio communications dispatcher. They were traveling on Highway 31 some 25 miles east of Pierre when they topped a hill and started down the other side. They both saw a bright red, bowl-shaped object hovering several feet above the ground on the north side of the road. It illuminated the ground around it with red light and was about the size of an automobile. Kelm slammed on his brakes and stopped within a block of the thing. Both men got out and watched it for several minutes. Then, it began to rise upward and away from them.

The men jumped back into their patrol car and tried to chase it, however, it always stayed about a mile ahead of them. For the next seven miles, the UFO swung back and forth across the roadway, "It wouldn't let us go by it," said Kelm. *Finally, they turned around and headed back to Pierre and the UFO followed (behind) them, "but never went ahead of them."*

Comments: Here is yet another cat-and-mouse game played out in a remote area and reported by two highly credible witnesses.

Reference: Gross, L.E. 1994. *The Fifth Horseman of the Apocalypse UFOs: A History 1956 November–December.* Fremont, Calif.: Privately published, 30-31.

Cross references: 182 (6); 187 (6); 188 (6); 192 (6); 199 (6); 204 (6)

Case No. 170

Date: 57-01-10
Local Time: 1630
Duration: estimated 5 minutes
Location: Stafford, England
No. Witnesses: 6
DB: p   CB: a   PB-1: a   PB-2: x   UFOD: O

Abstract: Mrs. Jesse Roestenberg and five others saw a brightly glowing cigar-shaped object in the daylight sky over their town at what seemed to be a high altitude—estimated to be over 20,000 feet. It moved slowly toward the southwest and left an orange trail behind it twice its own length. "A jet plane was approaching the object which then ascended vertically and vanished into the sky."

Comments: While this UFO's response to the approaching airplane is a very familiar feature in UFO literature, the orange-colored atmospheric trace is relatively uncommon.

Reference: Anon. 1957. Flying cigar at Stafford. *Flying Saucer Review* 3, no. 1 (January-February): 9.

Cross references: 172 (6)

Case No. 171

Date: 57-06-05
Local Time: 2230
Duration: 30+ minutes
Location: Warrensburg, Missouri

No. Witnesses: 3
DB: x   CB: e   PB-1: a   PB-2: x   UFOD: b

Abstract: Robert J. O'Brien, 27, his wife, and mother were in an automobile driving home on Highway 13. They noticed an intense orange glowing mass or ball-shaped object in the southeast sky. Wanting to get a better look at the odd light, he pulled over, stopped, and all three got out. Then it suddenly dimmed and in doing so made it possible for them to see "numerous small, light-blue, bodies" inside the sphere. O'Brien stated, "The objects moved away from the point of origin at a high rate of speed. The objects each took a different direction and angle of flight, some moved to the horizontal, some vertical. They circled, hovered and moved about that portion of the sky for some thirty minutes" (see Figure 37). Then, they finally disappeared from view.

Some time later while resuming their northbound trip toward Lexington, Robert's wife turned around and saw a "dark oval" following their car only about 100 feet behind and about thirty feet in the air. Its surface had some small points of light on it and as far as could be determined was totally silent. Mr. O'Brien "floored the gas pedal. It was no use however, since the *ebony blob matched the vehicle's speed without much effort. Robert then slowed down to see what would happen. The 'thing' dropped back in response, maintaining a constant distance.*"

283

Figure 37

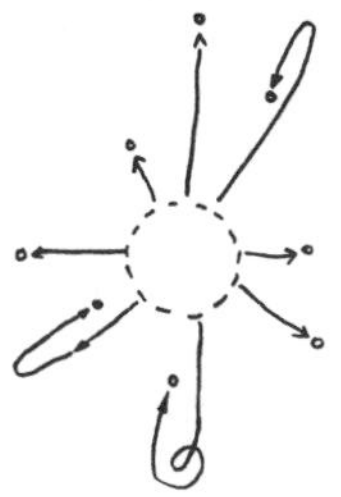

Sketch of UFO appearance and dynamics of June 5, 1957, sighting at Warrensburg, Missouri

Comments: Both the car's radio and Mr. O'Brien's wrist watch failed to work during the encounter.

References: Gross, L. 1995. *UFOs: A History 1957, May 24th–July 31st.* Fremont, Calif.: Privately published.
Smith, W. 1957. UFOCAT File 5 (June).

Cross references: 175 (6); 182 (6); 187 (6); 188 (6); 191 (6); 193 (6); 194 (6); 196 (6)

Case No. 172

Date: 57-06-21
Local Time: 1730
Duration: approximately 5 minutes
Location: Capel, Surrey, England
No. Witnesses: 1
DB: v   CB: a   PB-1: a   PB-2: x   UFOD: b

Abstract: Mr. A. N. Haskett was lying on his back looking up into the sky following a picnic tea. He saw a stationary point source of light high overhead. The witness was about 22 miles east-southeast of London's Heathrow International Airport. He noticed a DC-6 commercial airplane flying in the direction of the airport at from 5,000 to 7,000 feet altitude; it appeared to fly directly beneath the unknown light. *As soon as the aircraft had passed over the witness, the saucer made an extremely rapid ascent and vanished within seconds.*

Comments: Was the unknown object a meteorological balloon suddenly caught up in rising air currents from the airplane's engines or did it simply burst? There is not enough information to assess these or other possibilities. If it was a UFO, why did it wait to ascend only after the airplane had passed? Does this suggest that it did not care if it was seen from the airplane.

284

Reference: Anon. 1957. *Flying Saucer Review* 3, no. 5 (September-October): 7.

Cross references: 170 (6); 202 (6)

Case No. 173

Date: 58-06-end
Local Time: unknown
Duration: 2+ hours
Location: Via Anchieta Highway. 8 kilometers from
          Sao Paulo, Brazil
No. Witnesses: several
DB: x   CB: e   PB-1: a   PB-2: x   UFOD: 0

Abstract: This very bizarre case contains many details which seem to defy any reasonable explanation. A very large shining object "like a big steel house" was seen by Mr. and Mrs. Arlindo Alves and their young daughter, as well as by a large crowd of other motorists who had pulled off the highway leading to Sao Paulo to look at it. Everyone present noticed that it floated only a few feet above the ground for several hours! It also had windows and beings were seen inside the craft. No one could approach it due to the great amount of heat which it produced. The highway police on duty there called the Highway Engineers Department in Sao Paulo who sent a team to the site immediately. One of the engineers who arrived calculated the length of the UFO to be 143 meters while another took "an impressive photograph of the craft."[1]

Suddenly, the craft began to climb rapidly into the sky and soon disappeared from sight behind nearby hills. "Almost immediately after that, a jet aircraft of the Brazilian Air Force appeared, flying at low altitude (1,500 or 2,000 meters). Eyewitnesses concluded that the UFO had detected the approaching aircraft."

Comments: Of course, it cannot be known for sure whether the UFO's departure was caused by the approaching airplane, regardless of the conjecture of the witnesses. Nevertheless, the incident is included here under the premise that this was the case.

Note: 1. To the author's knowledge, this photograph has never been released publicly.

Reference: Buhler, W. 1983. More Early Brazilian Reports. *Flying Saucer Review* 28, no. 6 (August): 26.

Case No. 174

Date: 59-10-25
Local Time: unknown
Duration: approximately 45+ minutes
Location: West of Fort William
        (trans-Canada Highway), Canada
No. Witnesses: 4
DB: jj   CB: b   PB-1: a   PB-2: x   UFOD: b

Abstract: In this incident, an oval-shaped, glowing object followed a car for 30 miles along a barren stretch of wilderness. Four hunters—Victor Arnone, John Defilppo, Ray Disguiseppe, Douglas Robinson—were returning to the Lakehead when they noticed a white light which was spinning and maintained a fixed position relative to their car. It stayed about 40 feet above and ahead of them at times. At other times it would fall behind the car slightly. *When Robinson stopped, the object also stopped; when they drove off again, so did the object, holding the same separation distance.* It was silent as far as the men could tell. Then it unexpectedly veered off and disappeared.

Comments: This case sounds like a classic remotely piloted vehicle (RPV) sensing mission. I know of no silent RPVs shaped like this that existed in the late 1950s, however.

Reference: Anon. 1960. World Round-up: Canada: Hunters followed. *Flying Saucer Review* 6, no. 1 (January-February): 17.

Cross references: 175 (6); 176 (6); 177 (6); 180 (6); 184 (6); 191 (6); 196(6); 199 (6); 204 (6)

Case No. 175

Date: 59-11-03
Local Time: 0400
Duration: estimated 15+ minutes
Location: Statesville, North Carolina
No. Witnesses: 1
DB: cc   CB: e   PB-1: a   PB-2: x   UFOD: b

Abstract: A ball of blue light literally chased Mr. Robert L. James of Taylorville, North Carolina, as he drove along a deserted highway well after midnight. The UFO came to within several yards of his car a number of different times and also hovered directly above him "just above tree-top level." It traveled both fast and slow. Later, James said, *"I stopped [my car] six times and [the object] stopped six times."* James was "scared to death" during this encounter. Other witnesses in Hickory, and Salisbury, North Carolina, also saw a similar object.

Comments: Unfortunately, nothing is said about how the UFO disappeared or other details. This incident is placed in the indeterminate category since we don't know whether the driver signaled to the UFO in any overt manner other than continuing to drive his car and stopping periodically. Perhaps the UFO interpreted the various vehicle stops as some a kind of communication.

References: Anon. 1960. World Round-up: U.S.A.: Ball of light causes scare. *Flying Saucer Review* 6, no. 1 (January-February): 18.

APRO Bulletin, January 1960, 3.
New York Mirror, 4 November 1959.

Cross references: 171 (6); 191 (6); 196 (6); 199 (6); 201 (6); 204 (6)

Case No. 176

Date: 60-summer
Local Time: 2330
Duration: estimated 10+ minutes
Location: east of Crater Lake, Oregon
No. Witnesses: 3
DB: cc   CB: e   PB-1: a   PB-2: x   UFOD: 0

Abstract: Tom Page[1] and his two college roommates were on their way to Canada in a pickup truck. Mr. Page's first-hand account provides interesting information about his psychological responses to the two UFO which confronted them. This summary only concentrates on the CE-5 aspect of the case. Page was driving; his two friends were asleep in the back of the truck.

287

Alone, he had seen a white light ahead of him which suddenly shot straight up into the night sky. It could not have been an automobile. Somewhat later, he stopped and pulled over because the light (or a similar one) had reappeared and he wanted his friends to see it. One of his friends, Bob, got up and sat in the cab with Tom. He turned all of his lights off and then attempted to roll slowly down the hill toward the UFO ahead of them. *"But as the truck moved forward, the light backed up, keeping the same distance from us,"* Page said. Then Jim, the third roommate, knocked on the window and Tom stopped again to let him into the cab.[2]

The white light dimmed and then they could see a faint object in its place as wide as the roadway and shaped like an "upside-down cereal bowl." A separate, tiny red light was seen above the object. All three witnesses felt "a considerable amount of apprehension." *Once again, Page let the truck roll toward the object and it became much brighter again, moving slowly away from their truck.* Then, Page noticed another white light about 50 yards behind them so he stopped and all three got out of the cab. No sound could be heard of any kind. Page recounts what happened next.

"I was frightened, but also frustrated at not understanding what was going on. I started running toward the light [in front of the truck] to get closer to see what it was. *As I started running towards the light, it started moving toward me....*The light was still moving towards me so I raised my hands, not knowing what else to do. *The light dimmed, reversed direction and went back up and over the hill out of sight."*

Comments: This incident raises many interesting questions such as: why did the UFO maintain a fixed separation distance from the truck? Why did its illumination not produce a shadow? Why did the UFO react to Page when he raised his arms? Page and his two friends talked about what had happened for a while and then decided to tell no one except their parents about what they had seen. He had nothing to do with the subject for the next eleven years. He is currently a MUFON field investigator.

> Notes: 1. I have known Tom Page for many years and have every reason to accept this fascinating account as truthful and accurate. Page is a successful and popular high school science teacher in northern California. 2. The three young men noted that while the white light was very bright it did not illuminate their faces inside the truck nor did their heads cast normal shadows on the back of the cab. At this point the light was only about 20 to 50 yards away from them.

Reference: Page, T. 1987. Coming out of the closet. *MUFON UFO Journal*, no. 226 (February): 8, 9, 18.

Cross references: 171 (6); 188 (6); 196 (6)

Case No. 177

Date: 61-01-13
Local Time: 2045
Duration: approximately 45 minutes
Location: Virginia Airport (Durban), South Africa
No. Witnesses: 2
DB: r   CB: a   PB-1: a   PB-2: x   UFOD: O

Abstract: Mr. and Mrs. L. Saddler were driving past airport property when they first noticed lights in the night sky. They looked much like landing lights on an airplane, however, when they stopped the car to watch; *the lights also stopped.* They were point sources and looked like "large stars." Then, the lights moved in the direction of Berea at an estimated height of only 500 feet, hovered in one spot for another thirty-five minutes, and finally descended slowly and disappeared to the south of the fascinated witnesses.

Comments: The witness's descriptions suggest the presence of a single body containing many smaller lights. The implication is that the lights moved and stopped as a unit. Except for the long duration of this sighting this description is suspiciously similar to the appearance of lights on a very distant airplane.

References: Anon. 1961. World Round-up: South Africa: Saucer over coast. *Flying Saucer Review* 7, no. 3 (May-June): 22.

Sunday Tribune (Durban, So. Africa), 15 January 1961.

Cross references: 175 (6); 176 (6); 177 (6); 180 (6); 184 (6); 191 (6); 196 (6); 199 (6); 204 (6)

Case No. 178

Date: 61-10-21
Local Time: 0200
Duration: approximately 40+ minutes
Location: Route 60, New Mexico
No. Witnesses: 2
DB: jj   CB: d   PB-1: a   PB-2: x   UFOD: b

289

Abstract: This car-pacing incident extended over at least 30 miles and involved Mr. and Mrs. Richard DuBois of Westminster, California, who were returning home from a vacation. It was 2:00 A.M. and the highway was completely deserted of other traffic when the couple noticed an intensely bright white ball of light falling out of the sky in their direction. It slowed, passed in front of them, and "turned back and followed alongside." Then it streaked back up into the dark sky and was gone. Feeling somewhat more relieved, the couple reasoned that it was probably nothing more than an unusual reflection of moonlight.

Then, as suddenly as before, the light dropped downward again from nowhere at high speed, raced ahead of the car, "then slowed to let them catch up." As the UFO remained near them, they were even more frightened to see the object split into four smaller, glowing objects. All four maintained approximately the same distance from their car. Then, with great relief, the couple noticed the lights of a service station ahead and began to slow down. *As they did so the lights "flashed upward and went out of sight."*

Comments: This is a typical pacing incident where the aerial phenomena approaches at high speed and then decelerates to permit the car to catch up with it and then assumes pacing speed. These UFO motions require precise energy management (perhaps thrust vectoring), distance sensing, and navigation in three dimensional space. It is not likely that America had such a capability in the early 1960s.

Reference: Anon. 1962. World Round-up: U.S.A.: A 30-mile chase. *Flying Saucer Review* 8, no. 3 (May-June): 27.
UFO Investigator, January-February 1961.

Case No. 179

Date: 63-08-05
Local Time: early am
Duration: estimated 15+ minutes
Location: Wayne City, Illinois
No. Witnesses: 4+
DB: cc   CB: e   PB-1: a   PB-2: x   UFOD: b

Abstract: Ronnie Austin, 18, and Phyllis Bruce, 18, were followed for 10 miles as they were returning from a drive-in theater in Mount Vernon. A strange aerial light allegedly caused the car radio to "go crazy" and the engine to stall as it flew over them. The light was white and made a humming sound; it approached to within 100 feet and the air felt cooler when it did. When Ronnie got the engine started again he accelerated to 120 mph *with the UFO following close behind him for the rest of the distance home.*

Upon arriving home, both teenagers ran into Ronnie's house yelling for his dad to get a gun. At this point the UFO seemed to approach them so they closed the door, turned off all the lights, and phoned the police. When Deputy Sheriff Harry Lee arrived at the home he too saw the light in the distance.

Comments: Other witnesses included a police radio operator at Fairfield who sighted a light in the shape of a cross, Austin's grandmother, Wayne City Marshall George Sexton, and a neighbor.

References: Anon. 1963. World Round-up: USA: Morning star chases car. Flying Saucer Review 9, no. 6 (November-December): 24.
Dublin Evening Express, 6 August 1963.

Cross references: 169 (6); 182 (6); 185 (6); 193 (6); 203 (6)

Case No. 180

Date: 65-07-27
Local Time: 1940
Duration: 25 minutes
Location: 70 miles southeast of Carnarvon, Australia
No. Witnesses: 2
DB: jj   CB: e   PB-1: a   PB-2: x   UFOD: 0

Abstract: This event took place on the west coast of Australia on a winter evening. Mr. A. Kukla and Mrs. A. Lawrence were traveling together in a car on a rock hunting expedition. Suddenly, they saw a green light in the sky about 40 degrees above the horizon. It approached and then dove directly at them while also changing to an orange color and then a bright red. It had the shape of a "squashed football" when viewed from the side. Pulling over, Kukla turned his engine and headlights off. When he did so *the UFO "seemed to change course and hover in mid-air"* where it remained for almost thirty minutes, never changing again in shape or color. *The light circled their automobile slowly several times and then departed to the west.*

291

Comments: The remoteness of the witnesses would have made their car lights become the most prominent detail for many miles around. Was the aerial object on some type of reconnaissance mission?

Reference: Anon. 1966. World Round-up, Australia: UFO Buzzes car. *Flying Saucer Review* 12, no. 1, (January-February), back cover.

Cross references: 175 (6); 177 (6); 191 (6); 196 (6); 199 (6); 201 (6); 204 (6)

Case No. 181

Date: 67-03-08
Local Time: 0105
Duration: 30+ minutes
Location: Leominster, Massachusetts
No. Witnesses: 2
DB: q   CB: e   PB-1: a   PB-2: x   UFOD: a

Abstract: Mr. and Mrs. William L. Wallace were on their way home in their 1955 Cadillac traveling southeast on Lancaster Street. The night was dark, calm, and clear. Upon reaching St. Leo's Cemetery on Lancaster, they noticed a heavy patch of fog to the side of the road and ahead of them. It was unusual as they had not noticed any other fog previously; it was 1:05 A.M. when they drove directly through it. Upon emerging from the fog, Mr. Wallace noticed a bright glow to his left. After traveling about a minute more, he decided to go back to investigate what the light might be. It was then that they both noticed a bright light hovering motionless above the cemetery about 400 to 500 feet up.

292

Mr. Wallace pulled over and stopped, put the car in neutral and set the emergency brake. The motor was kept running. Then he rolled his window down to listen and finally opened his door and got out. "He pointed toward the UFO. *As he did so, his arm was pulled abruptly against the roof of the automobile. His car then stalled; the headlights went out, and the radio ceased playing.*" He was paralyzed, but his mind was still sharp. He said he felt a shock or numbness. When Mrs. Wallace, now very frightened, slide across the front seat and grabbed her husband's jacket, she noticed he did not move; he stood for about thirty seconds beside the car. He saw the UFO rocking back and forth and then the radio and lights came back on. The object moved upward and disappeared from sight, emitting a humming noise.

Comments: The main witness admits driving home after the sighting in a "slow and sluggish" manner. His reflexes seemed slower than normal so that he hit his own garage door as he drove into his driveway. The couple decided to drive back to the cemetery about ten minutes later but by then the fog was gone. The manner by which his arm fell to the roof of the car suggests a sudden loss of muscular strength to hold it up against gravity rather than a downward force acting only upon his arm. His body paralysis without mental impairment is not only significant but quite familiar in the UFO literature. Such symptoms suggest a focused stimulation at the brain stem or spinal level.

Reference: Keyhoe, D.E., and G. Lore 1969. *Strange Effects from UFOs: A NICAP Special Report.* Washington, D.C.: NICAP, 7-8.

Case No. 182

Date: 67-11-08
Local Time: night
Duration: estimated 5+ minutes
Location: near Sun Prairie, Wisconsin
No. Witnesses: 9
DB: v   CB: e   PB-1: a   PB-2: x   UFOD: c

Abstract: A UFO was hovering directly over the roadway about 100 yards ahead of their car, so Mr. James Dunn stopped and got out. Mrs. Dunn watched from the passenger seat. The object simply hung about "two house tops high" and was completely silent. The object seemed to be about the size of a Piper Cub airplane and was definitely solid, with no wings or other protrusions. Two closely spaced white lights were seen but no luminous projections or beams were visible.

293

Mr. Dunn got back in the car and began driving slowly. The UFO followed them. He accelerated to 90 to 95 mph yet, *the light "stayed with them at a constant distance."* [1] Upon reaching Main Street, they were relieved to see the UFO veer sharply to their left. Shortly, *the object reappeared and followed them to the home of Mr. Dunn's sister-in-law.* Mrs. Dunn ran inside and got her sister-in-law to come outside where they all saw the UFO. It left the area while phone calls were being made to Truax Air Force Base and the state police.

Comments: The capability of the UFO to leave the immediate location of the witness and later to return strongly suggests either a highly intelligent and planned response by the UFO, simple curiosity shown toward the humans, or both!

Note: 1. Here they noticed that the UFO had three or four small flashing red lights spaced around its circumference. The two white lights were no longer visible.

Reference: *APRO Bulletin,* September-October 1970, 8.

Cross references: 171 (6); 184 (6); 187 (6); 188 (6); 191 (6); 193 (6); 194 (6)

Case No. 183

Date: 67-12-03
Local Time: 0230
Duration: estimated 3+ minutes
Location: Ashland, Nebraska
No. Witnesses: 1
DB: v   CB: g   PB-1: a   PB-2: x   UFOD: b

Abstract: Officer H. L. Shirmer of the Ashland Police Department (and former Navy Vietnam veteran) was on patrol by himself after midnight. He was traveling southeast on Highway 6, intending to turn onto Highway 63 to the north. As he passed a low hill, his headlights illuminated something sitting on the road which he first thought was a truck.[1] Shirmer did not recognized the aluminum-colored object as any familiar surface vehicle. It looked like "two saucers fastened together lip-to-lip" and its circumference had many lights on it.

Shirmer drove to within 40 feet of the saucer shaped object "and could see red lights inside which began flashing as his lights illuminated it.[2] Then, the object rose until it was within 50 feet of the ground, an eerie loud "twanging" noise was heard, then it began to emit beeping sounds which became faster, louder and shriller." Then, a "brilliant orange and red beam" came out of the UFO toward the ground and it vanished upward in three seconds.

Comments: A measurement of 8 microroentgens (mr) radiation was found at the site soon after the event—normal background there is about 0.05 mr. While the officer's headlight illumination wasn't a deliberate act, it still caused a obvious response by the UFO. The witness also stated that he could not account for about thirty minutes of time.

Notes: 1. Later, he estimated the object to be about 26 feet wide and about 15 feet high. A square protrusion was seen centered on the bottom which gave off a continuous orange glow. 2. His engine died at this point and his flashlight also did not work.

Reference: Anon. 1967. Landed UAO Reported in Nebraska. *APRO Bulletin.* (November-December), 3-4.

Cross references: 186 (6)

Case No. 184

Date: 67-12-26
Local Time: 0400
Duration: 3+ minutes
Location: Aconquija, Argentina
No. Witnesses: 2
DB: v   CB: a   PB-1: a   PB-2: x   UFOD: o

Abstract: This car-pacing event took place in the early morning hours on the road to Aconquija and involved a newlywed couple. Upon reaching a particular hill called Dolla Chirca, they both saw a bright purple light near the ground which changed to green from time to time. They pulled over and stopped to watch the object and then got out of their car. According to the investigative report, the object "lifted up into the air. *Frightened, they got into the car and drove at high speed and the object followed them for a few minutes, then ascended into the air and out of sight in the distance.*"

Reference: Anon. 1968. Around the World. *APRO Bulletin*. (July-August): 7.

Cross references: 175 (6); 176 (6); 177 (6); 180 (6); 184 (6); 187 (6); 191 (6); 192 (6); 193 (6); 196 (6); 199 (6); 204 (6)

Case No. 185

Date: 68-01-20
Local Time: 2310
Duration: estimated 15+ minutes
Location: near Vermillion, South Dakota
No. Witnesses: 2+
DB: cc   CB: e   PB-1: a   PB-2: x   UFOD: b

Abstract: This car-pacing incident involved a large ball of orange and red fire which flickered in a nearby field and then developed into a frightening automobile chase at high speed. Mr. and Mrs. Robert Ballard had left his parent's home just after 11:00 P.M. Both Robert and his father, Howard, had previously seen the huge round, luminous object to the east of them and thought it was the moon. Robert's wife Lynn also saw it and thought it was the moon at that time. As the couple and their young son left the driveway, everyone saw the object emit a "real bright" flash; a dog also was "barking unusually loud and long." Ballard described the object as hovering first about 20 feet off the ground and then quickly dropping down vertically. It was spinning very fast and it looked solid.

At one point, the car happened to travel in the direction of the UFO so that its headlights shown in its direction. The Ballards then turned onto Highway 50. Robert noticed that the UFO was following them, so he accelerated to 60 mph; *the object suddenly "jumped or leaped" to a location right behind the fleeing car.* The couple estimated the diameter of the orange-red ball to be about 30 feet and "at times [it] had a white ring around it." *Later, it flew on ahead of them and then hovered about 3 feet above the ground at an intersection, as if challenging them to pass by.* Ballard had no choice. He passed it and started to climb a hill, scared of what it might do. As they neared the top of the hill, the UFO caught up with them but now remained at the height of the telephone poles, illuminating the roadway with white light. *Ballard sped up to about 100 mph and then 110 mph and the UFO did the same.* "I actually thought the craft was going to pick us up. It kept diving at my car, as though it was trying to grab us," Mrs. Ballard said.

Comments: Did the car's headlights somehow trigger this event since the object did not approach the group of four witnesses at the house before the couple departed?

References: Keyhoe, D.E., and G. Lore, Jr. 1969. *Strange Effects from UFOs: A NICAP Special Report.* Washington, D.C.: NICAP, 37-38.
The Vermillion Plaintalk, 25 January 1968.

Cross references: 187 (6)

Case No. 186

Date: 70-03-21
Local Time: 2015
Duration: estimated 10+ minutes
Location: Scranton, Pennsylvania
No. Witnesses: 2+
DB: q   CB: e   PB-1: a   PB-2: x   UFOD: 0

Abstract: This event involved two officers of the Scranton Police Department on duty at 8:15 P.M. They sighted a stationary light with an orange glow around it in the eastern sky; it looked unusual. Officer Sammes stopped the car in order to verify whether or not the light was moving. It hung at about 45 degrees above the horizon and appeared to be about one-quarter the angular diameter of the full moon (about 8 minutes arc). Officer Reina got out of the car. Soon, the UFO began to move upward and away from the two men until reaching the crest of a nearby hill where it came to a complete stop. Reina radioed officers in another car in the Providence section of Scranton to find out if they could also see the light from their position, they could not due to local hills. A third patrolman was reached by radio; he also saw the UFO.[1]

After the UFO traveled toward the south, it stopped again. *Reina "turned on the rotating red light on top of the patrol car and the object moved toward his position, becoming brighter and larger.* Reina said to Sames, jesting, "Get the shotgun out." Whether by coincidence or not, at this moment, *the object flashed a bright red light then receded into the distance and was out of sight within ten seconds.*

They started the engine and drove to Highway 81 and Davis Street at 8:25 p.m., pulled over and stopped again. They and another witness saw a light which may or may not have been the same object.

Comments: This is one of the few cases in which a red, rotating emergency vehicle light was used to signal the UFO. The bright red flash is also notable here; similar intense flashes of light from UFOs have been reported many other times.

> Note: 1. The UFO appeared increasingly smaller each time it changed positions as seen from Reina and Sammes' position. It isn't clear whether this was due to an actual distance effect or to a diminished object size.

Reference: *APRO Bulletin*, March-April 1970: 1, 3.

Cross references: 56-11-08(6); 183 (6)

297

Case No. 187

Date: 71-01-25
Local Time: unknown
Duration: estimated 10+ minutes
Location: Jackson, Ohio
No. Witnesses: 2
DB: q   CB: b   PB-1: a   PB-2: x   UFOD: c

Abstract: A young couple was driving along a country road near Jackson when they saw what looked like a moon rising up from the ground. Jay Chase, 19, claimed it rose quickly and took up a hovering position directly in front of his car. *He stopped and began to back away from the light which had now changed from orange to a purple color.* It began to follow them. Eventually he turned the car around and drove to a house nearby to try to alert its occupants, but nobody answered their call; they drove on into Jackson. Nothing is known about what happened later or how the UFO disappeared.
Reference: *APRO Bulletin*, January-February 1971: 7.

Cross references: 184 (6); 188 (6); 191 (6); 192 (6); 193 (6); 194 (6); 201 (6)

Case No. 188

Date: 72-02-26
Local Time: 2100
Duration: estimated 6+ minutes
Location: near Larned, Kansas
No. Witnesses: 1
DB: cc   CB: a   PB-1: a   PB-2: x   UFOD: b

Abstract: High school junior, Johnnie Beer, was driving home from town at about 9:00 P.M. when he encountered a strange light which descended right in front of his car, maintaining about 100 yards separation as he drove along. *"He said it frightened him and he slowed down and speeded up, but the object kept right in front of him. When Johnnie turned north at Riverside School, the object circled and traveled in front of him...[and] as he turned west to the farm, the light turned again and followed him into the farm yard where it hovered over the corral."*

The frightened young man jumped out of his car and ran for the house, closing the front door and pulling the drapes shut. When his parents arrived home thirty minutes later, the object was gone. The next morning it was found that the bottom two strands of wire on the corral fence "looked like they had been cut and twisted up in a pile in the corral, and several of our milk cows had burn spots on their backs," he said.

Comments: Was the UFO escorting Beer home? Was it curious to see how he might react to its presence? There is no way to answer such questions at this time.

References: Gribble, R. 1992. UFOs over space and time. *MUFON UFO Journal*, no. 286 (February): 15-16.

Tiller & Toiler (Larned, Kan.), 28 February 1972.

Cross references: 175 (6); 182 (6); 187 (6); 188 (6); 191 (6); 193 (6); 194 (6); 196 (6)

Case No. 189

Date: 72-02 to 07-mid
Local Time: 2300-early morning
Duration: estimated 3+ minutes
Location: Dighton, Kansas ¹
No. Witnesses: many
DB: v   CB: a   PB-1: a   PB-2: x   UFOD: 0

Abstract: This series of sightings occurred several times each month over several months and involved many residents of Dighton, population 1,500. The same nocturnal light would appear between about 11:00 P.M. and remain until the early morning hours about 10 miles west of town. Police Chief M. R. Shelton said the round object was a "red-orange and white light-bright as a cluster of lights on a football field." He mentioned that "the object would usually remain stationary until an investigating officer radioed another car about it. *It would then begin to move away....Everytime he would radio another car about it (the light)," it would depart.* The UFO was estimated to hover at from 300 to 500 feet up. Once he chased this UFO in his patrol car at speeds up to 100 mph and it would fly parallel with him or slightly ahead.

Comments: Officials at nearby Forbes Air Force Base in Topeka said that it could not have been one of their airplanes since they normally perform their infrared photography missions at much higher altitudes.

Notes: 1. Dighton is 76 miles west-northwest of Larned [cf. 188 (6)].

References: *APRO Bulletin*, July-August, 1972: 5.

Data-Net. 1971. *UFO Amateur Radio Network* 5, no. 6 (June).

Cross references: 190 (6)

299

Case No. 190

Date: 73-?-?
Local Time: night
Duration: estimated 10+ minutes
Location: Yakima Indian Reservation, Washington
No. Witnesses: 1 to 5
DB: q   CB: b   PB-1: a   PB-2: x   UFOD: b

Abstract: At the time of this sighting, Bill Vogel had been chief fire control officer on the Yakima Indian Reservation for thirty years. During these years he had sighted many UFOs in the relatively remote regions he was in charge of monitoring. He covered an area measuring about 50 by 70 miles which is located between Portland and Spokane. Most often, he sighted bright orange balls of light, pairs of bright, round lights like automobile headlights, and other odd luminous phenomena.

As author and researcher, Greg Long writes, "During the high-activity period, J. Allen Hynek Center for UFO Studies investigator David Akers of Seattle periodically visited the Reservation to observe and photograph NLs (nocturnal lights). However, the night time objects were seemingly aware of the approach of an outside observer. Vogel put it this way: 'Strange as it may seem, *when these objects were spotted, if there was any radio traffic about them at all (lookouts radioing a sighting), they would quickly disappear. As soon as it seemed they realized they were spotted by somebody who started talking about them, they'd vanish'"* (Long 1981, 4).

Vogel elaborated on this effect saying that he would radio to the various fire lookout observers to let them know that the field investigator was coming "so that they wouldn't be frightened if he pulled up there and just parked in the middle of the night." Vogel noticed that there might be a very high level of UFO activity from Monday through Wednesday but if Akers arrived on Thursday *"nothing would happen until [Akers] left."*

Long recounts how Akers changed his arrival and stake-out plans to try to "thwart" the UFOs. He would arrive with a minimum of publicity and/or stay for several nights. He was successful in getting some photographs the first night but no UFOs would appear on following nights.

Comments: This type of case could be easily explained by the deliberate use of monitoring technology by human beings who simply listened in to the radio "traffic." Still, one must ask why people would behave in this way?

According to Long, some of the photographs obtained by investigator Akers were of high quality showing "definite orange globes of light with yellow centers." A number of other highly interesting phenomena were seen by forest rangers which are not included here.

Reference: Long, G. 1981. Yakima Indian Reservation Sightings. *MUFON UFO Journal*, no. 166 (December): 3-7.

Cross references: 189 (6)

Case No. 191

Date: 73-03-27
Local Time: 0015
Duration: estimated 20 minutes
Location: Providence, Rhode Island
No. Witnesses: 1
DB: cc   CB: e   PB-1: a   PB-2: x   UFOD: b

Abstract: This rather typical car-pacing incident took place as the witness was driving home from Rhode Island Hospital in Providence. Traveling south on Route 95, he sighted an orange-red glowing object which seemed to be flying parallel with his own vehicle, traveling about 55 mph. It was estimated to be 200 to 300 feet in the air. Thinking that it might be some kind of reflection from his windshield, the witness pulled over, stopped, and got out. *The completely silent UFO also stopped ahead of him and pulsated in intensity. The witness got back inside his car and drove off with the UFO accompanying him another 15 miles.* Just before reaching the Route 138 Kingston Interchange, it veered away to the west and disappeared, much to his relief.

301

Reference: Anon. 1973. Flap over Rhode Island. *APRO Bulletin*, May-June, 9.

Cross references: 175 (6); 182 (6); 187 (6); 188 (6); 191 (6); 193 (6); 194 (6); 196 (6)

Case No. 192

Date: 74-06-13
Local Time: approximately 2345

Duration: 45+ minutes
Location: Gladstone, Tasmania
No. Witnesses: 6
DB: r   CB: b   PB-1: a   PB-2: x   UFOD: O

Abstract: This encounter involved a UFO which paced an automobile driven by Raymond Groves, 19. He saw a round pulsating light in the sky to the south which seemed to fade out from time to time, leaving a glow in the sky in its place. About every five minutes it would change from silver-white to red, green, orange, and then back to silver-white. Groves was curious and drove in its direction for about 1.5 miles and began to realize that the object was keeping pace with his car—it passed in front of Sugarloaf Hill, below the horizon.[1] At this time, Groves became very frightened and stopped his car, turned around, and sped back toward town. *"The object now appeared huge and close."* Passing *Sugarloaf Hill again, the UFO suddenly shot over the top of his car at about a mile from town and disappeared from sight."*

Raymond drove on to his uncle's house in Gladstone, very relieved that the close encounter was over. Then, he saw the light again in the eastern sky at about a 45 degree elevation and much smaller in apparent size than before. "Groves says he flashed his car lights at the UFO and both he and his Uncle saw it *pulse as if in response*." They continued to watch it for another ten minutes. He then drove the 8 miles home to Rushy Lagoon to the north, the UFO remained in the night sky and continued to flash different colors. His parents, brother, and grandmother also saw the strange light.

Note: 1. Later calculations showed that the UFO must have been at about 400 feet altitude and under 0.5 mile distance.

Reference: Roberts, W.K. et al. 1975. From the Tasmanian "Flap" of 1974 - Part I. *Flying Saucer Review* 21, no. 3/4 (November): 47-51.

Cross references: 193 (6); 204 (6)

Case No. 193

Date: 74-06-16
Local Time: 0500
Duration: estimated 20+ minutes

Location: Caceres, Spain
No. Witnesses: 1
DB: cc   CB: e   PB-1: a   PB-2: x   UFOD: O

Abstract: The witness, a 46-year-old farm laborer, sighted a bright object that illuminated the highway on which he was driving. Afraid, he sped up and the light followed him for several kilometers. At one point he saw "three tall figures standing inside the ship.[1] *When the observer extinguished the lights on his car, the object moved away. Then as he turned them on again, [the] UFO approached again at enormous speed [coming to] within 70 meters over his car. It followed him home and when he turned out his car lights, it slowly flew away.*"

Comments: Here is yet another official U.S. DAO intelligence report dated 1974 which deals with UFO. Some segments of our government are clearly interested in the subject of UFO and White House inquiries to the Department of Defense about them should be able to turn them up with ease!

Note: 1. It is unclear how the witness could see these figures while driving his car at high speed.

Reference: U. S. Department of Defense Intelligence Report 6 889 0174 74, dtd. 22 August 1974, from Madrid, by Capt. Richard T. Fox, acting defense attache.

303

Cross references: 192 (6)

Case No. 194

Date: 75-03-26
Local Time: approximately 2230
Duration: estimated 25+ minutes
Location: near Wakefield, Rhode Island
No. Witnesses: 2
DB: cc   CB: h   PB-1: a   PB-2: x   UFOD: b

Abstract: The UFO involved in this car-pacing incident displayed a high degree of intelligence with a seemingly specific but unknown objective. Mrs. M. and a friend drove out of the village of Wakefield (see point A; Figure 38) via Sugarloaf Hill onto Route 1 traveling west. At an unknown location, they sighted a large, crescent-shaped, red object in the northwest sky to the right-front of them. The brilliant star-like object was enormous, and paced their car at an altitude much lower than that of a light airplane. *The speed of the UFO seemed to vary but it kept pace with the automobile, remaining on its right side.* It then seemed to approach them and flew from their right to left side directly in front of their automobile. *Then, it reversed direction, passing in front of them again matching their forward velocity,* but now on their right side (near point B). When the two witnesses reached point C, the UFO, quickly moved to their left across Route 1 coming to a stop at point D. The two people felt relieved that they had left the object behind them.

Mrs. M. had to execute a 180-degree direction change on Route 1 in order to reach Matunuck Road where she lived. In other words, she had to drive beyond point C before she could use a cross-over access road to reach the east-bound lane of Route 1. Upon making this U-turn, the two witnesses noticed that *the UFO now visible ahead of them, suddenly changed from a deep red to a very intense silvery-white color. It seemed to have waited for the women's automobile to turn around and approach it.* Mrs. M. drove east and then turned south onto Matunuck Road.

Figure 38

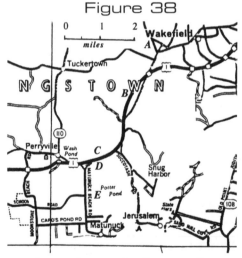

Chart of sighting area and approximate paths of motion of automobile and UFO seen on March 26, 1975, near Wakefield, Rhode Island

As soon as she passed the waiting, hovering object, it started to pace her car again. It then quickly flew across to the west side of the road and came to an abrupt halt over an open field opposite Mrs. M's. house (point E). There, it changed color from a silver-white to a pastel greenish glow. Both witnesses were, by now, terribly frightened; the driver careened into her driveway and both women ran inside and locked the door behind them. The UFO hovered just above treetop level until about 11:05 P.M. and then departed at a "moderate speed" to the southwest. By the time the UFO field investigator arrived at the scene at 11:15 P.M., the UFO was gone.

Comments: This fascinating case raises many questions which cannot be answered with only the present data. First, why would the UFO fly directly across in front of her car and then turn around and pass her again? Was it trying to be seen? Was it on some pre-assigned flight path and then changed its "mind?" Or was it on a deliberate mission to identify that particular automobile[1] and its occupants? Could the UFO have been part of a government-related surveillance project, perhaps searching for drugs or suspected terrorist activity? Maybe Mrs. M's automobile fit the description of a crime suspect's vehicle which might explain the sharp U-turn made by the object. Perhaps the controller of the UFO merely wanted to see if he could successfully intercept a ground vehicle selected at random, quickly identify its owner, and then perform a records check via radio for their address.

Second, what options did Mrs. M. have during this encounter? She could pull over and stop at any time. She could turn her car around and try to escape if the highway permitted this maneuver, or she could continue driving home. Third, what was the function of the color changes of the object? Were they but an artifact of its propulsion, guidance and/or navigation system? Others have commented on this possibility as well (McCampbell 1973).

The following night at about 8:45 P.M. another woman, Mrs. Sweet, and her daughter were driving home on the same general stretch of Route 1 and were paced in a very similar manner by a large, bright, silver-white object with a "silvery, stainless-steel, or chrome-like undercarriage beneath the object" (Todd 1976). It too anticipated where the driver was going to turn and shot ahead to wait at that location. Todd, the field investigator, also saw this same UFO and pursued it at speeds up to 60 mph, managing to take several photos of it which showed only a small rectangular area of light. It also crossed back and forth diagonally across the roadway from its north to south sides at times in front of him.[2] At no time could any sound be heard.

Notes: 1. Could the object have been a silent, unmanned remotely piloted vehicle with high resolution infra-red television camera? Its initial red hue might have been part of a ground illumination system and the later hue changes could have supported broad wavelength photography or television imagery. 2. The object was estimated to be about twice the size of a Cherokee light airplane or about 35 feet across.

Reference: McCampbell, J. 1976. *UFOLOGY.* Millbrae, Calif.: Celestial Arts.
Todd, D.R. 1976. Rhode Island car paced by UFO. *APRO Bulletin* 24, no. 8
  (February): 5.

Cross references: 21 (3); 182 (6); 188 (6); 191 (6); 193 (6) 204 (6)

Case No. 195

Date: 76-09-08
Local Time: unknown
Duration: estimated 15 minutes
Location: Sao Paulo, Brazil
No. Witnesses: 12
DB: w   CB: d   PB-1: a   PB-2: x   UFOD: 0

Abstract: This particular encounter came close to being an overt act of aggression against the UFO but, fortunately, stopped just short. A dense white cloud appeared close to the surface of Cajati Street in Sao Paulo and was witnessed by three men. As it hovered over one spot it began to dissipate, "revealing a flying disc" in its place. The men could not seem to move as they watched the object give off various colored beams of light. A neighbor was watching these events from a window in her house and telephoned the police. Soon, eight policemen arrived. They all drew their guns and one of the officers shouted, "It's the police!" *Suddenly, all of the policemen were paralyzed "like statues." Then the fog or vapor began to reappear around the object. It became increasingly dense and soon the disc was invisible while the cloud of vapor departed into the sky.*

Comments: While there is no obvious UFO response in this event, something happened to the humans present. How all eight policemen could have been simultaneously incapacitated is not known. Did the UFO take some pre-emptive action to prevent the officers from firing their weapons? Previous cases have been reported in which an apparently solid object "materialized" out of a vapor or clear air. However, we should not rush to assume that the idea of materialization is anything more than "becoming visible." If the UFO could have changed its emitted and/or reflected wavelengths by as little as 500 nm beyond the visible spectrum it would become invisible to the normal human eye. The interested reader should consult chapter six on the subject of "invisibility" in Haines (1980) for further comment on this particular subject.

Reference: *O Dia* (Rio de Janeiro, Brazil) 8 September 1976.

Case No. 196

Date: 76-11-?
Local Time: evening
Duration: estimated 5+ minutes
Location: Albacete, Spain
No. Witnesses: 2
DB: cc   CB: b   PB-1: a   PB-2: x   UFOD: b

Abstract: Mr. Martinez Sanchez and one of his daughters were driving home on the National Highway between Barrax and Albacete when he first sighted a very dark object in the air with many sparkling lights around it. He stopped his car and got out to see it better. *The perfectly silent object also stopped!* Becoming scared he got back inside, started the car and drove on down the road. *So did the UFO! "To test its movements Sanchez would slow down and the object would also slow down. Keeping always to a height of 10 to 12 meters, it followed them for the remaining 8 or 9 kilometers to Barrax."* When they arrived at the Tajo-Segura canal, the object   307 began to climb higher and higher. Sanchez described the object as having a round shape and about 5 meters in diameter. It also had a chain or "bars" within which he saw the many smaller lights.

> Note: This UFO was not a star or a planet because of its nearness to the witnesses, its obvious change in altitude as it departed, and its overall shape.

Reference: Anon. 1977. World round-up. *Flying Saucer Review* 22, no. 6 (April): 32.

Cross references: 175 (6); 182 (6); 187 (6); 188 (6); 191 (6); 193 (6); 194 (6); 196 (6)

Case No. 197

Date: 76-12-26 (approximately)
Local Time: night
Duration: approximately 22 minutes
Location: 25 miles northwest of Whitehorse, Yukon
No. Witnesses: 2
DB: i   CB: e   PB-1: a   PB-2: x   UFOD: b

Abstract: On their way north on the Alaska Highway to pick up a patient to take back to Whitehorse, ambulance attendants Tom Banks and Ken Schofield, saw an "extremely bright star" hovering in the dark night sky. "Then, all of a sudden, it was down in our area in no time," they said. It had an oval outline but its very high intensity prevented them from making out its exact shape. It paced their ambulance for twenty minutes, all the while darting back and forth and hovering at a low altitude. They successfully radioed Whitehorse about the presence of the object but also experienced vehicle electrical system interference (exact nature unspecified) each time the UFO came nearer.

"Eventually they got the [emergency] dome light to work and turned it on. *When it began flashing, the UFO flashed a bright light back at them for a few seconds and then skipped off over the trees and out of sight," they said.*

Comments: The connection between the bright flashes of their red emergency light and the UFO's response seems quite unambiguous. The two witnesses reported these events to local police later and said that the UFO made a head-on pass at them, darting to one side at the last moment. Here we find yet another instance of electrical system interference only while the UFO was nearby.

308

References: Anon. 1978. Spate of Sightings in Yukon. *The A.P.R.O. Bulletin* 26. no. 12 (June): 7.
Edmonton Journal, 10 January 1977.

Case No. 198

Date: 79-11-?
Local Time: evening
Duration: estimated 4+ minutes
Location: Roslin Glen, Scotland
No. Witnesses: 1
DB: v   CB: a   PB-1: a   PB-2: x   UFOD: O

Abstract: The witness to this event was Mr. Derek Scott Lauder, 29, who lived alone in a cabin about seven miles from Edinburgh. He had had numerous prior visual sightings of unusual phenomena [see 56 (3)] and liked to go for walks in the evening to possibly experience more. His account suggests that a CE-5 event took place. On this particular occasion he was in a broad glen adjacent to a castle (see Figure 39) and suddenly noticed "two beams of intense white light [that] shot out from near the castle.[1] They looked just like car headlights but no cars could be in that particular place. "I stood watching," he said, "as the black silhouette of a lens-shaped object could be seen to be emitting the beams of light....*I signaled at the object with the torchlight and the next thing I remember is seeing the object moving sedately away to my left, over the trees.*"[2] "I have never felt threatened by any of my experiences. On the contrary, I feel very safe knowing that 'They' are up there," he later wrote.

Figure 39

Original Watercolor Drawing of UFO Seen in November 1979 near Roslin Castle, Scotland
(Drawn by Derek Scott Lauder)

Comments: This UFO response was also witnessed by a British UFO investigator who traveled to Scotland to interview Lauder.

Notes: 1. He lived on the grounds of the historic Roslin Castle. 2. This point in the narrative has some similarity to cases involving missing time and should be researched further.

Reference: Lauder, D.S. 1996. Letter to author. 30 December.

Cross references: 56 (3)

Case No. 199

Date: 81-11-28
Local Time: night
Duration: estimated 20+ minutes
Location: Starks, Maine

No. Witnesses: 2+

DB: cc   CB: e   PB-1: a   PB-2: x   UFOD: O

Abstract: Deputy sheriff Harold Hendsbee and his wife Helen were driving home to Madison, Maine, on Route 43 when they saw a UFO hovering above a nearby hill. All they could see was an intense white light which illuminated the whole area. Hendsbee pulled over and stopped in order to watch it. It seemed to be only 20 to 30 feet in altitude and did not move. *When he pulled back onto the road to drive forward, the UFO* "*advanced toward us, then stopped.* I then backed up 30 to 40 yards and stopped."

He tried to get past the thing three separate times. On the second try *the UFO moved towards them and shown a very bright light through the windshield.* Mrs. Hendsbee said, "We had the feeling that it didn't want us to go up there [over the hill into the village of Starks]." After waiting twenty minutes they turned around and drove home on Route 148. Other eye witnesses also had seen the object.

Reference: Gribble, R. 1991. Looking Back. *MUFON UFO Journal*, no. 283 (November).

Cross references: 191 (6); 204 (6);

Case No. 200

Date: 85-01-16 to 26
Local Time: night
Duration: estimated 8+ minutes
Location: Hessdalen, Norway
No. Witnesses: several
DB: w   CB: e   PB-1: a   PB-2: x   UFOD: a

Abstract: This interesting event occurred in the valley of Hessdalen southwest of Trondheim in which hundreds of people had reported seeing very unusual aerial and ground light displays. The light shows began in November 1981[1] and continued to appear at least until February 1985 when Dr. J. Allen Hynek, founding director of the Center for UFO Studies in Illinois, traveled there to see them for himself. The first scientific expedition to investigate the colored lights occurred from the end of 1983 through the Spring of 1984 when a group of ufologists[2] succeeded in obtaining various equipment, including a spectrum analyzer, fluxgate magnetometer, seismograph, geiger counters, cameras, infrared night scope, and an Atlas 2000 radar set. At the end of their field research they concluded that "the phenomenon is capable of being registered on measuring instruments." However, they also had to conclude that more data was needed "before any conclusions can be drawn." An electrical gradient was discovered to exist in the same vicinity as the luminous display along with rather consistent winds and extremely cold dry air.

What did the phenomena look like? In the December to January 1982, period, three adults experienced a silent, bright, oval- or egg-shaped object hovering above a nearby hill at Finnsahogda (see Figure 40).

Figure 40

Sketch of UFO seen by several witnesses in early 1982 Near Hessdalen
(Drawing by Leif Havik)

When a group of 130 local eye-witnesses met on March 26, 1982 in the village of Alen, under the auspices of UFO-Norge, seventeen people present claimed

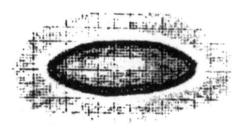

having seen a yellow spherical light, twelve saw "possible cigar-shaped" objects, eight observed "egg-shaped" objects, and six (4.6%) claimed to have seen "oblong" objects, each having one or two red lights. What the rest of the attendees saw wasn't disclosed. No particular physical or psychic events were reported. Many photographs were obtained of the typically small angular luminous source. Other witnesses reported seeing relatively large lights moving in waves. During daylight hours they appear as metallic balls or discs emitting a glow. Mr. Leif Havik and Mr. Age Moe watched in utter fascination on February 20, 1984, as a red light flew around the feet of Havik and then rose out of sight into the air.

In addition to the above events, a CE-5 incident occurred. Hynek writes, "On two occasions a laser light was directed toward a periodically flashing light.

Immediately the light doubled its flash period, only to resume its normal flashing pattern. As soon as the laser was again directed toward it, *the flashing doubled.* This occurred a total of eight out of nine times, straining coincidence and suggesting intelligent behavior" (Hynek 1985).

Comments: A variety of mostly silly explanations have been given for these lights, including: (a) atmospheric refraction at the interface between a layer of warm and cold air which produces a mirror-like image from other light sources (e.g., sun, moon);[3] (b) landing lights on airplanes approaching the airport at Vaernes, (c) superstitious behavior of the local residents combined with mental and/or physiological defects caused by centuries of in-breeding! No one yet has proposed any explanation for the light frequency doubling effect that reportedly occurred.

There are a number of important, unanswered questions which remain concerning the frequency doubling effect. Did the doubling continue only as long as the laser beam impinged on the light or for a shorter period? How stable was the laser beam aimed? Could the laser beam have alternately struck the light and then moved away as to trigger or "pump" its energy into the luminous ball at a higher frequency. How accurately was the pulse rate of the UFO measured; was it really a constant period or were there irregularities? How many witnesses were there to this particular phenomenon and what equipment was used to record it? These questions do not in any way challenge the veracity of the report made. They do attempt to identify more accurately the mechanism for the effect.

> Notes: 1. There is some evidence that unusual lights have been seen here since 1944. 2. "Project Hessdalen" was established on June 3, 1983, by representatives of UFO-Norge, UFO-Sverige, Foreningen for Psyko-biofysik, and several Finnish ufologists to carry out a concerted analysis of the phenomena. 3. Most of the sightings occurred around 7:30 P.M. and/or from 10:30 P.M. to 11:00 P.M.

References: Evans, H., and Spencer, J. 1987. *UFOs: 1947 - 1987.* London: Fortean Tomes, 88-92.

Hynek, J.A. 1985. *Tracking the Hessdalen Lights. International UFO Reporter* (March-April): 10-11.

Krogh, J. S. 1984. Hess-dalsrapporten, *NIVFO.*

Strand, E. 1984. Project Hessdalen. *UFO-Norge.*

UFO-NYT no. 4 and 5, 1985.

Case No. 201

Date: 87-11-11
Local Time: 1730
Duration: estimated 5+ minutes
Location: Delchamps (near Gulf Breeze), Florida
No. Witnesses: 2
DB: v   CB: a   PB-1: a   PB-2: x   UFOD: b

Abstract: This encounter involved a mother, 38, and her 15-year-old son who were in the family car on Highway 98 west, traveling to Delchamps. They sighted a "crown-shaped" object moving smoothly "to and fro like a feather" at tree-top level and keeping pace with their car. Its luminance lit up the interior of their car. *"Each time they talked about it, it would get brighter." She stopped the car and the UFO also stopped.* The next thing she remembers is watching it accelerate up and away into the dark night sky. They arrived home after 8:30 P.M., having lost track of more than two hours.

Comments: The original abstract indicates that all four members of this family have experienced many very unusual things before, the mother since age five. The ever-growing collection of evidence surrounding UFO continues to suggest a psychic component.

Reference: Current Case Log. 1992. *MUFON UFO Journal*, no. 294 (October): 19.

Cross references: 176 (6)

313

Case No. 202

Date: 90-06-?
Local Time: 2315
Duration: estimated 4+ minutes
Location: Baldim (Minas Gerais), Brazil
No. Witnesses: 1
DB: v   CB: d   PB-1: a   PB-2: x   UFOD: 0

Abstract: Mr. Sinval Santos, 25, was riding his bicycle home through a hilly area near Baldim after seeing his girlfriend. As he was repairing the chain on his bicycle, a bright white light shown down on him from overhead and he dashed off toward a nearby forest. *The light followed above him.* He ran into a barbed wire fence and was cut in many places. He got up, climbed through the fence, and hid in the woods as the light flew on past him about 25 feet up. He described the UFO as a "squashed sphere" with a bright yellow headlight in its middle and purple, green, and smaller dull yellow lights revolving around its circumference. He also heard the sound of air escaping from a balloon. As he hid there waiting, he saw an airplane approaching. "when the plane got close, *the UFO turned all of its lights off.*...After the plane passed over, I looked and saw the object far away," he said.

Reference: Pratt, B. 1996. *UFO Danger Zone.* Madison, Wis.: Horus House Press, 147-48.

Case No. 203

Date: 90-12-04
Local Time: 1930
Duration: 20 minutes
Location: 6 miles southwest of (Rio Grande do Norte), Brazil
No. Witnesses: 3
DB: cc   CB: e   PB-1: a   PB-2: x   UFOD: O

314

Abstract: Mr. Jose Garinberto Dantes and two of his fellow workers were in a water tank truck on a dirt road traveling toward Acari. Suddenly, they saw a 3-meter diameter ball of reddish-orange fire to the right side of the road in some bushes. *Dantes stopped the truck and everyone watched the object as it became only one-half its original diameter. As it did so it started to pulsate in brightness.* The men got out of the truck for about ten minutes but began to get scared. There was no truck radio interference and no sound was heard from the object. Everyone climbed back in and they drove off. After moving only about 30 meters one man cried, "It's going up!" Everyone saw it spinning and rising slowly. Dantes accelerated the truck to get away from the thing. It continued to zig-zag and then came back down 300 meters in front of them hovering only 3 to 4 meters over some bushes. It now looked oval-shaped and just over one meter in diameter.[1]

The UFO remained in this same location relative to the truck for about 2 miles, "fol-

lowing the contour of the ground, going up and down just above the height of the bushes." As the truck neared Acari, the UFO accelerated at a very high rate out of sight.

Comments: At no point during this pacing incident did the UFO approach very near to the truck as has occurred in many other cases. The lack of E-M effects in the truck is also notable.

> Note: 1. At this point its color changed to a "dead yellow," like it "had lost its color." The exact meaning of this is not clear but might refer to a significant decrease in luminance, color temperature, or both.

Reference: Pratt, B. 1996. *UFO Danger Zone.* Madison, Wis.: Horus House Press: 242-243, 293.

Cross references: 175 (6); 176 (6); 177 (6); 180 (6); 184 (6); 191 (6); 196 (6); 199 (6); 204 (6)

Case No. 204

Date: 92-04-18
Local Time: 2305
Duration: estimated 5+ minutes
Location: East Providence, Rhode Island
No. Witnesses: 1+
DB: cc   CB: a   PB-1: a   PB-2: x   UFOD: b

Abstract: This case is complex since the same 29-year-old woman was involved in two separate car-pacing events in the same part of town two weeks apart. During the second car-pacing encounter, she was driving by herself west on Cushman and Pawtucket Avenues. The triangular UFO was described as dark metal with seams or lines on the bottom surface in "distinctly staggered positions." It had a separate white, green, and blue light at the three corners and a small dome on the top front end. She estimated its size as over 30 by 30 by 30 feet. She felt it displayed a knowledge of exactly where she was going and felt that she was being watched. Then, the very low flying object[1] stopped so she stopped as well, about 50 to 75-five feet from it. *Then, she began driving again and it paced her, crossing the roadway several times directly in front of her. "When she slowed, it slowed."* She felt panicky and accelerated to well over the speed limit and the UFO traveled toward where she had first encountered it two weeks earlier.

Comments: The witness's mother claims that she saw several craft hovering over her daughter's home in April 1991 and also that the daughter had experienced some "missing time" at age 5 Of course, such evidence is circumstantial at best. At most, it points to a prolonged interest in the subject of UFOs within the family, perhaps leading to a predisposition to interpret ambiguous stimuli as UFO rather than as something else. The young woman said that she heard a distinct low frequency hum on her first encounter. This account is similar in many respects to that of Case No. 194 presented in this chapter.

Note: 1. It never flew higher than five to ten feet above the tops of buildings and trees and passed under a set of 100,000 volt power lines

Reference: Ware, D.M. 1993. Current Cases. *MUFON UFO Journal.* no. 298 (February): 14.

Cross references: 171 (6); 182 (6); 187 (6); 191 (6); 193 (6); 194 (6)

316

Case No. 205

Date: 92-06-27
Local Time: 0025
Duration: 10+ minutes
Location: Raeford, North Carolina
No. Witnesses: 2
DB: cc   CB: a   PB-1: a   PB-2: x   UFOD: a

Abstract: Two women were getting ready for bed when they heard a "roaring, train-like sound" outside their mobile home. The home shook and the sound seemed to travel past their home, seeming to come from a nearby field. Then it suddenly stopped. From a window they both saw a "curved row of about ten reddish-orange 'windows' close to the ground" about 450 feet away. They noted that their infrared-controlled light was off as was a neighbor's to the west. After calling the sheriff's office to report an airplane crash they went outside to see it better. *Upon turning their front porch light on, the UFO suddenly disappeared.*

Two infrared sensor-controlled lights came back on by themselves while three more porch lights farther away went off. None of the women's eight dogs barked at any time; their cats had disappeared as well.

Comments: A MUFON investigation team arrived on July 11 and found that a compass needle swung continuously 25 degrees to 30 degrees on each side of north inside a 15-foot-diameter area of grass that was matted down in the field where the UFO had been seen. The compass operated normally outside the circle. It is possible that the UFO emitted sufficient infrared radiation to trigger the porch light's sensor into responding as if it were daytime. This would turn the light off. Laboratory tests should be performed on this sensor to determine the minimum amount of radiation required to cause this to happen.

Reference: Current Case Log. 1992. *MUFON UFO Journal*, no. 294 (October): 18.

Cross references: 168 (6)

Case No. 206

Date: 93-08-10
Local Time: 0211
Duration: estimated 3+ minutes
Location: Norman, Oklahoma
No. Witnesses: 4+
DB: v   CB: d   PB-1: a   PB-2: x   UFOD: a

317

Abstract: Several college students were in their backyard watching for meteors as part of the Perseid shower of 1993. They noticed a tightly spaced formation of "blue-gray fluorescent" light patches flying across the sky toward the south. Each was separated from its neighbor by less than its own width. They thought the lights were at about 1,500 feet altitude and seemed to be flying very slowly at first. No sound could be heard from them and no navigation lights were seen. *"As one of the witnesses was saying 'no red lights,' a small red light appeared between the square panels!"* The young men had to run down the street in order to keep the objects in sight. But the objects turned to the east and zipped away rapidly, disappearing from sight in about ten seconds.

Comments: I have carried out a review of the literature on cases in which two or more objects are seen at the same time and place (*Project Delta*, LDA Press,

1994). This book contains scores of cases involving many identically shaped objects flying in very geometrically precise formation.

Reference: Current Case Log. 1994. *MUFON UFO Journal.* no. 311 (March): 15.

Figure 41

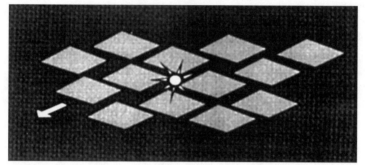

Reconstruction of UFO formation of August 10, 1993 in Norman, Oklahoma

Overview

Of the forty-three cases presented in this chapter, twenty-five (58.1%) involved some type of surface vehicle-pacing by the UFO. The UFO literature contains several times this many car pacing incidents. These CE-5 events were placed into the following general categories:

> Driver accelerates, decelerates, turns corners but UFO remains at a constant distance and relative position at all times in seven cases (29.2%).

> Driver stops, continues on. UFO also stops and continues on, pacing it in eight cases (33.3%).

> Driver accelerates to get away and UFO quickly comes within a very small distance from car in four cases (16.7%).

> Driver stops and backs up or turns around to get away and UFO maintains a constant distance from car, follows car in five cases (20.8%).

The remaining eighteen cases are heterogeneous in nature and include alleged UFO response to the use of radios, verbal command to the UFO, airplane pilot actions to evoke some response, etc.

The many cases in which the vehicle changes direction relative to a compass eliminates celestial bodies as explanations for the UFO. The relatively large angular size—i.e., apparent nearness—of the UFO also suggests other answers. The silent hovering and/or rapid directional changes of the UFO eliminate helicopters and lighter than air craft. The relatively long sighting durations eliminate most transient astronomical bodies as the cause of these CE-5 incidents.

References

Dodd, T. 1993. Mystery at Newstead Abbey. *UFO Magazine* 12, no. 4 (September-October): 21.

Fawcett, L., and B. J. Greenwood. 1984. *The UFO Cover-up*. New York: Prentice Hall Press, 134.

Haines, R.F., *Observing UFOs*. Nelson-Hall, Chicago, 1980.

Pratt, B. 1996. *UFO Danger Zone*. Madison, Wis.: Horus House Press.

Silberg, P.A. 1965. A Review of Ball Lightning. In *Problems of Atmosphere and Space Electricity*. Ed. S.C. Coroniti. Proceedings of the Third International Conference on Atmospheric and Space Electricity held at Montreux, Switzerland, May 5-10, 1963. New York: Elsevier Publishing Co., 438.

Slate, B.A. 1974. The Alien of Blount Island. *UFO Report* 1, no. 6: 32-35, 77-78.

PART II
Foreword

For nearly fifty years, it has been known that UFOs are real, and are probably interplanetary vehicles which are under intelligent control. In 1990, after an extensive survey of cases and the data related to UFOs, it occurred to me that a new category of UFO encounters was needed. The existing categories at that time, such as close encounters of the first, second, third, and fourth kind, were all passive, non-interactive categories. And yet we knew that these objects behaved intelligently, and on occasion, could interact in a voluntary manner with human observers. After reviewing a number of such cases, I decided that they were an important enough category to warrant their own classification: a close encounter of the fifth kind, or CE-5.

Simply put, a CE-5 is any UFO encounter which contains an element of voluntary human/UFO interaction or communication. It may be as simple as a driver of a car intentionally flashing his headlights at a UFO, and it stopping to return the signals, to as complex as a deliberate vectoring of a UFO to a landing site with a resulting near-landing event.

I have found that the CE-5 experience is not as uncommon as one might think. Dozens of people have related to me events during which a UFO reacts, signals, or takes some type of action as a direct consequence of a human attempting to communicate with it. Many times, these cases are hidden within the narrative of a seemingly routine case, such as a close encounter of the first kind, in which a UFO is seen within 500 feet—but which then interacts with the human observer's request to move in a certain direction, or to return the signaling of a flashlight.

In studying this aspect of the UFO/E.T. phenomenon, I found that a CE-5 could be the result of human efforts which utilized light, sound, and, remarkably, thought. That is, a UFO may respond to intentional human efforts to make contact which are mediated through visual stimuli, such as lights or lasers, through auditory stimuli, such as voices, music (and here I include radio wave signals) and through intentional and directed thought.

The inclusion of the last stimuli, thought, is important, even if conventional modern science would tend to dismiss it. While impossible to quantify at this juncture, the repeated accounts of UFOs interacting or responding to clearly directed thought may be giving us a glimpse into one of the most important scientific breakthroughs of our time: the interface between thought, mind, and machine.

Now this aspect of the phenomenon requires some very careful consideration—and a full measure of humility. The evidence would suggest that the UFOs may have the technological ability to receive clearly directed thought from a human, and respond to that thought. If true, then the communications telemetry of these objects have gone way beyond routine radio waves or microwave signals, and have crossed the great gulf between mind and matter.

I am reminded here of the fact that, as pointed out by Arthur C. Clarke and others, any sufficiently advanced life form capable of traveling through interstellar space to reach our solar system would possess technologies which would look like magic to us.

Could it be otherwise?

While much has been written on the possible propulsion and energy systems which might allow these objects to transfer through interstellar space, the matter of communicating through such vast distances has been left largely unaddressed. Surely, they are not using AT&T microwave systems.

The speed of light, a mere 186,000 miles per second, is simply too slow for effective and meaningful communication across lights years of space. If one's home planet is 50 light years away (a light year is the distance a beam of light traverses in one year) it would take 100 years for the simplest two-way conversation to occur using radio waves which travel at the speed of light! Could it be that these life forms have developed sub-electromagnetic communications systems which allow for nearly instantaneous communication across vast distances, by interacting with the non-local, non-linear aspect of thought and mind? That is, might they have developed technological systems which allow for accurate and reproducible up-links to thought and consciousness? It would appear that this indeed may be the case.

And, while this may all sound like magic to us in the late 1990s, perhaps we should remember that we have possessed real technology for only slightly more than 100 years, and in a universe over 15 billion years old, perhaps some other life form came this way one million years ago. Put another way, would not our current technology appear like magic to someone 500 years ago. How might a laser, a hologram, a TV, or a telephone appear to someone in the 1400s? It would all look like magic.

In founding the Center for the Study of Extraterrestrial Intelligence in 1990, I had as one of our chief goals that of exploring the possibilities of CE-5s.

That is, if indeed these objects are real and are intelligently controlled, how far might we be able to stretch the envelope of current understanding by deliberately engaging the UFOs in real-time field research? What might happen if we took the chief modalities for engaging in a CE-5—lights, sound, and thought—and attempted to systematically interact with UFOs? We have not been disappointed.

Dozens of unfunded, and admittedly ad hoc, CSETI research teams around the world have successfully engaged the UFOs in deliberate CE-5s, ranging from light signaling, to several near-landings with multiple witnesses present. While still highly experimental, our research results would suggest that the UFOs are under intelligent control, and are willing, from time to time, to interact voluntarily with humans attempting contact.

The real importance of the CE-5 category, then, may be the promise it holds for developing protocols for engaging in real-time field research, and, someday, for full-fledged contact and communication with whatever life forms and intelligences are behind the UFO.

In this book, Dr. Haines performs an important and enduring service to the area of UFO studies by collecting and reporting well-documented CE-5s. By doing so he provides us with excellent cases which we can study, and perhaps discern patterns of interactions which may help us in furthering our knowledge of how—and why—CE-5s occur.

From a scientific model perspective, the CE-5 is important because it may give us clues as to how to approach this subject in a way which can lead to real-time observations, something sorely lacking in UFO research. Retrospective studies—those which look into events and cases already past—can only yield so much. Ultimately, science and knowledge, must go forward in the here and now, not the past. The CE-5 category is important, because it gives us tools for exploring this fascinating topic in real-time, in the here and now.

Beyond this, the variety of CE-5 experiences may be a window on a new understanding of reality, one which bridges the immensity of space, time, matter, and mind.

And, ultimately, the CE-5 experience may lead to a time of open contact and communication with the life forms within the UFOs. And this in turn may lead to a time when human civilization becomes a truly cosmic, universal civilization. Perhaps sooner than we think.

Steven M. Greer, M.D.
Director, CSETI
Asheville, N.C.

> "UFO occupants are
> all things to all men."
>
> —Peter Hough

Chapter 7
Humanoid Responses to Overtly Friendly Human Behavior

(4 cases)

A Place to Begin

In many cases, science fiction and UFO literature contain fascinating if overlapping stories of beings from outer space who land on earth and carry out strange antics. The correspondence of themes and imagery has not gone unnoticed. Indeed, one may have spawned the other. But the important question is which spawned which? K.D. Randle (1996) expresses one side of the argument this way. "Long before the first alien abduction report appeared, the phenomenon was well established, fueled by speculation from science fiction and the popular press. All the familiar elements were reported—on the radio, on television, in movies, and in books—and were well-publicized. The evidence supporting the claim is overwhelming." A. Simon (1979) and others both before and after him commented on the possible cultural biasing which movies may have had on generating alien reports of various kinds. G. Nesom and R. Schatte (1996) list some twenty-seven early science fiction movies (1902 to 1964) in which aliens are portrayed graphically. N. Watson (1987) discusses the same general subject from the British point of view starting with the 1903 black and white film *A Trip to the Moon*, directed by Georges Melies. Watson maintains that aliens didn't really start "plaguing us" until after World War II. The 1945 thriller *The Purple Monster Strikes,* by Spencer Gordon Bennet and Fred Brannon, a fifteen-part serial film, is a case in point. "But it was not until the 'flying saucers' hit the newspaper headlines, that the cinema worked over-time to bring them into view" (333). Nevertheless, it must be remembered that both UFO and their alleged pilots have been reported in one form or another down through recorded history, long before movies and television existed.

Some maintain that Hollywood's portrayal of space aliens served as a metaphor for the perceived threat of communism back in the 1950s and 1960s when America and the Soviet Union were locked in the cold war. Whether aliens are portrayed as good or bad is usually determined by what's happening at the time, some writers contend. Others, working in the field of psychology, suggest that alien imagery is merely a projection of ourselves, a matte canvas on which we project our own fears—fears about possible galactic neighbors trillions of miles away as well as fears about human criminals who break into our homes. Nevertheless, it is a fact that aliens have been portrayed in many different ways by Hollywood and elsewhere so that the man on the street's image of an alien should probably represent some kind of average of them all. I suggest that no predominant type of alien should exist in mind or emotion.

Two movies released in America in 1951 portrayed E.T.s in almost opposite ways. Howard Hawks' *The Thing* thrust a hostile, Frankenstein-like monster upon the big screen. A group of scientists arrive at the scene of his ship's crash and attempt to capture him. While intelligent, this E.T. is bent on "purposeless, irrational destruction" (Jacobs 1975, 129). On the other hand, America was also treated to a "handsome benevolent" creature in Robert Wise's *The Day the Earth Stood Still.* While this very human humanoid tried to warn mankind of the dangers of nuclear power, "Earth people reacted with hostility and attempted to destroy him" and his invincible robotic sidekick (Ibid.). The 1953 movie *It Came from Outer Space* portrayed hostile earthlings confronting alien visitors after they crashed on Earth. Fortunately, the aliens managed to escape. The alien "star" in the movie *This Island Earth* (1955) was a half-human, half-insect creature with a bulging, exposed brain and perfectly round eyes. Interestingly, this particular creature has never reappeared in accounts of alien abduction!

The many truly weird looking space creatures encountered in Steven Spielberg's movie *E.T.: The Extraterrestrial* (1982) exhibit shapes, colors, appendages, and orifices which some anthropologists and biologists think probably come closer to reality than any other movie portrayal to date.

It is likely that, on the average, 50 percent of Americans think that space aliens are ugly and hostile and the other 50 percent think that they are pleasant, cute, and harmless creatures. In American culture, ugliness tends to be associated more with aggressiveness than does beauty. Ask yourself how you would react if you suddenly came across the creature shown in Figure 42 on a dark and lonely road? Compare this with encountering the creature in Figure 43 under the same circumstances. What does the UFO literature tell us about the responses of alleged aliens to various kinds of human behavior? This is what is addressed in Part II. Our focus is no longer on the craft but their alleged pilots!

Figure 42

E.T. from *This Island Earth* (1955)

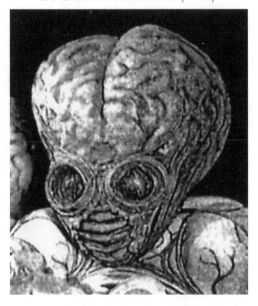

Figure 43

Creature from *E.T.:
The Extraterrestrial* (1982)

327

Distinguishing among different aliens

If alien life forms come in different sizes, shapes, colors, textures, intelligences, spiritual and moral persuasions, etc., we certainly don't yet understand how to distinguish between them. While several UFO investigators have suggested that there are many different alien races visiting Earth today (e.g., Pereira, Phenomenes Spatiaux, nos. 24, 25, 27-29; Huyghe 1996), reliable supporting evidence for this claim is virtually non-existent. Yet, if we automatically exclude Pereira's work and the evidence summarized by Huyghe, should we not also do the same for the rest of the evidence presented here and elsewhere in the literature? Indeed, much of the data is anecdotal in nature. Many come from only one witness; some sound like spooky campfire tales. And so I suggest that we suspend our judgment about the reliability of the evidence long enough to read the following accounts and consider what messages each may have for us. Then we can decide to conclude something. To prejudge this area of research without having even read it is scientifically dishonest. While the scientific method may not yet be adequate to deal with such narrative accounts, it still represents the single best starting point we have this side of theology.

It isn't clear whether the broad array of names given to the alleged visitors to earth from outer space is merely descriptive of what was perceived or is an attempt to describe our pre-existing beliefs concerning biological diversity in the universe. Indeed, these "saucer people" have been called by many names over the years. Some names we use seem to project something of our own nature or desires upon them like "ufonaut," "space man," "little men," "occupant," and "pilot;" while others like "alien," "alien life form (A.L.F.)," "demon," "space giant," and "reptile" seem to reflect our hidden fears of the unknown; they help us maintain separation between them and us. Other more neutral names such as "being," "extraterrestrial (E.T.)," "extra-terrestrial biological entity (E.B.E.)," "space creature," and "space being" could reflect our yearnings to understand the mysteries of space and our place in the vastness of the universe. The name "humanoid" clearly casts them in our own image, as if they must be human-like. Of course, this makes them far less formidable. And a name like "critter" places them in the status of a pet or vermin. "Angel," "entity," "little green men," "grey," "nordic," "Martian," and "Venusian" are names harder to explain psychologically. Nevertheless, it is still true today that some people want to keep them at a distance while others try to communicate with them as this chapter makes clear. What does all of this indicate about ourselves and about the apparent reality of these beings?

Communication

Part II of this book presents accounts in which the human tries to communicate with or signal to aliens in various ways. This chapter presents cases in which humans behave in a seemingly friendly manner. Of course, we immediately face the conundrum of what friendly behavior might mean to another race of beings. What are some kinds of overtly friendly behavior from our point of view? They include:

A wave	Mimicking behavior
A smile	A kiss
A slow, gentle embrace Slow, smooth body motions	A soft whistle Handing flowers or food
Doing something that attracts attention without causing injury to the other party	Setting out a plate of food or something to drink

But to a visiting alien these behaviors could be merely precursors to murder, like the kiss of Brutus or the "embrace" of a scorpion just before it strikes its victim. Only by prior experience with human beings could the alien know for sure that nothing harmful is represented by these overt behaviors. As will be seen, the kind of "replies," if that is what they are, are fascinating indeed. And so let us review some reports in which humans have claimed to have had peaceful motives in their hearts and have tried to act out these motives toward one or more aliens. We will learn as much about ourselves as about the alleged aliens. Indeed, merely compare the number of cases in chapter 7 versus chapter 8.

329

Case Files

Case No. 207

Date: 59-06-26 and 27
Local Time: evening
Duration: 3+ hours
Location: Boi-ani, Papua, New Guinea
No. Witnesses: several dozen
DB: dd   CB: e   PB-1: a   PB-2: x   UFOD: b

Abstract: (Incident #1; June 26; 5:45 P.M. to 9:45 P.M.) The Reverend William Gill, Anglican priest and headmaster of the mission at Papua, New Guinea, had just finished dinner and went outside at about 6:45 P.M. looking for the planet

Venus since it was very prominent at that time. After he located it, he noticed a "sparkling object…and very, very bright" located visually above Venus (Story 1980, 150). Then, it descended toward his location. When the object stopped, it subtended an angle of about 10 to 15 degrees arc. This incident puzzled Gill greatly; it forms the backdrop for the astounding CE-5 event which occurred the following night.

In a signed statement, the witnesses agreed that the object was circular, had a wide base and a narrower upper "deck," had something like legs beneath it, at times producing a shaft of blue light which shone upward into the sky at an angle of about 45 degrees, and that four humanoid figures appeared on top (see Figure 44). Some of the witnesses described seeing about four portholes or windows on the side (see Part B, Figure 45); Gill saw what appeared to be bright panels on the side but did not interpret them as portholes. Papuan teacher, Stephen G. Moi, saw "four dots beneath" the saucer and assumed that they were legs (Cruttwell 1961).

Figure 44

Sketch by Father Gill of UFO seen on June 26, 27, 1959, in Boi-ani, Papua, New Guinea

"As we watched it," Gill said, "men came out from this object and appeared on top of it, on what seemed to be a deck on top of the huge disk. There were four men in all, occasionally two, then one, then three, then four; we noted the various times the men appeared.… [1]

"Another peculiar thing was this shaft of blue light, which emanated from what appeared to be the center of the deck. The men appeared to be illuminated not only by this light reflected on them, but also by a sort of glow which completely surrounded them as well as the craft. The glow did not touch them, but there appeared to be a little space between their outline and the light."

Both the nearer object and another similar object moved erratically, changing directions, accelerating and then slowing, sometimes swinging to and fro like a clock's pendulum. One UFO then flew off toward the village of Wadobuna. So everyone ran down to the beach but the disk flew up and away over the mountains to the southwest, changing to a red color as it did so. It did not reappear that night.

Figure 45

A	B	C
(Stephen Gill Moi)	(Ananias Rarata)	(Dulcie F. Guyorobo)

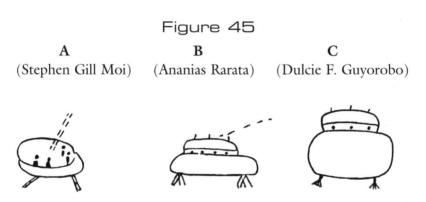

Sketches by three other eyewitnesses of UFO
seen on June 26, 27, 1959

(Incident #2; June 27)On the following evening at about 6:00 P.M., a similar looking object as before was seen in the still-bright sky by Annie Laurie Borewa, a medical assistant at the Papua hospital. Gill was called to come outside to see it along with several others. His subsequent report stated, "We watched figures appear on top, four of them. There is no doubt that they were human.[2] This is possibly the same object that I took to be the 'mother ship' last night. Two smaller UFOs were seen at the same time, stationary, one above the hills, west, and another overhead."

Two figures on top of the disc seemed to be doing something, "occasionally bending over and raising their arm as if adjusting or setting up something [not visible]. One figure seemed to be standing, looking down on us (a group of about a dozen)." This figure, he explained later, was standing with his hands on the "rail" looking over, "just as one will look over the rails of a ship."

"I stretched my arm above my head and waved. To our surprise the figure did the same. Ananias waved both arms over his head, then the two outside figures did the same. Ananias and myself began waving our arms and all four seemed to wave back. There seemed to be no doubt that our movements were answered. All the mission boys made audible gasps (of either joy or surprise, perhaps both)."

As darkness began to settle in, Gill sent one of the natives for a flashlight and directed a series of signals—"long dashes"—toward the UFO. *After a minute or two, the UFO wavered back and forth like a pendulum, in apparent acknowledgment. They waved and flashed signals again, and the UFO appeared to descend toward them, but stopped and came no closer.* After two or three more minutes, the figures disappeared." Then, at 6:25 P.M., two figures resumed their activity, and *the blue spotlight came on for a few seconds twice in succession.* By 7:45 P.M. the sky was totally overcast and no UFOs were visible. Thus ended the sightings (Story 1980, 150).[3]

Comments: The central figure in this classic, well-known case was Reverend William Booth Gill, an Anglican priest who directed a mission in Boi-ani, Papua, New Guinea. This event involves several intriguing details. The fact that a total of thirty-eight witnesses were present and that the beings on the craft waved back at Gill strongly suggests an intelligent response. The form of the beings was described as human. The response of the disc wavering back and forth when a flashlight was used to signal to it may or may not be coincidental. The precise visual details included in Gill's report lends further credibility to them. According to Gill (Gill 1971), the day after the second sighting, he had some students go out onto a playing field some distance from him. When they finally appeared about the same angular size as the figures onboard the disc he had them stop and then measured the distance to them; this distance was several hundred feet (assuming the beings were of the same height as average human height). Gill also had each witness go to his or her own room and write a report and make drawings of what had happened. These procedures point to a careful, educated man who took the entire incident very seriously. Indeed, he must have realized the scrutiny to which his report would be subjected.[4]

Is it possible that this aerial apparition was not extraterrestrial at all but a carefully contrived optical display perpetrated by humans for some unknown reason? Let us consider this matter for a moment. There is reason to believe that some people were very interested in determining the degree to which people can be fooled into believing something that is seemingly impossible (e.g., extraterrestrial life). The deliberate selection of a heterogeneous cultural group in a very remote location far from the probing and biased eyes of civilization might be an ideal social group and venue to use. The involvement of an educated leader like Gill might suggest the study of a cult-formative model. Perhaps someone wanted to determine whether such an aerial display would lead to a new religion or at least to formation of a cult. However, if this explanation is accurate then we would expect to find evidence of: (a) foreigners in and around the mission prior to this event, (b) some kind of advanced technology in the region to monitor the native's behavior, (c) written reports of the "experiment" turning up later, and (d) technical means of creating a craft that would support at least four human beings (at least 600 pounds) plus the weight of the vehicle (at least 15,000 pounds). I don't know of any such evidence. But what about the other side of this assertion? Why is it virtually impossible to accept this account as being the result of human activity?

First, the aerial object appeared metallic and solid to everyone on the ground and yet remained hovering over the clearing in total silence. Only a lighter-than-air ship can do this today. But what about other technologies? Even in the late 1990s, metallic-appearing holographic images cannot be projected into the air for many technical reasons. Second, the disk was seen by many people standing at different locations. Most large area holograms can only be seen when viewed from within a very limited area and even then they possess unnatural color fringe effects. Third, the disc was hovering silently below an overcast at about 2,000 feet altitude (at 7:10 P.M. to 7:20 P.M.) so that an upward light projection on the bottom of the cloud would appear very uneven, fragmented, and undulating. No such descriptions were given by the witnesses. Fourth, the "thin electric blue spotlight" which shown upward from the top of the disc is hard to explain. Where did the electrical power for this source come from? Either the spotlight and/or the glow from the object illuminated the cloud as it rose up through it.

It also may be mentioned that a colleague and unofficial chronicler of this event, the Reverend Norman E. G. Cruttwell, wrote another report regarding several scores of sightings from the New Guinea area for that same period of time (Cruttwell 1960).

Notes: 1. A family of three near Handbury, Staffordshire, England, saw a hovering UFO the size of a house in the afternoon of November 20, 1968, for about five minutes. There were "several humanoid figures walking across the bright top of the UFO," however, they did not wave or signal in any way to the captivated witnesses. The figures seemed to be inside a transparent dome (Keyhoe 1969). 2. Most likely, Gill was trying to describe the beings' outward appearances as human-like. At the time Gill said this, he was not at all convinced that the beings could be anything but human. Later, he came to a different point of view. 3. Later, Cruttwell wrote that, "Gill has come under a great deal of criticism and says that he sometimes wishes he had never seen the object......He himself is the first to say that, if anyone can give him a reasonable explanation of the phenomenon, he is prepared to accept it, but he cannot deny what he saw" (Cruttwell 1960). 4. In a review of this case published in 1977, the editor of the International UFO Reporter concluded that this sighting was generated either by an "extraordinary UFO as described" or by Venus distorted in size and shape by (amazing) atmospheric distortions! If the planet Venus, or any other planet, can produce this array of sensory impressions then mankind is in for some astounding sights indeed!

References: Cruttwell, N.E.G. 1960. What happened in Papua in 1959? *Flying Saucer Review* 6, no. 6 (November-December): 3-7.

Cruttwell, N.E.G. 1961. A letter from the Rev. N.E.G. Cruttwell. *Flying Saucer Review* 7, no. 6 (November-December): 29.

Gill, W.T. 1971. Interview by *Flying Saucer Review*, London, 28 January. I am indebted to Fred Beckman for making this recording available to me.

Hynek, J.A. 1972. *The UFO Experience*. New York: Ballantine Books, 167- 172.

Keyhoe, D.E., and G. Lore 1969. *Strange Effects from UFOs.* Washington, D.C.: NICAP, 30.

Papua/Father Gill Revisited. International *UFO Reporter* I, Vol. 2, no. 11 (November 1977): 4-7; (Part II) Vol. 2, no. 12 (December 1977):4-7.

Story, R.D. 1980. *The Encyclopedia of UFOs*. Garden City, New York: Doubleday & Co., 149 - 151.

Cross references: 30 (3); 59 (3); 81 (3); 212 (8)

Case No. 208

Date: 67-08-29
Local Time: 1030
Duration: estimated 5 minutes
Location: Cussac, (20 kilometers west-southwest of
          Saint-Flour), France
No. Witnesses: 2
DB: q   CB: d   PB-1: a   PB-2: x   UFOD: b

Abstract: Nine-year-old Anne Marie Depleuch and her older brother Francois, 13, were tending their father's herd of cows, with the help of their dog. Climbing over a rock wall, Francois saw four "children" whom he did not recognize. They were about 150 feet away and wore completely black shiny clothes. Standing near them on the ground was a very intense "silvery" sphere;[1] it was difficult to view directly due to its brightness. One creature was bending over as if studying something on the ground. A second held something in his hands reflecting sun like a mirror. He seemed to be waving to his companions.

Figure 46

Drawing of spherical UFO and creatures seen on August 28, 1967,
at Cussac, France (Courtesy of *Flying Saucer Review*)

Francois called out, "Are you coming to play with us?" *At that moment, the creatures, who did not seem interested in the children, seemed to realize that they were being watched. The first figure "flew up" vertically and dived head first into the top of the sphere. The second followed him in the same manner and the third one, after standing up, did the same....The fourth figure rose off the ground, came down again, appeared to pick up something, then he took off again and caught up with the sphere which, during this time, had begun to rise"* (*APRO Bulletin*). The last being then entered the object as had the other three.

The intensity of the sphere increased greatly as it rose. The children also reported hearing a "low, fairly sharp hissing noise...mingled with a noise like a breeze blowing."[2] Then, all noise stopped and the UFO accelerated out of sight toward the northwest.

Comments: The beings were estimated to be from 3.5 to 4 feet tall (but of differing heights). While their heads seemed to be uncovered, the children did not see any border between the black body covering and their skin. Their arms were long and slender. And, while their heads were of "normal" proportion to their bodies (relative to human beings), their skulls were pointed and the chin very marked. Other details are left to the interested reader to research.

This incident contains numerous high strangeness elements typical of our general topic. The four beings seem to be able to levitate into the air. Their proximity to the spherical object should be noted as it may have provided some kind of local, anti-gravity field effect for them. The possibility of controlled robots comes to mind which are pre-programmed to take evasive maneuvers if necessary. The reported return back to the ground by one creature to retrieve something left behind is also an interesting detail which the children probably would not have thought up by themselves.

Pratt also recounts a somewhat similar case in which a number of short creatures jumped up and floated back down to the ground. It occurred early in 1988 near the village of Mossoro in western Rio Grande do Norte, Brazil. In that case a dull, silvery, oval UFO of 4.5 to 5 meters diameter and 3meters thick sat on the ground on three legs. The creatures "would go down and do something on the ground and then go back up [to the craft]" (Pratt 1996, 221).

Notes: 1. The sphere was about 6.5 feet diameter and stood on thin legs off the ground. No markings or openings of any kind could be seen. 2. Neither witness felt any breeze during this time but did smell sulfur before the object departed.

References: *APRO Bulletin*, July-August 1968, 1-4.

Mesnard, J., & C. Pavy. 1968. Encounter with Devils. *Flying Saucer Review* 14, no. 5 (September-October): 7-9.

Pratt, B. 1996. *UFO Danger Zone*. Madison, Wis.: Horus House Press.

Cross references: 21 (3); 126 (4); 131 (4); 195(6)

Case No. 209

Date: 73-10-17
Local Time: night
Duration: estimated 3 minutes
Location: Falkville, Alabama
No. Witnesses: 1
DB: v   CB: a   PB-1: a   PB-2: x   UFOD: O

Abstract: This encounter involved police officer Jeff Greenhaw, 23, of Falkville. He was on his way to the home of a woman who had called about a landing of a "spaceship" in a field behind her house. After looking around her property and finding nothing, he drove along a side road. It was then that he saw a silver-suited being near the road. Thinking that it was someone in a theatrical costume, he decided to play along. He pulled over, stopped, and called out to the being, "Howdy stranger." *The figure did not respond but continued to approach the young officer.* He then grabbed his Polaroid camera and started taking some pictures of the "person." By now, the two were only ten feet apart and the being stopped. Greenshaw became alarmed at this point. He jumped back into his vehicle and turned on its red flashing light. *"At the sign of the light, the strange creature started running down the road, with Greenshaw in pursuit."* The policeman went off the road when his tires lost their traction and dust swirled up and around the car. When the dust settled the figure was gone. He admitted later that, "It scared me to death."

Comments: Fowler gives no further details of this case nor of the whereabouts of the photographs. We have seen other cases in which creatures approach and then runs away from the witness.

Reference: Fowler, R.E. 1974. *UFOs: Interplanetary Visitors.* New York: Bantam Books, 310-11.

Cross references: 21 (3); 126 (4); 195(6)

337

Case No. 210

Date: 82-07-?
Local Time: 0200
Duration: estimated 5+ minutes
Location: Sainte-Dorothee (near Laval), Quebec, Canada
No. Witnesses: 4
DB: v   CB: d   PB-1: a   PB-2: x   UFOD: 0

Abstract: Sixteen-year-olds Stephane Lebeau, Michel, and Francois Cousineau, and twenty-five-year-old Denise Labre all watched as a UFO the size of a helicopter hovered and also maneuvered silently at about 200 feet in the air early one morning. An intense beam of white light came from the object and illuminated the ground. It had multi-colored flashing lights around its circumference; their flash rate gave the impression of motion known in psychology as the Phi effect. Suddenly, the object emitted a "dull sound" and vanished from sight, moving behind nearby buildings.

The group returned to their tent, wondering what they had seen. Then they heard a sound like "electronic beeps." They got a small flashlight and went back outside and shone it into the darkness, a face was caught in its beam. Francois then arrived with a more powerful spotlight. Its beam fully illuminated an entity who was from 5 to 6 feet tall, "with a huge brown head, and orange eyes, bigger than ours." *The four youths also heard a sound in the nearby field of maize and then felt a whirlwind of dust swirling around them.* They also experienced intense stomach pains and then ran terrified into a nearby house.

Comments: A deliberate prank? The product of overactive imaginations? Perhaps the result of drug consumption? This account raises many troublesome questions. Yet, taken at face value, this incident is not all that unusual when compared with many others which have been reported over the years. The nature of the humanoid's or UFO's CE-5 response is not particularly clear. It is included here as it might relate to other similar cases.

References: Anon. 1982. World Round-up. *Flying Saucer Review* 28, no. 2 (November): 28, Nov. 1982.
La Presse (Quebec), 24 July 1982.

Overview

These four cases have shown that all of the visitors: (1) were associated with some type of craft or physical object; (2) reacted, sooner or later, to the human behavior in some way—by waving back (friendly response: case 1); quickly ducking inside their craft (escape response: case 2); approaching, stopping, then running away (approach-avoidance response: case 3); and by producing a sound, a whirlwind, and stomach pains in the witnesses (unclear response: case 4); (3) never harmed the humans; and (4) did not communicate any significant information except that they exist perceptually. Other patterns are presented in chapter 10.

References

Hough, P. 1987. UFO Occupants. In *UFOs 1947-1987: The 40-Year Search for an Explanation*, eds. H. Evans and J. Spencer, 126-131. London: Fortean Tomes.

Huyghe, P. 1996. *The Field Guide to Extraterrestrials*. New York: Avon Books.

Jacobs, D. 1975. *The UFO Controversy in America*. Bloomington, Indiana: Indiana University Press.

Nesom, G., and R. Schatte. 1996. 27 early sci-fi movies. *Houston Sky*, no. 11 (June-October): 12.

Randle, K.D. 1996. Does pop culture affect our view of alien abduction? *Houston Sky*, no. 11 (June-October): 1, 3-5, 7-8.

Simon, A. 1979. The zeitgeist of the UFO phenomenon. In *UFO Phenomena and the Behavioral Scientist*, ed. R.F. Haines, Chpt. 3. Metuchen, New Jersey: The Scarecrow Press.

Watson, N. 1987. The day flying saucers invaded the cinema. In *UFOs 1947–1987: The 40-Year Search for an Explanation*, eds. H. Evans and J. Spencer, 333-37. London: Fortean Tomes.

> "Super smart...does not mean super nice."
>
> —Loren E. Gross

Chapter 8

Humanoid Responses to Overtly Hostile Human Behavior

(26 cases)

If we were to ask the average person about aggression between people and space aliens, he or she would probably think immediately about lurid stories of grotesque monsters doing terrible things to humans. Indeed, there are documented accounts about this subject in literature (Clark 1967; Creighton 1977; Fawcett 1986; Holiday 1978; Michel 1958; Pratt 1996). However, the opposite situation is far more likely to be the case. This is the subject of the present chapter.

I suggest that humans are the direct or indirect cause of alleged aggressive alien responses in almost every encounter. If this can be proven to be true the implications are enormous. It is impossible to control the aggressive tendencies of human beings, and therein lies a potential danger for the future, whenever alien life forms may visit Earth.

One possibility is that our visitors are basically peaceful beings and do not present a clear and present danger to mankind. But, of course, the opposite might also be true! The highly advanced intelligence and technology seemingly displayed by UFOs (Haines 1994; Hall 1964; Hynek 1972; McDonald 1967) only shows that they are super smart, not necessarily super nice. As Loren Gross has correctly pointed out, "Super-smart UFO pilots no doubt consider violence stupid and counterproductive in almost every case" (Gross).

The well-known French ufologist, Amé´ Michel, came to the conclusion that the visitors never displayed fear under any circumstances (1958). Of course, the claimed behavior of some aliens does not prove the intent of the majority.

Do they follow one programmed behavioral response in all confrontational situations, or do they possess a controlled response capacity that is dependent upon the encounter circumstances? And why do so many close encounters with humans seem to take place under the control of the aliens? Are they trying to provoke a human response of some kind, like a psychologist who stimulates a rat in a cage to see how it will respond?

As was true in our review of cases about UFO responses to overtly hostile human behavior (chapter 4), there are a number of reported cases of hostile human behavior shown toward the alleged humanoids as well. Assuming for the time being that these space beings exist, this unfortunate occurrence is usually motivated out of fear. As George Fawcett has put it, "when confronting UFO occupants (ufonauts), bone rattling fear and blind panic become the order of the day or the night, depending on when such incidents took place" (1986, 2). The interested reader should consult his interesting review of human responses to UFO and humanoid close encounters. Other times the human is motivated out of simple curiosity. But whatever one's motive, it is a stupid and dangerous thing for humans to do. If one is wise, one does not pick a fight with someone who is very likely more powerful than oneself. And just because we do not yet know if these alleged humanoids are friendly or hostile does not give us the right to shoot at them.

While people have the right to defend themselves, they do not have the right to provoke an in-kind hostile response. Indeed, many holy books teach that we must not kill each other, since only God has the power of giving and taking life. Do alien beings fire back at humans? Yes, as the following case illustrates. Do they react in-kind or worse? Sometimes. What if they share a common genetic makeup with us—a far more problematical issue. Have they developed an extraordinary level of self control that is far beyond ours. These and other questions are raised by the cases presented here.

Case A. 63-10-13 The following terrifying encounter reportedly took place near Monte Maiz, Argentina, and clearly illustrates a provoked act of human aggression. Most of the details of this incident are presented later in this chapter. Sr. Eugenio Douglas, a truck driver, was on his way home to Venado Tuerto, Province of Santa Fe, at 3:30 A.M. when he was flash-blinded. He collapsed and ended up in a ditch in a dazed condition. Upon regaining his faculties, he got out and saw an oval or round object sitting in the highway ahead of him. It was at least 30 feet high. Then, through the driving rain, he saw a doorway appear in the side of the object and three huge "beings in human form" emerged; each was from 13 to 16 feet tall. Glowing helmets covered their heads but he could not make out their faces. They noticed Douglas standing beside his truck and "shone a beam of [many] red lights at him...of tremendous power, which seemed to irradiate his skin as they struck him. The beams were burning him, and in his confusion and terror he drew his revolver and fired three or four shots at them, then fled in panic back down the road toward Monte Maiz." He turned to look back at them and saw that the object was taking off and flying toward him. It dove at him several times, each time he felt the burning, prickling sensations as before. Upon entering the outskirts of Monte Maiz, the mud-covered truck driver was met by a number of townsfolk who also saw a "vivid red light in the sky" (Creighton 1965).

What constitutes provocation?

It is one thing to shoot at a wild animal that is threatening your life. It is quite another to shoot at an apparently living, intelligent being that is minding its own business and is not harming you in any way. Yet, the important issue remains, what really constitutes provocation from a cosmic point of view, and how do we know for sure that a humanoid is intelligent? In the final analysis, our judicial system will not decide such issues even though it may try to. Indeed, judges have no concept at all of the ethics, morals, and mores of alien life forms, if they exist. Juries and attorneys are likewise as biased. Provocation to alien visitors may be little more than the blink of an eye or turning one's back toward them or getting in the way of their carrying out an objective. If the UFO literature is to be believed, it looks as if humans have not prevented them from their "appointed rounds." The following cases illustrate this.

Case B. 54-10-? The following event reportedly happened at a rock quarry at Marcilly-sur-Vienne, France, in October 1954. A group of men were working at the quarry when there appeared a "small helmeted and booted entity, carrying a sort of ray-pistol. He fired "at the foreman of the group and then at six other men nearby." They all saw a "luminous and paralyzing ray" (Wilkins 1955). Unfortunately, nothing more is known of this alleged incident. But it does illustrate the problem of trying to decide what constitutes provocation. Were some facts left out of this story? Did the foreman pick up a steel bar and approach the creature with it in a menacing fashion? Did the men slowly surround the being as if to capture him? Did the visitor first signal some sign of greeting to the workers and then try to communicate with them using a luminous projector such as the ray pistol?

Provocation between human beings is not as simple to determine as one might believe. For example, behavior I might tolerate from another person without making any response on a clear sunny day on the sidewalk of a busy city street is very likely different from what I would tolerate from the same person at that same place in the dark, early morning hours with no one else around. Certainly the immediate physical and social situation is important. Yet, provocation usually implies a threatening act taken toward another person or their property. In the following cases do you think the humans involved were justified in using weapons against the aliens?

Case C. 54-11-28 This incident, as reported, involved an apparently unprovoked attack upon two young men by several highly unusual creatures on a lonely road at 2:00 A.M. near Caracas, Venezuela. Gustavo Gonzalez and Jose Ponce were traveling in a van in the early morning hours. At one point, they saw a spherical, self-luminous object hovering over the road ahead of them, so they stopped, not willing to get any closer to it. But eventually, Gonzales got out and walked toward the object. Suddenly, "he was set upon by a slight, dwarf-like, bristly creature, who sent him sprawling with a casual push." The dark figure, partly outlined by light from the distant object, then leapt on top of the terrified man, the creature's eyes "glowing" in the dark. Gonzales drew out his knife and jabbed forward at the creature but its blade *"glanced off the creature's hairy body as if from a rock."*

Then, a second creature appeared and dazzled Gonzales' eyes with an intense light. Finally, Jose Ponce jumped out of the van to lend assistance. He noticed two more beings coming out of the nearby bushes carrying rocks. The account ends abruptly by indicating that, "All four then leapt effortlessly into the hovering craft and were gone." Unfortunately, part of the narrative is missing, particularly what Ponce might have done to cause the four creatures to leave so abruptly. A physician who was in the same area allegedly saw all of this happen and arrived to lend assistance. He discovered a long, deep scratch on Gonzales' side (Brookesmith 1984).

It is most probable that since it is the human being who ultimately tells the story it is he or she who may add details that support his or her position and reputation afterwards. Are some facts left out that could be damaging later? In case B, which took place amid hundreds of other equally bizarre events reported from across Western Europe in 1954, the quarry workers were brave to disclose as much as they did. They certainly could expect to face laughter and derision from their coworkers, employers, police, and the general public. This fact probably adds to the credibility of their story. In the next case there were signs of a violent struggle left on the body of one of the men. Were these placed there deliberately? If so for what reason? Case No. 217 in this chapter contains several details similar to those of this case.

Case D. 76-09-09 Forty-year-old Hermelindo Da Silva was on his way home at 2:00 A.M. from his roadside bar near Vargem Grande, north of Belo Horizonte, Brazil. His dog walked beside him. The distance to his house was only about 70 yards. As Pratt (1996) states, "the sky suddenly lit up as bright as day." A 10-foot-diameter object hung over his head and was "surrounded by yellow and violet hues (and had)...a dark hole in the bottom." Suddenly, the light went out and Hermelindo ran to the back door of a small building for safety. Hermelindo was unable to unlock the door because he was shaking so badly. Then, he was blinded when the light came back on again. He grabbed a long pole and thrust it upward and hit something solid. He said, "I got a slight shock. Then the light turned off again." A faint buzzing noise also stopped when the light went out, yet a "hissing-like" sound could still be heard. Something struck his shoulder and he fell to the ground. Now terribly afraid, he got up and ran toward his house, the UFO following him only 25 feet up. I will not recount the other bizarre details except to point out that a 4-foot-tall creature—who looked like a human being but without a visible face—descended from the UFO; Hermelindo wrapped both his arms around the being in a powerful bear hug. *"He felt like metal," the witness said later. "I tried to throw him to the ground," and the being "shuddered." I let go when I became scared.*

Once again, we can ask if this witness was provoked into hitting the UFO or grabbing the humanoid? From the safety and comfort of our arm chairs it is easy to say no. But had we been in his place what would we have done? (Pratt 1996)

Case E. 77-03-13 In the following encounter, the witness ran away so fast that he wasn't certain whether the figure responded or not. Was he provoked into striking out? This incident took place on a dark roadway just outside of Brawdy, West Wales, at about 9:00 P.M. Stephen Taylor, 17, was returning on foot to his home when he sighted a light in the sky; but he didn't think much about it at first. After stopping by to see some friends he continued on. A nearby security fence where he was walking separated the public road from a NATO installation at Brawdy, near Pen-y-cwm.

Then, the young man noticed a dark, dome-shaped object that blocked the lights of a nearby farmhouse he had seen many times before from this same spot. The object should not have been there; its body seemed to be 40 or 50 feet across with a dim glow coming from its underside. As he approached it (out of simple curiosity), he eventually realized that it was a definite object that had landed there just outside the security fence of the American base. Indeed, a sign on the fence made it clear to stay out of the facility and that the installation came under the purview of the Official Secrets Act of England.

Stephen stopped and leaned against the gate and lit a cigarette. Immediately, he heard a noise "like someone stepping on dry leaves," and he looked away to his right. Standing near to him was a figure. Later he told a reporter from the *Western Telegraph,* "I was so frightened that I just took a swing at it and ran. I don't know whether I hit it. I never looked back, but kept on running till I got home." The figure stood over 6 feet tall and was wearing a semi-transparent, silver suit. It had very high cheek bones "and large eyes like a fish's—round and sort of glazed," the witness said. While the figure was humanoid in its overall form, it also possessed very unusual features as well. For instance, where its mouth should have been was a small "box-like device" and a "thick, dark tube leading from this over its shoulder." Stephen's keen perceptive capability is remarkable considering the darkness and brevity of the encounter. A similar strange creature had been sighted on April 24—six weeks previously—at Ripperston Farm by Billy and Pauline Coombs and by Mark Marston on April 15 (Paget 1979; *Flying Saucer Review* 1977).

What constitutes murder?

Murder is the unlawful killing of another human being with malice aforethought. But while hundreds of library shelves are filled with law books on this subject there is almost nothing written about the murder of beings that are not fully human in form or intellect. Premeditation is a key factor in murder.

Case F. 65-01-end Fowler recounts an interesting series of events that took place early in 1965 in the Brands Flat area (Augusta County) of the Shenandoah Valley (1974). Several people there had armed themselves because creatures from outer space were supposed to have landed. A group of people were out looking for the creatures and Sheriff John E. Kent was concerned. Things were getting "completely out of hand" and were "dangerous to county residents," according to Kent (*Arkansas Gazette*, 30 January 1965). His public position was that even though they might find "little green men from space," they still had "no right to mow them down." Finally, the Justice of the Peace in Fredericksburg requested a legal ruling by the Attorney General, Robert Y. Button. Button replied in due course that, "There apparently is no state law making it unlawful to shoot little green men who might land in the state from outer space!"

Interestingly, nothing at all was said about possible danger to the alleged visitors from space! This frame of mind underscores man's egocentric nature and an unconscious denial that such creatures could even exist. Kent went on, "Anyone can go out at night and see reflections in the sky. But anyone carrying firearms in the county without good reason will be dealt with according to the law." To the many Americans who firmly believe in extraterrestrial beings and are fearful of them, this is reason enough to carry a firearm. Nevertheless, shooting and killing an alien becomes a moral issue with very grave implications, implications that have not eluded many science fiction writers over the years.

347

The main issue of U.S. law in regard to murder of an E.T. is whether or not our laws were written to include nonterrestrials, regardless of their intelligence, physiology, appearance, personality, or other characteristics. Certainly, foreigners visiting America from other nations are covered under U.S. laws in this respect by virtue of their very humanity. But is an E.T. similarly covered? This will be an issue for our Supreme Court to decide when the time comes. Is a space being a "human being" under earthly law? What exactly qualifies one for being a person? Certainly not just one's planet of birth.

Quoting an article in the newsletter of the Civilian Research Interplanetary Flying Objects Organization, Derek Dempster writes:

> Flying saucers have turned from surveillance to aggressive action....And any such action constitutes a state of interplanetary war....It is a war without precedence, without land armies and minus the conventional bombs and rockets used to destroy cities. Such a war, so phantasmal, could well exist under a tight censorship

and no one save a few military heads would know of it as a fact or of its scope. Pilots scrambled to intercept UFOs might also be ignorant of this fact even though they were participating perilously in defense action, while innocently in the conspiracy. (1955)

There are some people around the world who still hold such a belief as that expressed back in 1955. They are actively seeking ways to mobilize Earth's military might and scientific technology to defend against an anticipated invasion from outer space. The conspiratorial-minded individual probably can find some minor evidence to support such claims. Yet, most conspiratorialists are not very logical or thorough in their research. They often overlook the much larger mass of evidence pointing toward a non-hostile alien response to human provocation. The American movie *Independence Day* (1997) portrayed the alien "visitors" as aggressively hostile, seeking their own ruthless objectives through the use of highly advanced technology. This kind of provocative image of alien life forms plays directly into the hands of the conspiratorialist. This genre of movie will help solidify pre-existing attitudes and beliefs as it entertains us.

The Brazilian scene

348

This discussion would not be complete without acknowledging the informative book *UFO Danger Zone* researched and written by Bob Pratt. In his introduction he states, "The terror comes from UFOs...and the aliens that operate them. These aliens are not our space brothers coming here to help us, as so many people want to believe. These are non-human creatures that for years have been tormenting and terrorizing human beings, hurting many and killing some" (1996, xii). Later he writes, "to my knowledge it is only in Brazil—not neighboring countries or the United States or any other nation in the world—that UFOs have been so overtly hostile. No one knows why" (6).

The abduction transcription project

Under the direction of Dan Wright, the Mutual UFO Network (MUFON) has been collecting audio tapes of interviews conducted between so-called "alien abduction" researchers and alleged alien abductees. As of early 1995, a total of 215 separate cases have been transcribed and recorded in a computer for subsequent sorting and analyses. But, does this data base contain any mention of abductee aggression shown toward the visitors and potential captors? In a letter to me dated January 5, 1995, he wrote, "There are only two cases among the 215 thus far indexed in which a human [abductee] reportedly acted aggressively toward his/her captor(s). Both [cases G and H] involved hitting or grabbing. No instances are recorded of firing a weapon, throwing a projectile or engaging in some other form of aggression."

Case G. (Undated) The following excerpt is from a male witness who divulged his story while under hypnosis. "Well,… when I stopped [the car], I was sitting there and somebody opened the door…and I looked and there was this hand reaches in, not a human hand, and it's unbuttoning the seat belt. For a moment there I caught myself, and as I'm stepping out I see there are two greys. And I grabbed both of them by the neck. 'See, you little bastards, you're not (unintelligible to hypnotist during the interview).' I was angry. Y'know, 'leave me alone!' I grabbed one, well, both of them by the neck, and had picked them up. Very light.

"*And someone must…somebody must have come behind me and poked me with something, because it's like I lost all the feeling in my left hand.* One fell and, as I was going down, threw the other one up against the truck.…*And the next thing, I'm being carried into this opening off the side of the road.* [There were] probably three or four of them. They had me up in the air, just lifted up, and I was like, I couldn't move.…And one was saying, '*Calm down or we are going to rip your head off.*'" Under these circumstances anyone would struggle to protect themselves. Apparently, the creatures used some means to subdue the man other than this rather human-sounding verbal threat.

Case H. (Undated) This event occurred to a female witness whose story was obtained while she was under hypnosis. "There are others around now. I don't know where they came from and nobody is saying anything, but I get the idea I'm supposed to get in that tub. I don't want to. Oh, I got an arm out and I think I hit someone. Oh, my god, I can't move."

The investigator asked, "Are you in the tub now?" To which the woman replied, "No, I'm still in the hallway and facing back toward the way I came. And I did manage to hit one of them, and he's sitting on the ground. He got up and I feel kind of bad because I might have hurt him. They want to hurt me. I can't really move. I'm in the water."

It is interesting to note that only two out of the 215 alleged abductions (1 percent) in the MUFON study involved human aggression. To these few cases must be added the following, which come from many different countries none of which were included in MUFON's abduction transcription project.

Case Files

Case No. 211

Date: 1877-05-15
Local Time: unknown
Duration: estimated 5 minutes
Location: Aldershot, Hampshire, England
No. Witnesses: 2+
DB: v   CB: d   PB-1: a   PB-2: x   UFOD: O

Abstract: Humanoids were seen floating over the heads of several guards at Aldershot. The men drew their weapons and fired at the creatures who wore tight-fitting clothing and helmets that shined. Their shots were without effect. *Then, a blue beam was directed at the guards and stunned them.*

350

Comments: This historic account took place near the beginning of science fiction literature and long before man possessed beams of light that can stun people. It is not known whether the rounds struck the creatures or if the subsequent blue beam of light was a direct result of the gun shots. The story is, indeed, very creative for its time.

References: Bougard, M. 1976. *Soucoupes Volantes.* Brussels: SOBEPS.
Flying Saucer Review 61, no. 3.
Vallee, J. 1969. *Passport to Magonia.* Chicago: Contemporary Books.

Cross references: 98 (4); 124 (4); 223 (8); 230 (8)

Case No. 212

Date: 47-08-14
Local Time: morning
Duration: estimated 5+ minutes
Location: Villa Santina, Italy

No. Witnesses: 1+

DB: q   CB: d   PB-1: a   PB-2: x   UFOD: b

Abstract: A Professor R. L. Johannis, geologist and anthropologist, was near the village of Friuli in northern Italy one morning when he saw a red object sitting on the rocky bank of the Chiarso river. It was shaped like a lens and had a low symmetrical protrusion like a dome on its top surface. He estimated it to be about 10 meters in diameter and 6 meters thick. He also noted a brightly polished metal antenna sticking out of its upper surface.

Then he noticed what he first thought were two boys and yelled out to them, pointing to the disc. He continued on towards them and, at a distance of about 20 meters or so, he stopped, "horrified at what he saw....The two boys were dwarfs, the like of which I had never seen or even imagined," he said. He continued shouting at them, "demanding to know who they were and where they came from. But he made the mistake of waving at them with his alpinist's pick,[1] and I supposed the strangers felt threatened," he said later.

As Holzer writes, *"One of them raised his right hand to his belt and from the center of the belt there came something that seemed as though it might be a thin puff of smoke.* I now think it was a ray or something of the sort. Anyway, before I had time to move to do anything, *I found myself laid out full length on the ground. My pick shot out of my hand, as though snatched by an invisible force"* (Ibid., 110).

The two beings came over to within two meters of Johannis and picked up the pick "that was longer than they were."[2] As he finally managed to sit upright the two creatures had returned to the landed object, climbed into it, and flew away.[3]

Notes: 1. Also known as an ice axe, this wooden handled tool is used in winter mountain climbing; it usually has a pointed metal tip at one end and various shaped metal endings on each of the other two ends of a T-shaped split. 2. The witness estimated the height of the "boys" at about three feet. 3. The witness inquired in the nearby village the next day whether anyone had seen anything unusual and located two people who had seen a "red glow being carried aloft by the wind" and "a red ball rising at great speed and vanishing into the clear sky."

Reference: Lorenzen, C., & J. 1967. *Flying Saucer Occupants.* New York: Signet Books, 87.

Cross references: 30 (3); 81 (3)

Case No. 213

Date: 54-02-06
Local Time: unknown
Duration: unknown
Location: Torrent, Argentina
No. Witnesses: several
DB: kk   CB: a   PB-1: a   PB-2: x   UFOD: 0

Abstract: "Martians" tried to abduct people from the village of Torrent. The townspeople "drove them off the first time." The humanoids returned on February 6, 1954, and were met by shotgun fire "from the well-prepared villagers." *The gunfire had no effect on the beings who flew away and never returned.*

Comments: This rather cryptic account contains little detail other than that it allegedly occurred. Were it not for other similar stories that are well documented it would be more difficult to include this one.

Reference: *Saga Magazine* 1981, 52.

Case No. 214

Date: 54-09-10
Local Time: approximately 2235
Duration: estimated 5 minutes
Location: Quarouble, France
No. Witnesses: 1
DB: p   CB: d   PB-1: a   PB-2: x   UFOD: b

Abstract: Mr. Marius Dewilde, 34, was reading while his wife and child were in bed when he heard his dog bark. With his flashlight in hand, he went outside to investigate, thinking that there might be a prowler. He had a small garden in front of his house with a fence surrounding it; a railroad track bordered his property. As his eyes became accustomed to the darkness, he made out an "ill defined shape to his left, on or near the railway line." His first impression was that it was a truck belonging to one of the farmers nearby. Then he noticed that his dog was approaching him slowly, walking on her belly as if afraid or in pain. It was then that he heard something on his right side.

As he turned in the direction of the sound, his flashlight illuminated two "odd creatures." They seemed to shuffle forward on very short legs and were not over one meter tall. Each wore something like a diver's suit with a large helmet. While he was surprised at first, he still thought that they were intruders who should not be there so he ran toward his gate "with the intention of cutting off the interlopers from the path. *He was about 2 meters from them when a blinding beam of light, the color of magnesium flares, issued from an opening in the side of the dark shape.*" He was rendered completely immobile and could not speak. But he could still see and hear normally. The two beings simply walked by in front of him no more than one meter away toward the formless mass near the tracks.

Immediately, as the white beam went out, he regained use of his muscles and he started running toward the beings but he was unsuccessful in catching them. A door in the side of the object closed behind the beings and then he heard a whistling sound. The UFO rose up "like a helicopter;" it gave off a white vaporous cloud. Then, at an altitude of about 30 meters, the object changed directions to a more horizontal flight path moving toward the east.[1]

Comments: The short creatures, UFO on the ground, paralyzing ray of light, and rapid neuromuscular recovery are all familiar elements in close encounter narratives. Stories such as this one raise significant questions concerning the nature of the paralyzing beam's energy and the seeming indifference of the creatures to the presence of the human being. How was the ray of white light aimed? This account suggests that the human was under continuous surveillance by the object, which then acted to protect the two creatures. The witness's boldness is commendable, if unusual.

353

Note: 1. The witness woke his wife and then ran to the local police station to report what had happened. At first, the police did not believe his story, but a later investigation was performed by the Territorial Security Department and the air police.

Reference: Brookesmith, P. ed. 1984, *The UFO Casebook*. London: Orbis Publishing, 26-27.

Case No. 215

Date: 54-09-27
Local Time: 2030
Duration: estimated 5 minutes

Location: Premanon, France
No. Witnesses: 4
DB: v   CB: a   PB-1: a   PB-2: x   UFOD: c

Abstract: Four children playing in a farmyard saw two humanoid creatures. A 12-year-old boy got his bow and arrow and sent a rubber-tipped projectile toward one of them. It had no apparent effect. *However, when the boy approached the stranger, the boy was "flung to the ground as if by an ice-cold invisible force."*

Comments: Again, it is unfortunate that no more data is available on this interesting case. Numerous anecdotal reports suggest that children respond to humanoids with less fear and aggression than do adults. Is this because children have not yet been conditioned by society to fear strange appearing beings?

Reference: Maney, C.A. 1959. Aime Michel's study of the Straight Line Mystery. *Flying Saucer Review* 5, no. 6 (November-December): 10-14.

Case No. 216

354

Date: 54-11-11[1]
Local Time: 1945
Duration: 18 minutes
Location: Isola, (near La Spezia) Italy
No. Witnesses: 1
DB: kk   CB: e   PB-1: a   PB-2: x   UFOD: O

Abstract: A bright, cigar-shaped UFO approached a farmer's field and then landed. The owner hid from sight and watched it in fear; he saw three "dwarfs" emerge from the object wearing what he thought were metallic diving suits. The three beings went over to cages in which the farmer kept rabbits and he ran to his house for his shotgun. He could hear them talking to each other in an unknown tongue.

"As he attempted to fire [at the "little fellows"], two strange things happened. *The shotgun failed to fire when he pulled the trigger, and the gun suddenly felt so heavy in his hands that he had to drop it."*[2] *He felt paralyzed. He couldn't move or speak.*

"The little humanoids seized the rabbits and jumped into their craft. As it rose up, the farmer regained his shotgun and fired but it was too late. The UFO streaked away at high speed."

Comments: This instance of temporary paralysis is not only interesting but quite common in UFO literature. Whether or not the beings anticipated the farmer's aggressive behavior by causing his shotgun to malfunction (quite unlikely) or causing his trigger finger to "malfunction" cannot be known with certainty.

Notes: 1. Holzer gives the date as November 14. 2. Holzer refers to this effect as a "projected magnetism."

References: Holzer, H. 1976. *The Ufonauts.* Greenwich, Conn.: Fawcett Publishing, 94. Lorenzen, J. & C. 1967. *Flying Saucer Occupants.* New York: Signet Books, 99. Mauge, C. 1976. *Recontres de Lyons.* Privately published, 1988.
Official UFO, February 1976.

Cross references: 181 (3)

Case No. 217

Date: 54-12-10
Local Time: night
Duration: estimated 5 minutes
Location: near Carora, Venezuela
No. Witnesses: 2
DB: v   CB: b   PB-1: a   PB-2: x   UFOD: c

Abstract: Two young men were night hunting near Carora. Upon reaching the Trans-Andean Highway, both Mr. Lorenzo Flores and Mr. Jesus Gomez noticed some lights in the sky and then a strange washbasin-shaped (or "giant clam") object hovering just above the ground next to the roadway. They estimated it to be 9 or 10 feet across; flames came from its bottom. *They were then amazed to see four short, about 3-feet-tall, dark creatures climb out of it and move in their direction. The creatures selected Gomez, surrounded him, and tried to drag him toward their ship.* But Flores had a unloaded shotgun, so he used it to hit the creatures with its butt. He found that the creatures were very strong and also were "hard." As the man struck one of the beings with his gun, the gunstock broke in two. Breaking free, the two men ran away as fast as they could, finally arriving at the police station. Both had deep scratches, bad bruises, and were hysterical. Their clothes were badly torn.

Comments: Carora is about 240 miles west of Caracas, where an encounter reportedly took place two weeks earlier. It has been suggested that the men were so poor that the deliberate breaking of a valuable shotgun would be unthinkable. Upon returning to the site, the police found "signs of a fight" but no other evidence to support the men's story.

References: Brookesmith, P., ed. 1984. *The UFO Casebook.* London: Orbis Publishing, 84. Gross, L.E. 1991. *The Fifth Horseman of the Apocalypse-UFOs: A History 1954 November December.* Fremont, Calif.: Published privately, 51.

Case No. 218

Date: 55-08-21
Local Time: 1900-2200
Duration: 20+ minutes
Location: Kelly (near Hopkinsville), Kentucky
No. Witnesses: 11
356   DB: cc   CB: e   PB-1: a   PB-2: x   UFOD: a

Abstract: This story is both famous and complex, and only the facts relevant to this book will be recounted here. It took place over several hours on a hot, dark, and clear night in August. A glowing circular or oval-shaped UFO was seen at about 7:00 P.M. by Billy Ray Taylor, a family friend of the Sutton family and owner of the farmhouse. Taylor was visiting the family at the time and had gone outside to the well for water. The object he observed had an "exhaust all the colors of the rainbow" and flew across the sky before dropping into a 40-foot-deep gully near the edge of the property. The "flying saucer" had "settled down back of the barn," he said. At first, the Suttons just laughed at his story.

About 7:30 P.M., everyone heard a dog barking violently, so Billy Ray and Lucky Sutton went to the back door to find out what was wrong. They saw "a strange glow approaching the farmhouse from the fields." As it neared the house they made out, "a glowing three-and-one-half-foot-tall creature with a round, oversized head. The eyes were large and glowed with a yellowish light; the arms were long, extended nearly to the ground, and ended in large hands with talons. The entire creature seemed made of silver metal. As the creature approached, its hands were raised over its head as if it were being held up" (Story 1980, 191).

Each man grabbed a gun and remained inside the back door, waiting. One had a .22-caliber[1] rifle and the other a 20-gauge shotgun.[2] The creature continued to approach and, at about 20 feet from the doorway, both men opened fire directly at it. *"The entity flipped over backward[3] and then scurried off into the darkness."*[4]

As the two men eventually returned to the living room, they saw a similar-looking creature at a side window. *Both immediately fired their guns at it through a screen. The creature flipped over as before and ran away into the darkness.* They were sure that they had hit it this time and walked outside to find the body. With Billy Ray in the lead, the men walked through the front door. The rest of the family stayed inside the hallway and watched in horror as a taloned hand and arm reached down and touched Billy Ray's hair from above just as he was going to step down onto the dirt. Those inside the house pulled him back inside while Lucky ran out into the yard. *He "turned, and fired pointblank at the creature, knocking it off the roof.*

"There was another creature in the maple tree close-by." *Both men fired at it and watched as it was knocked off a limb. It began to float gently down[5] to the ground and run off rapidly into the darkness.* Another creature appeared around the side of the house "almost directly in front of the group. Lucky fired his shotgun at point-blank range and the result was the same: no effect. *A sound was heard as the bullets struck, as if a metal bucket had been hit, but the creature scurried off unhurt"* (Ibid.).

The creatures continued to peer in the windows in intervals until about 11:00 P.M.; the children were becoming hysterical. So the men decided to get all eleven people present into two cars and leave for town. They drove at high speed to the local police station and demanded help. The obvious fright and excitement of everyone was not ignored. The police returned to the property with the Sutton family and Billy Ray but could find no trace of the creatures. The police left and things seemed to quiet down again; their presence had helped provide a sense of security.

However, all of the witnesses claimed that the creatures returned after the police left; the available accounts are unclear about how long the creatures remained or what the witnesses did. The interested reader should study the remainder of this fascinating story presented in the references cited below.

Comments: When not moving, the creatures appeared to be standing erect, but when running, they seemed bent over, their long arms almost touching the ground. Two sources (Davis and Bloecher 1978; Story 1980, 191) commented that the entities were able to float through the air.[5] The outer surface of the creatures seemed to glow in the dark, becoming brighter when they were shot at or shouted at.

Despite all of the shotgun blasts, none of the creatures reacted in an aggressive way. If true, this raises some fundamental questions concerning their defense mechanisms, their psychological makeup, or both. This case is included in the U.S. Air Force Project Bluebook files as "Unidentified."

Notes. 1. Typical muzzle velocity : 1,100 fps; mass : 50 - 60 grains (1 grain : 0.00225 ounce); muzzle energy : 720 ft-lb. 2. Typical muzzle velocity : 2,680 fps; mass of shot : 45 grains; muzzle energy : 2,000 ft-lbs. 3. I disregard the possibility that the entity flipped over voluntarily or just prior to being hit by the projectiles (e.g., a Kung Fu-like motion). If we assume: (a) the creature's body possessed mass and was rigid, (b) the pellets did not pass through it (i.e., inelastic collision), (c) there was little or no friction between the creature's feet and the ground, and (d) a coefficient of restitution=1, then it can be shown that a shotgun blast to the upper torso of a three-and-on-half-foot-tall object will cause it to flip over backward within a second or more which is nearly what was claimed. 4. The shot was not fatal! In a signed statement given to Bud Ledwith, a local engineer and radio announcer who investigated the event, one of the two men said that the sound was "just like I'd shot into a bucket" (Hynek 1972). 5. The statement that the creature floated downward is one of the stranger elements of this account. It suggests that it possessed a very low density. Yet, its body seemed to absorb the kinetic energy of the shotgun pellets and be pushed backward by them.

References: Davis, I., and T. Bloecher 1978. *Close Encounter at Kelly and Others of 1955.* Evanston, Ill.: Center for UFO Studies.

Gross, L.E., *The Fifth Horseman of the Apocalypse-UFOs: A History 1955: July–September 15th.* Fremont, Calif.: Published privately, 54-59.

Hynek, J.A. 1972. *The UFO Experience: A Scientific Inquiry.* New York: Ballantine Books.

Story, R., ed. 1980, *The Encyclopedia of UFOs.* Garden City, New York: Doubleday & Co., 190-92.

Cross references: 211 (8); 221 (8)

Case No. 219

Date: 57-10-15
Local Time: 2130
Duration: estimated 10+ minutes [1]

Location: Sao Francisco de Sales, (Minas Gerais) Brazil
No. Witnesses: 1
DB: cc   CB: a   PB-1: a   PB-2: x   UFOD: a

Abstract: Certainly the most publicized human abduction case in history, the Antonio Villas Boas encounter contains a wealth of information for those interested in psychological themes. It may be of relevance here as well. However, it must be approached very carefully because of its high strangeness and the fact that it is a single-witness case.

On two separate nights before the date of the claimed abduction, Antonio and his brother had seen both an unidentified light in the farmyard near their house (October 5) and then a huge, round, dazzling light that hovered over his field (October 14). It was while Antonio, 23, was by himself plowing his field the following night that he sighted a "large red star" (almost egg-shaped) approaching him (see Figure 47). It stopped above him at an altitude of about 50 yards and illuminated the entire field as if it were noon.[2]

The UFO was only a few meters away from his tractor and was settling down to the ground. Antonio could no longer control his fear; he leaped from the machine[3] and began running across the field toward his house. He ran only a short distance, however, when someone or something grabbed his arm. He turned to see "a strangely garbed individual whose helmeted head reached only to Villa Boas' shoulder. *[Boas] hit out at his assailant, who was knocked flying, but he was quickly grabbed by three other humanoids who lifted him from the ground as he struggled and shouted.*"

359

Figure 47

A   B
(Side View)   (Top View)

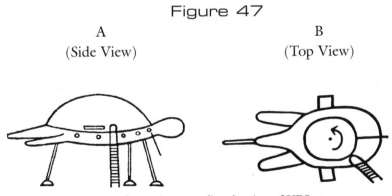

Approximate outline drawing of UFO
seen by the witness on October 15, 1957, in Brazil

Boas continued to shout as the creatures carried him toward the domed object which had now landed. Later, he remarked that his speech seemed to arouse their surprise or curiosity because *all three beings "stopped and peered attentively at my face (each time) I spoke."* Dr. Fontes, the investigator, maintained that there was "no communication or understanding between him and the 'space beings'…he was captured by force after a tremendous fight with four of the aliens, and carried to their craft on their shoulders—still fighting desperately" (1958, 2). As he was hoisted up into the craft again, he resisted by hanging onto a handrail.

He found himself standing inside a square room. He was escorted into an oval-shaped room with a metal rod or column extending from floor to ceiling.[4] It was here that Villa Boas heard the beings communicating with one another in a weird pattern of yelps and barks much like dogs make. Then five creatures advanced toward him and forcibly took all of his clothes off; he again tried to resist, but to no avail. *They peered at him carefully each time he shouted out at them.*[5]

He later claimed to have experienced apparent biological decontamination and possible blood-typing procedures, and sexual relations with a beautiful woman as a result of the abduction. Suffice it to say that this story has provided a larger than usual target to shoot at by the skeptically inclined for over thirty years. This case has even separated the "deeply committed true believer" from the open-minded UFO investigator. It is recounted here only for its CE-5 aspects.

Comments: Villas Boas was "set up" for this encounter by at least two prior detailed sightings confirmed by his brother. The fact that the aerial phenomenon reappeared in the same locale over at least a ten-day period is not only significant in itself but also not uncommon. The fact that the young man tried to resist capture a number of different times suggests that he maintained not only self-will but some degree of physiological self-control. Many other incidents have been reported where the human being possessed neither will-power or physical strength after being captured. And the fact that he recalls seeing a creature falling backward because of his blow is significant for it suggests it possessed a hard surface with possibly some amount of mass. Energy was imparted from his hands and arms to another physical object. The Antonio Villas Boas encounter cannot be dismissed lightly, nor should it be. It may have served as a prototype event to be copied by others, yet the alleged response of these aliens corresponds only slightly with later cases. Why would later reports of abduction not incorporate more of the details reported by Villas Boas?

Notes: 1. This estimate begins when the witness first saw the light over his field to the time he found himself inside the square room. 2. Summer sunlight on a clear day at mid-latitudes yields about two calories per square centimeter per minute. The amount of electrical energy required to illuminate a large field with an equivalent of noon sunlight using current light sources would be truly enormous (on the order of 100 megawatts or more), depending on various factors. 3. His gasoline engine and headlights both stopped working at this time. 4. He saw small light sources high up in the metallic room which is unusual in the UFO lore. Most often, wall surfaces are seen to glow with an even luminance. The vertical column is similar to that reported in the Enrique Castillo Rincon case of 1973 (Castillo Rincon 1995). 5. Apparently, the creatures were as intrigued at human speech as Villas Boas was at theirs. Their responses suggest a non-familiarity with human speech.

References: Castillo Rincon, E. 1995. *OVNI-Gran Alborado Humana* 1, Venezuela: Editorial Norte y Sur, C.A.

Fontes, O. ltr. to A. Mebane, 23 November 1958, *The Fifth Horseman of the Apocalypse-UFOs: A History 1957: Oct. 1st-Nov. 2nd*, ed. L.E. Gross, (Fremont, Calif.: Privately published, 1997), 42-45.

Martins, J. 1971. *Domingo Illustrado*. Rio de Janeiro, 10 October.

Lorenzen, C., and J. 1976. *Encounters with UFO Occupants*. New York: APRO/Berkley Medallion Books, 61.

Cross references: 221 (8); 229 (8); 233 (8)

Case No. 220

Date: 63-10-12
Local Time: 0330
Duration: estimated 5+ minutes
Location: between Monte Maiz & Isla Verde, Argentina
No. Witnesses: 1
DB: cc   CB: h   PB-1: a   PB-2: x   UFOD: c

Abstract: A truck driver, Mr. Douglas, was driving his truck toward Isla Verde (Cordoba Province), in a violent storm in the early morning hours when he was blinded by a beam of bright white light coming down from the sky in front of him on the roadway. It caused him to drive into a ditch, his face was burning. When he got out of the cab he saw a huge metallic oval disc he estimated to be 30 feet thick. Subsequently, he noticed a door open in its side and robots came out. He reached for a revolver and fired four shots at them and then started to run away. *The very tall robots quickly returned to the object which rose from the ground and began following him, circling over his truck several times.* He felt his body burn each time it passed overhead. *Then, he felt another burning sensation from a ray of energy that struck him.* It was similar to the feeling he had earlier when the truck had first stalled in the ditch.

Comments: Douglas later discovered that the truck's electrical wiring had burned out. Investigators also found an 18-inch-long footprint in the ground the following day. A physician at Monte Maiz police station examined Douglas and discovered "curious lesions" of an unknown cause; he also was found to be of "perfectly sound mind."

References: Lorenzen, C., and J. 1976. *Encounters with UFO Occupants.* New York: APRO/Berkley Medallion Books, 153.
Le Maine Libre, 21 October 1963.
Anon. 1964. World Round-up. *Flying Saucer Review* 10, no. 1 (January-February): 29-30.

Cross references: 211 (8); 223 (8); 233 (8)

Case No. 221

Date: 64-09-04
Local Time: late afternoon
Duration: estimated 2+ hours
Location: Cisco Grove (Placer County), California [1]
No. Witnesses: 1
DB: jj   CB: h   PB-1: a   PB-2: x   UFOD: a

Abstract: The only witness to this classic case was Mr. Donald S., a stable, married young adult who worked in a local factory. He was bow-and-arrow hunting with two friends, Tim T. and Vincent A., but became separated from them late one afternoon. [2] He realized that he would have to spend the night outdoors alone. Climbing the ridge he was on, he located a tree, and climbed up for safety. Then, he waited for sunrise. Two hours later, he saw a light "like a flashlight" approaching his location at a low altitude and bobbing up and down. He climbed down, thinking that it was a "helicopter from the ranger station."

He quickly set three fires on top of large rocks beneath the tree and began waving his arms, trying to attract the attention of the helicopter pilot. *Suddenly, the light turned abruptly and headed in his direction, coming from the northwest.* When the light was only 50 or 60 yards away, the witness became afraid since he heard no sound at all. The light seemed to hover between two trees and it was then that Mr. S thought it might be a UFO, "just a little tiny one."

He wrote later, "I just threw my bow up in the tree and got up there. I had camouflage clothing on from head to toe." He watched silently as the light traveled in a large half-circle to the east and over a canyon rim to the south. During this time he made out three glowing, shimmering rectangular panels on the object, oriented vertically in a "stepped-down" configuration. They "were spaced evenly and were illuminated like crumpled aluminum foil, shimmering." Soon the object came within about 50 feet of him and stopped. It remained suspended in the air for about five minutes.

Then, "something came out of the second (panel), and all I could see was a kind of flash. Something went straight down the hill....It went pretty fast....I just saw a dark object shoot right down and there was a flash when it came out." It seemed to disappear in the underbrush. At first he thought that the object dropped to the ground; all he could see was a "part of a dome on top and just a little light flashing on it." Later, he noticed that the object was still hovering silently in the sky. Soon afterward, he heard crashing sounds from the underbrush nearby and then saw a figure clothed in a "light-colored, silver, or whitish-looking uniform" with bellows at the elbows and knees. Its eyes were dark and large, the size of silver dollars and similar in appearance to welder's goggles." Then a second, identically shaped being joined the first and they both approached the base of the tree in which Mr. S. was hiding. They stood there looking up at him.

As the thrashing sounds continued in the distant underbrush, Mr. S. saw a third creature. Its eyes were reddish-orange and gave off their own light "just like two flashlights hooked together." Their reddish light illuminated a square metallic face, its mouth square and hinged. This robot joined the two other beings under the tree staring up silently at the terrified witness.

Now, afraid he would be captured by them, Mr. S. lit a book of matches and threw it down in their direction. *They backed up," he said,* "so then I started

363

going crazy with fire. I lit my hat and threw it at them. *As it flared up they backed a good 50 to 75 feet away."* The fire dwindled, and they began approaching him again so he began tearing his clothes off and setting them on fire until all he had on were his Levis and a T-shirt. "I tried shooting the robot with my bow, as he was the only one that was doing anything against me. The other two just stood and looked. The bow I've got has the velocity of a rifle at that close range, 12 feet or so.[3] So I just pulled it back as far as I could, and hit him the first time. *It was just like a big arc flash. It just flashed up real bright.* I only had three arrows with me, so that was all I shot. I shot him three times, and all three times it *pushed him back a little bit,[4] with just a big, bright flash,"* he said. *The other two humanoids scattered a little with every arrow shot.*

The robot allegedly emitted a white vapor from its mouth which rose up to where Mr. S. was now tied by his belt to the trunk. It caused him to black out temporarily. This happened several times; each time he awoke he saw the two humanoids starting to climb up the tree.[5] They tried to "boost each other up the tree." *Each time, he shook the tree and they got down again. A second robot, identical to the first, then arrived at the base of the tree and sent another cloud of white vapor upward, having the same effect as before.*

Throughout the night, Mr. S. threw branches and coins and other things down at them to try to "distract" them. *Once, when he threw his water canteen at them, one of the beings picked it up and "appeared to examine it, then threw it away."* The entities disappeared in a spectacular light show of "arc flashes" which passed back and forth between them and lit up the entire area beneath the tree. Apparently, this process produced a kind of fog which "rolled upwards" and prevented him from seeing them anymore. Then he fell unconscious. Upon awakening he was hanging by his belt alone. All of the creatures were gone.

Comments: This case is so bizarre that it could be taken for a result of a bad drug trip, except for the fact that other known cases containing similar elements exist. Let us accept the story as true for the time being to see what might be learned.

Mr. S. gave the blunted arrowhead to investigator Victor W. Killick who turned it over to the U.S. Air Force for analysis. Mr. S. gave the other two arrowheads to representatives of the National Investigations Committee on Aerial Phenomena (NICAP) for analysis by a lab in Pennsylvania. Nothing of interest was found. The Air Force sent their arrow tip to the University of Colorado UFO Study Project for analysis. It was never returned. The Air Force continued to research the event covertly, according to Clark (1996).

> Notes: 1. This spot is 2.5 miles south of Interstate 80 near the Loch Laven lakes and about 23 miles west of the north tip of Lake Tahoe. 2. Vincent A. also claimed to have seen an unusual "flying light" sometime that night. He found Donald wandering back toward their camp, weak, exhausted, and tired. 3. A typical hunting arrow travels from 240 to 260 feet per second. 4. This fact suggests that the robot: (a) had a solid surface which the arrow struck, and bounced off, (b) probably had relatively little mass, and (c) motion dynamics are somewhat similar to that of Case No. 218. Considering the weight and likely velocity of the arrow, if the robot was pushed backward over the period of a second, an estimated 125 ft-lbs. force would have been imparted to the robot. 5. The inability of the humanoids and robot to simply levitate up the tree is strange considering all of their other advanced capabilities.

References: Clark, J. 1996. *High Strangeness UFOs From 1960 Through 1979 3.* Detroit, Mich.: Omnigraphics. 64-68.

Hynek, J.A. 1977. *The Hynek UFO Report.* New York: Dell, 210.

Keyhoe, D.E., and G. Lore 1969. *Strange Effects from UFOs: A NICAP Special Report.* Washington, D.C.: NICAP, 17-23.

Cross references: 214 (8); 218 (8); 222 (8)

Case No. 222

Date: 65-04-03
Local Time: 2320
Duration: estimated 8+ minutes
Location: Fortaleza, Brazil
No. Witnesses: 2
DB: q   CB: b   PB-1: a   PB-2: x   UFOD: a

Abstract: Francisco Muller and Jose Araujo, both salesmen, were traveling together toward Fortaleza. A light source like a deep-red lamp shone a ray down over their car from one side and soon a flying object had descended to a point just above them. The car began shaking and its lights went out as Muller steered it off the road; he claimed that he had no control over the car. The two men watched in amazement as the UFO landed in front of them. Two "robots" emerged from the object and began walking toward the car. Muller and Araujo got out, one with a rifle. *Just as one of the men fired at the object, a "terrible explo-*

sion" occurred, leaving both men blinded. Their sight returned after ten to fifteen minutes. By then, the craft and the beings were gone. Upon getting into their car and driving to a gas station, they learned that an employee there also had seen the same object.

Comments: The visual "explosion" has all the hallmarks of intense visual flash-blindness where the retina's sensitive carpet of photosensors is so completely bleached by the impinging light energy that vision is impossible for many minutes. This could occur, for example, if one were standing near where a lightning bolt strikes at night with dark-adapted vision and wide open pupils. Based upon the data presented elsewhere (Haines 1970), it is possible to estimate the luminous energy of the light flash within an order of magnitude. Assuming that the men could see only dim detail as their vision recovered and visual recovery time was longer than one minute, then the luminance of the flash would have had to be at least 100,000 foot lamberts and very likely two or three times as much (249, Ibid.).

Reference: Pratt, B. 1996. *UFO Danger Zone.* Madison, Wis.: Horus House Press.

366   **Cross references:** 211 (8); 214 (8)

Case No. 223

    Date: 67-08-13
    Local Time: 1600
    Duration: estimated 5 minutes
    Location: between Crixas and Pilar de Goias, Brazil
    No. Witnesses: 2
    DB: v   CB: e   PB-1: a   PB-2: x   UFOD: a

Abstract: Mr. and Mrs. Inacio de Souza were returning home from an outing when they noticed a "strange, basin-shaped object"[1] about 35 meters in diameter hovering just over the runway of a private airfield nearby.[2] They also saw three creatures who looked like they were naked because of their tight-fitting yellow suits. About as tall as young children, the beings had no hair and seemed to be playing in some unspecified manner.

When the three creatures noticed the de Souzas, they began running toward the couple. Very afraid, Inacio told his wife to run inside their house as he got his

.44-caliber carbine rifle and began to fire at the closest advancing figure. *"At the moment that Inacio began to fire, a jet of green-colored light erupted from the disc and struck him full in the chest,[3] knocking him to the ground." The three beings were seen running back to the craft. It subsequently rose vertically into the air making a sound like a swarm of bees and disappeared from sight.*

Three days later, the farm's owner landed at the private airstrip on his property and discovered that Inacio was very ill. He felt nausea, tingling, and numbness throughout his body. His employer immediately flew Inacio to a hospital in Goinna. There, the doctor found burn marks covering his torso. Blood chemistry tests showed that he was suffering from leukemia and "malignant alterations of the blood." During the next two months Inacio lost weight and experienced great pain. Yellowish-white spots covered his body. He died from the consequences of leukemia two months later.[4]

Comments: Mrs. de Souza verified her husband's account in all respects. Many of these physiological symptoms correspond to those experienced by Pfc. Francis Wall in Case No. 98.

Notes: 1. The term tub ("tina") and washbasin ("bacia de maos") have been used by many other witnesses in Brazil. 2. De Souza was a tenant farmer and caretaker for a wealthy man who commuted by private airplane to meetings. 3. Brookesmith's account indicates that de Souza was struck on the head and a shoulder. 4. Many people believed that de Souza had passed away from a high dosage of ionizing radiation.

References: *Flying Saucer Review* 15, no. 2.
Lorenzen, J., & C. 1976. *Encounters with UFO Occupants.* New York: Berkley Books, 159-160.

Cross references: 98 (4); 211 (8); 233 (8)

Case No. 224

Date: 67-11-28
Local Time: 0230
Duration: estimated 5+ minutes
Location: near Americana, Brazil
No. Witnesses: 1
DB: w   CB: d   PB-1: a   PB-2: x   UFOD: 0

Abstract: A patrol officer was on duty driving on the Anhanguera Highway at Km 124 when he saw a UFO that at first he though was an airplane flying without lights. It was truly immense, "as high as a fifteen-story building and elongated." It spun about a vertical axis and seemed to be made out of aluminum with large rivets on its "fuselage." Its lower section had "hatchways."

The witness "took out his revolver, ran across the road, and hid himself in a depression of the terrain in such a way that he could not be seen, as it was filled with bushes. The object stopped some 30 meters from the ground at a distance of about 50 meters from the patrolman. He could not decide whether to fire [at it] and was extremely nervous."[1] Then, a kind of elevator appeared which extended down from the bottom of the object. It only had a floor, two sides, and a back. There were two figures inside the structure, "illuminated by dull light." They both wore tight-fitting clothing that looked like scuba suits. Their belts were about three-inches high and seemed to emit brilliant lights from different locations.

The frightened witness next heard one of the beings say to him, along with some gestures, "put your gun away...do not be afraid....You are an intelligent and a fearless man...we will be coming back again." The patrolman wanted to run but his legs would not support his weight.[2] Then, the huge UFO took off and disappeared into the night sky. The witness submitted an official report to the Commandant of the Highway Police in Sao Paulo.

Comments: This incident is interesting because the officer encountered the UFO at the same location on two different nights. During his first encounter, the craft had given off "powerful lights...which looked like those from electric welding torches...but millions of times brighter." He also recalled getting a headache from a very loud humming sound coming from the direction of the object. Intense ultraviolet radiation can also produce headache symptoms.

Notes: 1. The officer had seen the same object two nights before at the same spot. 2. Whether his paralysis was externally or internally derived is not clear. He said that his legs felt "very weak."

Reference: Two 1967 Brazilian Landing Cases. *APRO Bulletin*, Pg. 5, September-October 1968.

Case No. 225

Date: 73-10-25
Local Time: 2100
Duration: 20 minutes
Location: near Greensburg, Pennsylvania
No. Witnesses: 15+
DB: q   CB: a   PB-1: a   PB-2: x   UFOD: b

Abstract: At about 9:00 P.M., Stephen Pulaski, a farmer, sighted a bright red mass of light about one hundred feet across hovering above his field. The object was dome-shaped and sounded like a lawnmower. He and a group of other farmers and neighbors saw two tall beings near a fence. One was 7 and the other 8-feet-tall.[1] Both were completely covered with long, dark-gray hair and had green-yellow eyes. Their arms hung down almost to the ground. The astonished witnesses heard the creatures make whining sounds "like that of a baby crying." Pulaski fired a rifle bullet over their heads and then panicked and fired three more rounds directly into the taller figure. *Both figures "slowly turned around and walked back towards the woods."* Soon after, the self-luminous object took off into the dark night sky.

369

Note: 1. Their height was judged in relation to a nearby fence post of known height.

References: *MUFON UFO Journal*, no. 171.
Flying Saucer Review 20, no. 1, 1974, 3.

Cross references: 213 (8); 218 (8); 234 (8)

Case No. 226

Date: 74-10-25
Local Time: 1600-1830
Duration: 2 hours 30 minutes
Location: south of Rawlins, Wyoming
No. Witnesses: 1
DB: kk   CB: a   PB-1: a   PB-2: x   UFOD: O

Abstract: Mr. Carl Higdon was hunting elk at the northern edge of Medicine Bow National Forest at about 4:00 P.M. He had just walked over a hill and noticed five elk standing ahead of him. Higdon said, "I raised my rifle and fired, but the bullet only went about 50 feet and dropped." He walked over to where the round had fallen on the ground, picked it up, and placed it in his canteen pouch. As he did this, he heard a twig snap to his right side and looked in that direction. He saw "sort of a man standing there." The man was six foot two inches tall, weighed about 180 pounds, and wore a black suit, black shoes, a belt with a star in the middle and a yellow emblem below it. He was quite bowlegged. His chin was missing and the top of head seemed slanted. We shall not recount all of the other details of this alleged abduction here except to mention that *the man asked Higdon if he was hungry.* When Higdon answered yes, the man tossed him some pills; Higdon swallowed one without hesitation.[1]

Comments: While this incident does not fit precisely into the subject matter of this chapter, it is included because of the fact that a high-powered rifle was aimed and fired at some type of creatures, originally thought to be elk. It isn't known for sure if they were elk or something else (e.g., screen images of alien life forms or mental projections of animals rather than humanoids). The reported fact that the round traveled only 50 feet might be explained on the basis of a faulty charge of powder, particularly if Higdon made his own rounds. The statement that he saw the bullet leave his rifle and then located the lead on the ground is remarkable indeed.

This now famous account raises far more questions than it settles. For example, was the cause of the rifle misfire merely coincidental with the other bizarre events, or was it caused by an outside agent or force? We have already reviewed numerous cases where humans were kept from firing a weapon at strange, non-human creatures. Was there any period of amnesia or unconsciousness before or after the rifle fired? Perhaps Higdon experienced a concussion caused by the blow from a cartridge misfire. Where would the strange man dressed in a black suit come from out in the forest? Why would Higdon accept and swallow an unknown pill from a stranger so readily?

Note: 1. A similar detail is reported in Bigorne 1974.

Reference: Bigorne, J. et al. 1974. Forced Feeding by UFO Entities. Lumieres dans la Nuit, no. 139 (November); *Flying Saucer Review* 21, no. 6 (December).

Sprinkle, R. L. 1979. Investigation of the alleged UFO experience of Carl Higdon. In *UFO Phenomena and the Behavioral Scientist*, ed. R.F. Haines, Metuchen, N.J.: The Scarecrow Press.

Case No. 227

Date: 75-01-05
Local Time: 0330
Duration: unknown
Location: Near Bahia Blanca, Argentina
No. Witnesses: 1
DB: kk   CB: d   PB-1: a   PB-2: x   UFOD: 0

Abstract: This challenging incident involved a reported abduction of Mr. Carlos Alberto Diaz, 28, a waiter walking home after work. The sky was very dark and overcast, it would rain later. *He was several blocks from home when he was momentarily blinded by a very bright beam of light.* He thought at first it was a flash of lightning except that the light wasn't straight but "broken" and no thunder was heard. *Then, he found himself totally paralyzed yet his vision continued to function normally.*[1]

Diaz claimed he was "drawn and absorbed upwards" within a beam from above him. He passed out and awakened inside a spherical-shaped room, completely smooth and shining,[2] but he saw no lights. He lay on the floor on his right side, fully lucid. Three human-like beings "came sliding in," he said. Their faces and heads possessed a greenish hue. Their heads had no features at all and were about one-half the size of an adult human's head. Their arms were "grotesque and impressive, as they had no hands or fingers to them." *The creatures surrounded him on the floor and began pulling out large tufts of his long hair; this seemed to cause them great pleasure.*[3] This took place in the midst of a tremendous struggle, with Diaz putting up a stubborn resistance, in the course of which he was able to feel the tactual softness of the covering of their bodies, described by him as "rubber."

371

The beings continued to pull his hair out several more times, including hair on his chest, after which his eyesight began to fail. He became unconscious. Later, he awoke lying on the grass, disoriented and dazzled by bright sunshine on him. At first, he felt fine but then experienced nausea. He tried to vomit; subsequently he was taken to the Railway Hospital at Retiro, near Buenos Aires.[4]

Comments: This extraordinary report by Romaniuk (1975) provides numerous details concerning Diaz' apparent change of personality, motivation, and other interesting characteristics of the creatures and the inside of the enclosure. It is interesting to note that Diaz' physical behavior trying to defend himself did not in any way prohibit his captors from achieving their rather odd goal. It should be noted that Banchs (1976) disputes the validity of this case. This alleged event might have been patterned after a very similar event that allegedly took place in 1973 in the same location.

Notes: 1. His watch stopped working at 3:50 A.M. 2. The witnesses estimated the ceiling height to be about 3 meters and the room to be about 2.5 meters in diameter. 3. The beings jumped up and down and waved their arms about holding his hair. 4. A man in a passing car helped Diaz to the hospital, they arrived there at 8:00 A.M. on January 5, 1975.

References: *APRO Bulletin* 26, no. 2.

Banchs, R.E. 1976. Las Evidencias de los OVNIS. ed. Rudolfo Alonso, Argentina.

Romaniuk, P. 1975. The case of the "Greenmen." *Flying Saucer Review* 21, no. 3/4 (November-December): 39-42.

Case No. 228

372

Date: 75-10-27
Local Time: approximately 0230
Duration: estimated 3 hours
Location: near Norway, Maine
No. Witnesses: 2
DB: kk   CB: e   PB-1: a   PB-2: x   UFOD: O

Abstract: Two young men who were sharing a house trailer in the town of Norway reported a bizarre array of experiences involving premonitions, automatic car control, a strange cylindrical aerial object with associated electromagnetic effects, two aerial discs which put on an "aerobatic display," perceptual distortions of space and time, dream-like imagery, unconsciousness produced by a light beam, an abduction—complete with a physical examination and a tour of the craft—and various other unnerving experiences. It is the hypnotically recalled events just prior to the examination of one of the young men that is reviewed here.[1]

The first room on board the craft in which David Stephens, 21, was taken had curved walls. He was joined there by a "non-human being" whose hands had three webbed fingers and a thumb. Its head was mushroom-shaped; large, slanted white eyes were seen. It had no mouth and only a small nose. It wore a black robe almost reaching to the floor and full sleeves. Stephens was taken into another room where four similar beings waited. It resembled a hospital room with an operating table. "When the beings tried to persuade him to lie on the table for a medical examination, David became violent and struck one of the beings. *They did not retaliate, and eventually he relented.*" His clothes were removed and he was given a thorough exam with a box-like device. "The creatures seemed to want to be friendly," he said. When he returned to the car, his friend Paul didn't realize that he had been gone.

Comments: The non-retaliation of the creatures is very interesting. Still, they appeared to have achieved their intended ends using other means. Why the other young man apparently was not also involved is unclear. Classic UFO cases are characterized by high strangeness elements and frequent paranormal psychic and psychological experiences. By these criteria, this case ranks as a classic. The obvious similarities with the following case should be noted. It would be valuable to know when and where this case was first publicized.

Note: 1. A total of eight hypnosis sessions were administered by Dr. Herbert Hopkins from December 1975 to March 1976.

References: *Flying Saucer Review* 22, no. 2.

373

Lorenzen, C., & J. 1977. *Abducted*. New York: APRO/Berkley Books, 70.

Cross references: 219 (8); 229 (8)

Case No. 229

Date: 75-11-05
Local Time: 1817
Duration: estimated 6 minutes [1]
Location: 9 miles south of Heber, Arizona
No. Witnesses: 7
DB: q   CB: a   PB-1: a   PB-2: x   UFOD: a

Abstract: The primary participant in this alleged abduction, Travis Walton, has had a full length Hollywood movie made of the event[2] and has also written a book which describes the many intriguing details[3]. I will only present the human aggression aspect of this incident here. Walton claimed to have been taken aboard a large UFO after having been struck by a beam of light originating from it—several of his fellow workers observed this. He awoke with a bright light shining in his eyes and feeling panicky "because," as he said later, "there was a terrible pain in my head and chest.

"My mind cleared a little and I thought I was in [a] hospital. I was on a table on my back, and as I focused I saw three figures. It was weird....They looked like a crowd of well-developed fetuses to me, about five feet tall, and they wore <u>orange</u>, tan-brown robes, loose fitting...<u>with no emblems</u>.[4] They made no sound...and their eyes were very large" (Bowen 1976, 4).

Walton noticed <u>a thin cylinder about 18 inches long and transparent which was lying on a nearby table top. He grabbed it and raised it in the air to threaten the creatures.</u> *As he did so, all three creatures backed up away from him. Then they turned and left the room.* Then, another being, who seemed human, and who wore blue garb and a helmet, led him to another big, bright room" (Walton 1996).

Comments: Once again, violent human behavior did not appear either to evoke an in-kind response on the part of the strange beings nor to be particularly effective in self-protection. Their behavior seems to indicate that they wanted to protect Walton, their "specimen." If true, perhaps they only wanted an uninjured human being for further studies.

Note: 1. This estimated duration refers only from the time the glowing light was seen to the time the workers in the truck sped away from where Walton was last seen. The entire event lasted much longer. 2. *Fire in the Sky: The Walton Experience.* New York: Marlowe and Co., 1996; also see Price (1997). 3. *Fire in the Sky.* 4. Underlined text is corrections and additions made by Walton during September 15, 1996 interview.

References: Anon. 1975. The Travis Walton Case. *APRO Bulletin* 24, no. 5 (November): 1-5.

Anon. 1975a. The Travis Walton Case. *APRO Bulletin* 24, no. 9 and 10 (December): 1, 3.

Bowen, C. 1976. The Snowflake story: A commentary. *Flying Saucer Review* 21, no. 5 (February): 3-6.

Lorenzen, C., & J. 1977. *Abducted.* New York: APRO/Berkley Books, 80.

Price, G. 1977. *The Travis Walton UFO Abduction Case.* http://www.primenet.com/~bdzeiler/essays/walton.htm

Walton, Travis. 1996. Interview by author, 15 September, New Hampshire.

Cross references: 219 (8); 229 (8)

Case No. 230

Date: 76-09-03
Local Time: 1900
Duration: estimated 5+ minutes
Location: Serra do Mouro, Novo Trento, Brazil
No. Witnesses: 1
DB: v   CB: h   PB-1: a   PB-2: x   UFOD: b

Abstract: Mr. Joao Romeu Klein, 19, was on his way home after visiting a friend when he sighted a disc-shaped object coming toward him in the sky from the south. An approximate sketch of the three meters diameter object is shown here (see Figure 48); its bottom half was more flattened than the top and rotated in a counter-clockwise direction. The upper part of the object had a small light on its top which changed color with the speed of the object. The light appeared red at high speed, varying through orange, yellow, pale green, and turning a dim white when it was almost stationary. He said it was possible that the body of the craft was grey.

The UFO flew over his head at an altitude of only 5 meters and then stopped just 375 ahead of him. Then, it projected a slightly diverging ray of intense red light downward from the center of its bottom. He saw three short—approximately one-meter-tall—beings slowly descending in this beam to the ground. Standing side by side, they blocked the pathway. When he edged nearer to them, they "spread out their arms across the road, indicating that they would forbid him passage."

Determined to defy them, he seized his big knife…and threw it straight at them, but the knife seemed to encounter an invisible obstacle and was deflected.…The middle being at once reacted by pointing a sort of "rod" at him. From the rod came a beam of bluish, almost white, light, which struck Klein on his left thigh.[1] He became unconscious and was found by neighbors soon afterwards. Once inside his own home, he regained consciousness.

Comments: Doctors at Azambuja hospital at Brusque examined the young man's leg and found no obvious signs of trauma other than that his left leg muscles were totally rigid "rendering any movement of the limb impossible." This condition of complete tetanus is highly unusual and suggests that both sets of muscles (flexors and extensors) were in full and continuous contraction for perhaps several hours. Let us take a closer look at this particular aspect.

Figure 48

Drawing of UFO seen on September 3, 1976
(Courtesy of *Flying Saucer Review*)

The neural fibers which control these large upper and lower leg muscle groups exit from the spine at the second, third, and fourth lumbar level on the front of the leg and at the fourth and fifth lumbar and first, second, and third sacral level at the back of the leg. Interestingly, the front "femoral nerve" bundle lies several inches beneath the skin while the rear "sciatic nerve" bundle lies perhaps three to four more inches almost directly behind it. So what?

376   May we assume that the white beam of light was not the direct cause of the tetanus effect but, rather, microwave energy within the small beam which would penetrate deep into body tissue? Might this microwave energy have been pulsed in synchrony with the specific frequency response of the neural membranes to cause them to "free-fire" continuously over several hours or more? I know of no known existing terrestrial technology that can achieve this effect.

The ability of Klein to move only his right leg suggests that the energy from the beam somehow produced a relatively long-term activation effect (not an inhibition or blocking of transmission) of the nervous signals at a local level in the left leg.[2] Had it affected the spine itself both legs probably would have been affected. Happily, Klein's condition improved to normal "in due course." This suggests that no permanent physiological damage was done by the energy. It is unlikely that the tetanus effect had psychological causes.

Notes: 1. The other two creatures also wore similar "weapons" which they did not use. These rods were attached to their belts on the right side. 2. Human muscle tetanus can be produced through electrical stimulation or by introducing various toxins. However, both techniques typically do not last over long periods of time. While electrical stimulation can influence individual muscle groups toxins typically act more systemically, within the entire circulatory system and therefore would affect both legs.

Reference: W. Buhler 1982. Extraterrestrial dwarfs attack farm worker. *Flying Saucer Review* 28, no. 1 (August): 5-8.

Cross references: 211 (8); 214 (8); 233 (8)

Case No. 231

Date: 76-11-12
Local Time: 0145
Duration: approximately 35+ minutes
Location: Talavera la Real air force base, Spain
No. Witnesses: 3
DB: kk   CB: d   PB-1: x   PB-2: a   UFOD: 0

Abstract: This interesting case was reported by the Spanish researcher Juan Jose Benitez, a highly competent field investigator. Two air force base sentries, Jose Maria Trejo and Juan Carrizosa Lujan were on duty guarding the fuel depot at the air base. Each was in his own sentry enclosure about 60 meters apart when a strange noise was heard.

377

The noise first reminded them of radio interference and changed to an "acute, penetrating whistle,...so piercing that it hurt our ears." After about five minutes, the noise stopped and everything became quiet again. Then, it began once again but nearer the sentry box of Trejo and he called out for Juan Carrizosa to come over and help him search the area for its source. Both men carried automatic Z-62 military rifles. The terrible noise stopped again for five minutes. The men stated later, "It seemed our ear drums were going to be ruptured." It began a third time for about five minutes. As the two men looked around, they both noticed a bright flare-like light high in the sky which illuminated a wide area of ground around them. Then, the piercing noise stopped for good, and the light faded out for about twenty seconds. To their relief, a third sentry arrived, Jose Hidalgo, who had a guard dog with him. Hidalgo asked if they had seen the bright glow, which they had.

One of the men ran to a nearby hut where he sounded a general alert and other soldiers and a Corporal Pavon were soon at the site. The corporal ordered a search of the area. Walking about 300 yards along the edge of an adobe wall surrounding the base, they saw and heard nothing. As they approached a partially built sentry box, they felt a "whirlwind" on them and each loaded his rifle. The whirlwind seemed confined to one spot. They heard the sounds of breaking branches in an area of eucalyptus trees nearby and, alarmed, they let the dog go and it charged away into the darkness.[1]

Some time later, they saw the dog returning to them in an unusual manner, staggering "as though seasick." The dog was sent back into the grove of trees four or five times and each time it returned in the same way. His ears seemed to be hurting him and he whimpered. The highly trained German Shepherd began circling, a sign the dog had been trained to alert his handlers to danger.

As the dog started to snarl, the men shouted into the darkness as loud as possible. They could still see or hear nothing. Trejo claimed he saw a greenish light in his visual periphery and turned quickly toward it. He saw very tall—estimated 3 meters—human figure only 15 meters from him. It "seemed to consist entirely of small points of light." The outline of the humanoid was lined with brighter lights and its head was covered with some kind of helmet. His body was thick (in the shape of a bobbin) and his arms long. The soldier stood there paralyzed for some moments. As Benitez writes, *"[Trejo] had his rifle at the ready, all set to shoot, but when he did decide to do so, he felt as though totally bound and shackled. He was quite unable to fire. Then he started to feel...a sensation of general weakness."*[2] His vision and hearing were normal but he felt as if he was falling downward slowly. All Trejo could yell was, "Down! They'll kill us!" He found himself conscious of what was happening around him but face down on the grass.[3]

It was then that the two other soldiers opened fire at the "apparition." *They fired from forty to fifty rounds total almost point blank and the being simply faded from view like a TV set that is turned off.* As the two men ran to help Trejo, the loud piercing whistle began again, emanating from the grove of trees, and lasting only ten or fifteen seconds before becoming silent.[4]

Trejo experienced severe headaches which would begin just prior to a loss of vision. They would always begin at the base of the skull in back and then appear in the frontal lobes. The other serious physiological effects which Trejo experienced several times over the following fifteen days are not recounted here.

Comments: This air base, with combined civil airport and jet flying school, is located several kilometers inland from Badajoz, Spain, on the Portuguese border. It is not at all clear where the other two men were standing when Trejo experienced his confrontation with the being nor why they did not also experience the same physiological symptoms that Trejo experienced. This "isolation of effect" suggests the use of a collimated beam of energy rather than an angularly broad beam.

This case is intriguing from other points of view as well, not the least of which is the simultaneous disappearance of the greenish form as the bullets were fired at it. And, why did Trejo became blind several times and for so long after this encounter. Psychosomatic blindness may possibly be indicated, however, the migraine-like symptoms suggest a more traumatic, physical cause.

> Notes: 1. The investigator made it clear that prior to this time the dog had been quiet and under control. The three men never heard the dog bark at any time. 2. There appears to be some kind of correspondence between volition (will power) and the witness's weakness and paralysis which prevented him from carrying out his act of aggression. We do not understand the nature of such a relationship, however, it may be related to an unconscious self-protection mechanism. 3. Here the vision of Trejo began to fail, "as though everything was being blotted out." These symptoms are similar to those of retinal ischemia which can occur for many different reasons and which can occur over a ten second period or longer. During acute cerebral ischemia auditory perception persists for some seconds after the person has become blind yet before unconsciousness has set in. Benitez discovered that it was only when Trejo tried to fire his rifle that he began to fall down. He said, "It seemed as though that 'being' had guessed my intentions." 4. Trejo's chest felt painful upon getting up even though he had not fallen or been struck there by his rifle. The pain lasted for only fifteen to twenty minutes.

379

References: Benitez, J. J. 1978. Encounter at Talavera. *Flying Saucer Review* 23, no. 5 (February): 3-6.

Lumieres dans la Nuit, no. 187, August/September 1979.

Cross references: 30 (3): 211 (8); 216 (8)

Case No. 232

Date: 77-07-12
Local Time: 2100
Duration: estimated 3 minutes
Location: Quebradillas, Puerto Rico
No. Witnesses: 1
DB: v   CB: d   PB-1: a   PB-2: x   UFOD: O

Abstract: A strange-looking, 3.5-foot-tall humanoid was seen by Sr. Adrian de Olmos Ordonez from the balcony of his home. It appeared out of the darkness from a farm opposite his house. It walked normally and bent down to pass beneath a barbed wire fence. It approached the mercury street lamp pole in front of his house "cautiously." The being was only about 30 feet away.

Because the being's dress looked so strange,[1] the witness called to his daughter to bring him a pencil and paper (in order to make a sketch) and also for her to turn on the living room light inside. But she turned on the wrong light switch, suddenly illuminating the outside balcony instead. *"The creature by then almost got to the lamp standard, but when it saw the light turned on it took fright.* I saw it looking upwards at the street lamp, and my personal opinion is that it was seeking something, energy, or electricity, or something like that...[it carried in its right hand]...a small shiny object.[2] The light from our house was reflected off the glass of the creature's helmet, which shone with a tremendous brightness." *The being ran back towards the barbed wire fence immediately after the balcony light went on.* It passed under the wire and then stopped. It placed both hands on the front of its belt and something like a hiking pack it carried on its back suddenly became luminous and emitted a noise similar to an electric drill. Then, the creature began to rise into the air and disappeared toward the nearby trees. During the following ten minutes, the witness, his family, and some neighbors watched lights move from tree to tree on the nearby farm.

Comments: While this is not strictly a human aggression case, it is included here for comparison with others where the beings are attacked in some way. Interestingly, the sudden appearance of the balcony light seemed to be an aversive stimulus to the visitor. Is it possible that he didn't want to be seen? Of course, neither the witness nor the investigator could know whether the being felt fright when the light went on; this was imputed to it from their point of view.

Notes: 1. The being wore a green garment which looked as if it was filled with air (it didn't fit tightly). The helmet was tall and pointed on top with a tiny light or flame at its tip. 2. It gave off light, was the size of a matchbox, and had a sharp point on it.

References: *El Vocero,* 13 July 1977.

Lamarche, S. R. 1978. A "Flying" humanoid in Puerto Rico. *Flying Saucer Review* 23, no. 6 (April): 9-12.

MUFON UFO Journal, no. 124, 1978.

Case No. 233

Date: 77-09-15
Local Time: approximately 0225
Duration: estimated 1 hour 30 minutes
Location: Pacienca (near Rio de Janeiro), Brazil
No. Witnesses: 1
DB: dd   CB: a   PB-1: a   PB-2: x   UFOD: a

Abstract: Antonio Bogado La Rubia, 33, a bus driver, was walking to a bus stop to go to work. As he past a deserted field, he was surprised to see an object in the sky that looked like "an enormous hat." He felt paralyzed and two men grabbed him. He found himself floating around inside a disc. He saw many scores of similar beings there, all were about five feet five inches tall and wore metal helmets that looked like football helmets with a wide brim. La Rubia saw blue flashes coming from mirror-like sections that were cut into their helmets. The beings seemed like metallic robots yet he could hear them breathing.

La Rubia felt panic but he couldn't yell out. At one point during his abduction he shouted, "Who are you? What do you want from me?" As an article in the *International UFO Reporter* states, *"His shouting seemed to mortally affect the occupants, who fell back. Just as suddenly, Antonio lost his voice, and was paralyzed by a beam of light directed at him.* It looked as if they wanted to roast me alive. That beam of light turned me red, burning like fire, and I was overcome as it lessened." He felt very dizzy and still burning when he awoke beside the roadway. A passerby told him it was 3:55 A.M.[1] La Rubia continued on to work and then to its infirmary where he was examined by the nurse and a psychologist. He was in a "pitiful emotional state, crying like a baby…not to mention his variable temperature, swollen state, loose bowels, and vomiting." After examinations conducted at the Rocha Faria Hospital, La Rubia was found to have hypertension of the thyroid and flooding of the cerebral ventricle."[2]

Comments: For a long period of time after the claimed event, the witness was abnormally thirsty and a chain smoker.[3] The *IUR* article questions whether the witness may have been hit on the head from behind, leading to these traumatic symptoms and mental recollections. If true, then we should expect to find at least one other case of cerebral concussion in which the victim describes an alien abduction with many intricate, linked details. I know of none.

381

Notes: 1. Nothing is said about why the passerby did not give any more aid. 2. A concussion of the head or ruptured aneurysm could produce these symptoms. It isn't known whether he hallucinated during this time. 3. Both suggest a problem with secretion of the antiduretic hormone (ADH) produced by the posterior pituitary gland. Nicotine increases ADH levels and is compatible with counteracting sensations of thirst, perhaps due to the dehydration associated with vomiting and diarrhea.

References: Anon. 1977. *International UFO Reporter* 2, no. 11 (November): 2, 8.

Foreign Forum. *APRO Bulletin* 26, no. 4.

News World (New York), 10 October 1977.

O Dia (Rio de Janeiro), 4 October 1977.

Cross references: 219 (8); 221 (8); 229 (8); 230 (8)

Case No. 234

Date: 78-01-18
Local Time: 0300-0500
Duration: unknown
Location: McGuire Air Force Base, New Jersey
No. Witnesses: 2+
DB: kk   CB: e   PB-1: a   PB-2: x   UFOD: c

Abstract: Several UFO were sighted above the Fort Dix air field and near the adjacent Fort Dix Army Camp. An unnamed air force security policeman was on duty at the air base in the early morning hours and wrote the following account to UFO investigator Len Stringfield.

"I am a security policeman and was on routine patrol at the time. New Jersey State Police, and Ft. Dix MP's were running code in the direction of Brownsville, New Jersey. A state trooper then entered Gate 5 at the rear of the base requesting assistance and permission to enter. I was dispatched, and the trooper wanted access to the runway area which led to the very back of the air field and connected with a heavily wooded area which is part of the Dix training area. He informed me that a Ft. Dix MP was pursuing a low-flying object which then hovered over his car." The object was described as oval in shape and glowing with a blueish green color. His radio transmission was cut off. "At that time in front of his police car, appeared a thing, about 4 feet tall, greyish brown, fat head, long arms, and slender body." He panicked and fired five rounds from his .45-caliber directly into the thing, and one round into the object above. *The object then fled straight up and joined with eleven others high in the sky.* "This we all saw but didn't know the details at the time. *Anyway, the thing ran into the woods towards our fence line and they wanted to look for it.* By this time several patrols were involved."

We found the body of the thing near the runway. It had apparently climbed the fence and died while running. It was all of a sudden hush-hush and no one was allowed near the area. "We roped off the area and Air Force Office of Scientific Investigation (OSI) came out and took over. That was the last I saw of it. There was a bad stench coming from it too. Like ammonia smelling but it wasn't constant in the air." That same day, a team arrived from Wright-Patterson Air Force Base in a C141 and went into the area. "They crated it in a wooden box, sprayed something over it, and then put it into a bigger metal container. They loaded it in the plane and took off. That was it, nothing more said, no report made and we were all told not to have anything to say about it or we would be court martialed."

The witness continued, "I will be getting out of the air force in about two months. Do not disclose my name as I could get into trouble. I am interested in pursuing this and other matters if you need help."

Comments: It is not clear how this eye witness knew about the details of the crating of the alleged body which allegedly took place after personnel from Air Force OSI arrived and took charge. Did they tell him about their procedures, did they permit him to be present? This is not very likely. Until such questions are answered satisfactorily this story must be placed in a "hold for further details" category.

> Note: 1. At the top of this letter are the words: DEPARTMENT OF THE AIR FORCE, _th Security Police Squadron, APO San Francisco 96239 as if this was printed on the stationery.

References: Good, T. 1990. *The UFO Report,* 189.

MAJIC 6.DOC computer file received from Dale Goudie, Seattle, Washington, 1995.

MUFON UFO Journal, no. 230, 1987: 7-8.

Case No. 235

Date: 78-03-mid
Local Time: 2050
Duration: 1 hour 20+ minutes
Location: Kent, Ohio (rural area)
No. Witnesses: 2
DB: jj   CB: e   PB-1: a   PB-2: x   UFOD: b

Abstract: Two young men, Jim and Ned, had just set up their camp for the night. The air was cold when, at about 8:50 P.M., they noticed a large, bright orange object with "an iridescent glow to it" some distance away hovering over the lake. It appeared like a fluorescent light, pulsating from very dim orange to very bright. The globe jumped vertically upward and then angled back down about 40 feet at a time. A yellow ray of light shot out of it irregularly; it did not pulsate. Jim recalled, "It would dart over then all of a sudden stop. And Ned got his flashlight out—and...he flashed his light on and off at it. And that seemed right there to catch the attention of whatever this thing was. *And the beam retracted—came right back up to the ship.* It was weird—when you see a flashlight shut off it is just off but this thing it would retract back up to the ship."

384

The two young men saw this yellowish beam strike the ground almost deliberately, hitting a certain spot and then another. "This beam seemed to be under a force of power where it would not spread out, [it was] a straight beam, but it did get a little bigger at the bottom. You could see the end of it come back up into the ship.[1] The orange globe moved around the edge of the lake slowly approaching their campsite. *Then, the UFO came down to the ground "like it was going to land" just as an airplane flew overhead.* It was followed by a second airplane and, when they both had left the area, the UFO rose up out of nearby woods.[2]

As the UFO approached, its beam illuminated one of the witnesses for about ten to twelve seconds. It left him paralyzed. "I couldn't bat an eye," he said later. He did not remember thinking about anything at the time but does recall staring.[3] Neither experienced any fright or other emotion.

Jim dove into the tent for some protection, and the light "came back down and hit" Ned. Now looking out of a small window in the tent, Jim saw Ned's hair, face, and upper torso illuminated "like God had decided to come down."[4] Jim also recalls hearing a very mild pulsating "whoo whoo" noise coming from above him. The whole tent was illuminated with a bright orange light for thirty-five to forty seconds and then the UFO moved away into the night sky. "We were in a state of shock...[inside our tent for the next ten minutes]...but [we were not scared]. I want to emphasize that. It was very weird, sort of a tranquil feeling.

"I think under normal conditions we would have gotten out of there," Jim said. "After about ten minutes, we got our fishing poles and went down to the lake to fish. Nobody's going to believe this. We were two buddies and the conversations was like joking, 'Hey that was neat wasn't it! Wow did you see that thing!' It was like we were having a great old time. We were not scared that's the important part."[5]

The two men walked down to the lake's edge to fish for catfish, standing in total darkness. Jim had just put on a leather holster and .22-caliber revolver. After about forty-five minutes, they heard "heavy stomping through the woods" behind them near the campsite. It seemed to be coming toward them "at a fairly good pace." They swung around to face the noises, aiming their flashlights in the same direction. "I said I had a gun and anybody out there should talk because I would shoot....After I had talked a minute....man, there it was, it had come out from behind a tree. *He fired all six shots directly at the creature and it kept on coming."* The being was about 4.5 feet tall but bent over. It was estimated to be over 6 feet tall if it had stood upright, "A humongous thing and it weighed in at about 350 lbs., I believe, and it was solid whatever it was. And it was no bear." *The creature moved quickly behind a tree.*[6]

385

The two men ran back to their tent and zipped the door shut. Jim reloaded his revolver. Then they heard a "pop-pop" noise outside around their tent, like something beating on a hollow tree; this thumping lasted for five or six more minutes. The two men were terrified. And, to add to their fear, the orange sphere came back over the tent for another four or five minutes. They made a run for their car; the UFO had departed back some distance in the direction from which it had come. The men fired two shots as they came out of the tent. As they raced toward their car they could hear the rapid footsteps behind them and Jim cried, "Oh my God, they are comin' after us!"

The men reached their car, jumped in, and raced home. Later, when they looked in a mirror, they saw that their faces and the backs of their hands and arms were a shiny red. Their palms were not, however. "It was all very swollen and red and had little pimples all over my face," Jim said. "The next day we had runny bowels, but that same night I felt nausea, a queasy sickness, but couldn't vomit. My skin felt hot and tingly. These things lasted three or four days." He also passed some blood in his stool. Ned said he had much the same symptoms but didn't pass any blood. Jim's urination caused a burning sensation and was a yellow-orange color. He was extremely dehydrated for several more days. Both slept longer than normal for the next several weeks. Their psychological state was described as that of depression, exhaustion, and listlessness. They dreamed bizarre dreams after the incident, one of which was of having been abducted up into a ship.

Comments: This nightmarish account is filled with science fiction and symbolic content that would make for a good grade-B thriller movie. Our chief interest is in the response of the UFO and alleged occupants to the behaviors of the two young men, however. If a craft can cross the vast stretches of interstellar space, it can probably detect light from a flashlight in dark countryside at night. Indeed, U.S. military night field search operations routinely use infrared spotting scopes and even more sensitive sensors. It did not appear that the creature was harmed by the bullets fired. Although, it isn't known if it was actually hit. And, aside from a small part of the dream-like narrative of the alleged abduction on-board the craft, nothing harmful happened to the two young men. Apparently gun fire did not provoke an in-kind response. If the light ray actually had the powers claimed, the UFO did not need to fire any other type of weapon at the men.

Notes: 1. This is known in UFO literature as the "truncated beam" where a collimated light ray comes to a visible end in mid-air for some unknown reason. 2. Is it possible that the approaching airplanes led to the UFO's avoidance response? Similar responses are presented elsewhere in this book. 3. This clearly indicates that the witness was not unconscious. Paralysis of large, voluntary muscles is common during abduction events, e.g., controlling head rotation and nodding, shoulder and chest muscles, arm and leg movement. Muscles which are controlled by the autonomic nervous system, e.g., those involved in breathing, are not affected in most abduction cases. Significantly, small muscle fibers like those used to blink, turn, and close the eyes also are usually not affected. I suspect that the mechanism for this paralysis occurs within the axons which control the muscles and not the muscles themselves. 4. The men said that they felt scared only when the beam of light was not on them, then "you were not scared anymore....I felt no fear, pain, emotion—I felt nothing. I felt very aware of what was happening yet at peace...there seemed to be no time." 5. This is a rather classic example of psychological denial which is found in numerous other abduction and close encounter cases. 6. Jim felt he had hit the creature with .22 hollow point magnum rounds at least five times and had aimed at the neck-shoulder area. He claimed to be a good shot.

Reference: Worley, D. 1989. Field investigator's report, 22 November.

Cross references: 218 (8); 225 (8)

Case No. 236

Date: 78-12-06
Local Time: 2040
Duration: estimated 4 minutes
Location: Fronteira, Brazil
No. Witnesses: 1
DB: v   CB: d   PB-1: a   PB-2: x   UFOD: O

Abstract: This incident took place inside a security area at the Brazilian hydro-electric station at Marimbonda near Fronteira and involved night watchman, Mr. Jesus Antunes Moreira.[1] The witness was inside a guardhouse located on top of a nearby dam; his visibility was very good except for the darkness of night and the falling rain when this encounter took place.

At about 8:40 P.M., he noticed that the surface of the water behind the dam was being illuminated from above so he stepped outside to find out where the light was coming from. He saw an object about 5 meters long by 3 meters high moving slowly "slightly above the level of the horizon, about 200 meters away." It was moving in his direction across the Rio Grande near a large waterfall known as Cachoreira do Marimbonda. It seemed to be coming down as if to land on the dam or the powerhouse. The object did not resemble a helicopter.

Moreira began walking out along the concrete top of the dam toward the UFO which just flew above him in the opposite direction, descending to about 1.5 meters above the base of the earthen part of the dam. He described the surface of the object as being a light grey color with a 2-meter-high door in its side and a small window in its upper section. A small platform seemed to extend around its circumference. Only about 7 meters from the astonished witness, the window in the side of the UFO opened and "in it there appeared a face in many respects very like a human face. Then the main door opened, and from it came three beings dressed in blue overalls with a metallic sheen. They were all very tall—two meters maybe—and with quite long, black, smooth hair" (Buhler 1979, 19). At this point, we turn to Buhler's own wonderful account of the witness's description:

> "In an absolutely natural sort of fashion, they spoke to me, in
> some unknown language. I replied, in my confusion, that I would

'go to one of the telephones that are strung out along the 300-meter-wide top of the dam, one of them gestured to me to step back. At this stage, I began to get scared, and I felt for my revolver, which I was wearing over my rain-cape, with the idea of firing a warning shot should it be necessary. *And indeed I did try to shoot, but the revolver jammed, and would not fire"* (Ibid.).

After the encounter, Moreira was examined by Dr. Sergio Bandeira at the Hydroelectric Plant's first aid station and presented "attacks of nausea," acted like he was in a dazed state, and experienced photophobia for about an hour following his encounter. His recovery of visual sensitivity was "slow and gradual."[2] He was examined later by a medical team at Sao Jose do Rio Preto where he showed no abnormal symptoms.

Comments: Here is another instance where the person's willful response to fire a weapon at living beings was thwarted. If the humanoids prevented Moreira's gun from firing, how did they accomplish it? Did they react because of what they saw or because they knew what was going through his mind?

388

Notes: 1. The witness was described by co-workers as a conscientious, hard-working individual. 2. Moreira left his job at the dam after this event and went to work at Barretos.

References: Buhler, W.K. 1979. Conversation with Entities at Marimbonda. *Flying Saucer Review* 25, no. 3 (May-June): 18-19.

Pacheco, V. 1979. *Ultima Hora* (Sao Paulo), 2 March 1979.

Overview

Here are a few preliminary observations about the twenty-six cases presented in this chapter. As expected, they raise far more questions than they lay to rest.

Entity responses to being fired at...Of the ten incidents presented here:
- The entity reacted immediately in some way in six cases (60%).
- The entity reacted by firing something back in three cases (30%).
- The entity never moved or reacted at all in one case (10%).

Entity responses to being fought with...Of the 4 incidents presented here:
- The entity was knocked backwards in one case (25%).
- The entity pulled human's hair out in one case (25%).
- The entity gave human physical exam in one case (25%).
- The entity caused gun to break when hit in one case (25%).

Entity responses at being threatened in some other way. Of the five
incidents presented here:

- Human shouted/waved ice pick and entity touched his belt,
human suddenly fell down/paralyzed, pick taken out of
his hand by alien in one case (20%).

- Human drew a gun, aimed (didn't fire) and entity allegedly
made gun feel very heavy and unable to fire in one case (20%).

- Human aimed flashlight at entity and was hit by blinding beam
of light in one case   (20%).

- Human threw a knife at entity and it hit an invisible barrier in
one case (20%).

- Human raised object above his head and entity backed up in
one case (20%).

389

Other investigators have also remarked about the speed with which the
visitors react to any attitude or movement that could be interpreted as a threat.
Michel (1958) reported six instances of temporary paralysis and one case of
blindness and paralysis in humans.

When aliens behave with impunity at being fired at or take an extraordinarily
long time to respond to signals or hover over heavily guarded U.S. nuclear
missile silos does this imply that they have no fear of us? Does it suggest that
contact with humans has a low priority to them? If aliens snub humans does it
indicate that humans have little or no rights as far as the aliens are concerned?
These and related questions must be addressed.

References

Anon. 1977. West Wales round-up. *Flying Saucer Review* 23, no. 1: 6.

Anon. 1979. *Project World Authority for Spatial Affairs* (W.A.S.A.), Intercontinental U.F.O. New York: Galactic Spacecraft Research and Analytic Network., New York, 1979.

Brookesmith, P., ed. 1984. *The UFO Casebook.* London: Orbis Publishing, 84.

Castillo Rincon, E. 1995., *OVNI - Gran Alborada Humana,* Vol. 1. Venezuela: Editorial Norte y Sur, C.A. Venezuela, 1995.

Clark, J. 1967., Why UFOs are hostile. *Flying Saucer Review,* 13, no. 6 (November-December): 18-20.

Creighton, G. 1965. Argentina 1963-64, *Flying Saucer Review* 11, no. 6 (November-December): 14-17.

Creighton, G. 1977., UFO, Occupants and sex in Columbia. *Flying Saucer Review* 23, no. 1: 14-16, 18.

David, L. 1989., *Air Force Considers Planetary Defense.*

Dempster, Derek. 1955. Interplanetary War Claim. *Flying Saucer Review.* November-December.

Fawcett, G.D. 1986. *Human reactions to UFOs worldwide (1940-1983): A four year study project.* Published privately, 1986.

Fowler, R.E. 1974., UFOs: *Interplanetary Visitors.* New York: Bantam Books., New York, 1974.

Gross, L. 1997. Interview with author, July 19.

Haines, R.F. 1990. *Observing UFOs.* Chicago: Nelson-Hall, Publishing.

Haines, R.F. 1994. *Project Delta.* Los Altos, Calif.: LDA Press.

Hall, R. 1964. *The UFO Evidence.* Washington, D.C.: NICAP.

Holiday, F.W. 1978. Did humanoids kill these men? *Flying Saucer Review* 24, no. 1 (June): 20-22.

Hynek, J.A. 1972. *The UFO Experience: A Scientific Inquiry.* New York: Ballantine Books.

McDonald, J.E. 1967. *UFOs: Extraterrestrial Probes? Astronautics and Aeronautics* 5 (August): 19-20.

Michel, A. 1958. *Flying Saucers and the Straight-line Mystery.* New York: Criterian Books.

Paget, P. 1979. *The Welsh Triangle.* Granada, London: A Panther Book, 100-102.

Pratt, B. 1996. *UFO Danger Zone.* Madison, Wis.: Horus House Press.

Vallee, J. 1969. *Passport to Magonia: on UFOs, Folklore, and Parallel Worlds.* Chicago: Contemporary Books.

Wilkins, H.T. 1955. *Flying Saucers Uncensored,* 56-7.

"I think, therefore I am."

—Plato

"At least I think so!"

—Haines

Chapter 9

Humanoid Responses to Human Thought and Miscellaneous Behavior

(6 cases)

If Earth is being visited by extraterrestrial beings, can they read our minds or communicate with us via thought alone (telepathy)? Do they possess a higher native intelligence than humans? Are they similar to humans in how they process ideas and abstract thought? Do they possess deep intuition and/or wisdom? Perhaps they possess the ability to control things at a distance (telekinesis). These and many other questions have been raised over the years. Indeed, there are scores of books about people who claim to have been in telepathic—i.e., purely mental contact—with extraterrestrial beings.

As with the previous chapters in Part II, there are great methodological challenges associated with establishing the validity of the human behaviors that are described. We cannot yet measure thoughts. Nor can we be certain that a witness is telling the truth or was even recalling the thought accurately and without distortion. Nevertheless, let us suspend our disbelief and contemplate the following stories. Let us see what might be learned while holding the "evidence" with a loose grip.

The following cases are drawn from my own research files as well as those of other colleagues. These accounts have not been published before. Controversy swirls around these kinds of incidents, as indeed it should, considering the bizarre claims that are made. Nevertheless, we find an odd sort of correspondence from case to case, a remarkably similar correspondence of general themes as well as many details. The interested reader should consult Bullard (1987, 1994) and Pritchard (1994) for more information on this important matter.

These cases are grouped into three general categories for the sake of convenience. The stories that follow are fascinating as much for what they do not say as what they do.

Case Files
Human thought-provoked response(s)

Case No. 237

Date: 75-11-?
Local Time: 2115
Duration: approximately 2+ hours
Location: 22 miles west of Sonora, California
No. Witnesses: 2
DB: x   CB: a   PB-1: a   PB-2: x   UFOD: 0

Abstract: Two young women, Kathy, 19, and Susan, 20, were returning to Sonora from Modesto about 45 miles away on Highway 108/120 in Kathy's Ford Pinto. Kathy was driving. It was a cold night with thick ground fog in low lying valleys. Passing through the town of Oakdale, California, at 11:00 P.M., the two women were enjoying casual conversation. UFOs were far from their minds and conversation. It was Susan, sitting in the right seat, who first sighted a luminous globe of golden-white light in the dark sky through the left-front portion of the windshield. She pointed it out to Kathy who glanced at it quickly and exclaimed, "that's not the moon, the moon is over to the right!"[1] The object moved toward them, and Susan became very excited, pleading with Kathy to pull over and stop so Susan could get out of the car to see it better. Later, Susan admitted thinking to the object that it should come over and land.[2] Kathy noticed that the car "slowed down all by itself....It's like I wasn't driving the car anymore....It was like the car knew right where it was gonna go." Susan got out and ran forward beneath the large glowing UFO while Kathy remained inside clutching the steering wheel in fear.

The remainder of the alleged physical abduction events will not be repeated here so that we can concentrate on several telepathic events which took place inside the underground cavern where both women said they were taken.[3]

I had an opportunity to ask more than one hundred questions to both women during many independent hypnosis sessions. The following narrative is from Kathy on August 20, 1988, during the third and last stage of the "Three Stage Technique" (Haines 1989) during which I am permitted to ask questions of the person directly.

Haines: I want you to tell me every single detail you can about what you see, what you hear, and what you feel on your body, sensations of touch and so forth, what you smell, what you say out loud, and what people say back to you. It's just as if you hold a conversation. For instance, tell me in your own words the conversation that's going on. Is that clear?

Kathy: Uh huh. (yes)

Haines: Go from the time you are just set down on the ground at the entrance of that place to the time when you enter the round room. (I had previously learned from Kathy that this room was round.)

Kathy: This beam lowers me down. And I'm still laid out and I still can't, I still don't have any control of my body....I can't move anything....And there's two of them that are on each side of me as I'm put down and then there are two more that appear at my feet or my legs...and...the one, the one to the right seems to be tellin' the other two, or the other three, what to do. He seems to be the one responsible for me. And then...um, he says to the other one, and then, like I said, they're not talking, they're just talking telepathically....As we get inside the cave the other one asks, he says, "Do you know why you're here?" And then, then he told me, "You don't have to talk." And I said, "No." I didn't say it, I just thought it. And then he said, "Well, I don't want you to be afraid." And that's when he reached out and touched my arm, my forearm.

In the above telepathic "conversation" Kathy was convinced that she and the being carried on a human-like conversation over time. Questions were asked and answered, commands were made and followed. Later in the same hypnosis session, the following interchange took place.

Haines:	When he transmitted messages to you what language were they in?
Kathy:	English.
Haines:	Always?
Kathy:	Uh huh. (yes)
Haines:	OK. When you talked to him did you move your lips and your tongue and your vocal cords?
Kathy:	I tried but I couldn't.
Haines:	Uh huh, so how did you communicate?
Kathy:	He told me that I just had to think, and he would understand what I said.
Haines:	And did that always work?
Kathy:	Yeah, but I didn't think very much.

Susan provided me with the following narrative under hypnosis on July 11, 1988. She is describing a 4.5-foot-tall male being who accompanied her at all times inside the cave where she was taken with Kathy.

Haines:	OK, very good. I'd like you now to look right at that person you've been describing and tell me what he looks like.
Susan:	His mouth is just a little place…it doesn't move. And his ears are real small.…And he has no hair. And he has round large eyes, nice eyes.
Haines:	Does he ever open his mouth?

Susan: No.

Haines: When he talks to you does his mouth move?

Susan: No, no mouth.

Haines: No mouth?

Susan: Just a mouth, but no mouth.

Haines: Like a line or something?

Susan: Tiny...no teeth, no mouth.

Haines: When he talks to you, what language does he use?

Susan: My language.

Haines: And when you talk to him do you move your mouth or just...

Susan: Oh no. I talk to him.

Haines: I see. Do you ever think to him and have him understand you?

Susan: Oh yeah. Oh yeah, that has happened.

At one point, Kathy said she was lying on her back on a table inside a round room whose walls glowed evenly without any visible light sources. A door was located on her left side and a cabinet or control console on her right, both located about 20 feet or more away. A being wearing a khaki-green robe-like uniform stood on her left side and she pleaded with him mentally, "Oh don't hurt me." She felt he was communicating with her telepathically in English, as far as she could tell.

She said the short figure (who appeared almost human) passed his open hand back and forth above her left leg from thigh to ankle several times as she lay on her back. She felt a tingling sensation as he did so. He did not perform any operating procedures that she can recall. Kathy had experienced pain in her left knee for many years prior to this event and, soon after the event, it went away for some unknown reason and has not returned to this day (1998). Whether or not he was complying with her thought cannot be determined.

Similar accounts of so-called mental telepathy are given by many other people who claim to have been abducted by extraterrestrial beings. Whether these creatures are terrestrial, extra-terrestrial, or purely mental constructs remains to be proven.

Comments: This incident is but one of many in which the human being thinks thoughts in the presence of one or more aliens. This particular multiple-witness case is important since Susan provided separate and independent corroboration for many of the alleged events both above and underground, before, during, and after the abduction. How she could know about all of these details[4] without actually having experienced them is hard to explain.

> Notes: 1. The moon was almost full on this night. 2. I was told that the UFO did not land but remained suspended in the air over the car about 150 feet. I found some evidence pointing to earlier bizarre encounters in Susan's childhood (starting at about age 6 and occurring several other times) which would support her curiosity and relative lack of fear shown toward the approaching object. 3. I performed numerous regressive hypnosis sessions on each woman independently and did not disclose anything that the other had said. I asked each one the identical set of 104 questions and carefully noted their answers; these answer sets were then rated by several colleagues using a special correspondence rating scale. These ratings indicated a remarkably high degree of agreement on many questions. 4. The two women said they didn't talk about the incident the next day. Indeed, they just wanted to forget about it, which they did for over eleven years! When they finally did meet to discuss what happened to them that night, Kathy had the foresight to tape-record their entire meeting. It provided me with valuable evidence showing that only a few of their shared abduction experiences were discussed at that time. I referred to this recording to plan my questions—to be asked of each woman under hypnosis—so that only information that was not shared that day was discussed under hypnosis.

Reference: Author's file.

Case No. 238

Date: 87-09-14
Local Time: after midnight
Duration: unknown
Location: Hercules, California
No. Witnesses: 2+
DB: kk   CB: a   PB-1: a   PB-2: x   UFOD: b

Abstract: This account was obtained immediately following a hypnotic regression session conducted on May 24, 1989, in San Francisco using conscious recall techniques. Diane, a young mother, claimed to have had numerous nighttime visits in her home by several short aliens. Each of her four children told her of their own meetings with little people the same nights as Diane was visited.[1] She had presented no untoward physiological or psychological symptoms over the eight-month period of these counseling sessions. In short, she had adjusted very well to her bizarre experiences. The following narrative seems to display a CE-5 type event that took place telepathically.

Haines: Do you know anything about height or anything else about…

Diane: They're probably about our height. I got the feeling they were proportionally about our size. And…I remember seeing…sitting in a space ship….next to one, one night…..And I asked him to hold, asked to hold his hand, because I was scared of him, and he looked at me and I looked at him and *we both thought, well* (she laughs) *"repulsive!"*

Comments: The alleged two-way mental conversation qualifies as a CE-5 event, even though the being's response could have been imputed to him by Diane.

> Note: 1. I have several drawings made by these children on the morning after an alleged visit. They show a disk-shaped object and several beings with detailed, pear-shaped heads with large, dark, wrap-around eyes. According to Diane, they had never drawn such pictures before and told her that these were the "friends who came to play with us last night."

Reference: Author's file.

Case No. 239

Date: 88-11-29
Local Time: midnight to 0300
Duration: estimated 30 minutes
Location: Sherman Oaks, California
No. Witnesses: 2
DB: dd   CB: a   PB-1: a   PB-2: x   UFOD: b

Abstract: This narrative was obtained on November 29, 1988, by telephone with Maria, 42, a divorced mother of two boys. The younger son Billy, 8, was involved in several alleged encounters [1] while the older son had moved out of the apartment, allegedly for reasons of independence and because of the many weird things that had happened there. Maria said that she was asleep at about 2:00 A.M. the previous night when "they" arrived again. "There was a being, evidently…assigned to keep me absolutely frozen to the spot….Right there! And he seemed most unhappy with this task….He was standing like a sentry….I tried to open my eyes twice.…My eyes were slammed shut, and I couldn't turn. It had become very quiet behind me[2]…because I was aware of two other beings there by [Billy], and when it became quiet, I had assumed something had changed or maybe they had left. And I wanted to see, and I could not move, I wasn't permitted to turn, and, on two occasions…it was like a little of the paralysis was lifted and I opened my eyes, *they were slammed shut* and that's that! Restraint was like put on me twice as hard and I became very angry. And, you know, telepathically, of course, I said to this being, "Why are you not letting me open my eyes? Why aren't you letting me move? After all, I've seen you already. After all I've been through! Why are you doing this to me? And the being said to me…you know, with tremendous affirmation, *he said, 'Because you need to be reprepared,'*…almost bordering on anger.…Like frustration more than anger. I wouldn't say anger.

"He said, 'You have to be reprepared.' And I said, 'Reprepared? What do you mean by reprepared?' He said, 'Well, you haven't done anything you were told to do, and tonight, when you were told to change out of those clothes into your other pajamas, you didn't listen.'"

Comments: The CE-5 aspects of this case involved both physiological control of Maria's eyelids, and an enigmatic reply by the visitor to her thoughts. The UFO/alien abduction literature is literally filled with verbal and mental conversations between humans and various entities. Such two-way communications qualify as CE-5 events as long as the alien is said to respond to a prior human verbalization. I learned later that Maria very likely "heard" his voice in her mind, not acoustically.

> Notes: 1. I worked with Maria over an eighteen-month period. She claimed to have had over fifty separate visits, one every eleven days on the average. This event took place near the middle of this sequence. 2. On this night, Maria's son Billy was asleep on her left side in her king-sized bed.

Reference: Author's file.

Human speech-provoked response(s)

Case No. 240

Date: 79-09-02
Local Time: afternoon

Duration: unknown
Location: Livermore, Colorado
No. Witnesses: 1
DB: o   CB: a   PB-1: a   PB-2: x   UFOD: b

Abstract: Shirley P., a school teacher on vacation, and her dog Spank were taking an afternoon walk on her mother's small farm near Livermore. She suddenly noticed an odd-looking, dumbbell-shaped object hovering silently in the sky above her. It looked like polished chrome and had no seams or other markings. She soon found herself inside it looking down at the roof of her mother's ranch house, a totally new and fascinating experience for her. Fearing for Spank, the witness asked about him. The following hypnotic narrative was obtained on July 21, 1984, with Shirley in a very deep trance.

Haines:	All right. Tell me what else happens now.... Forget that I'm here. I want you just to tell me everything that happens from now on.
Shirley:	He asked me if I wanted to...to go and see some more, and I said, "Okay." And then I said that I wanted to be sure that Spank was going to be all right. And he said, "Don't worry...we'll take care of Spank. He'll be fine." He said, "He'll be right where you left him when you came in." And I said, "Okay."

401

Comments: As far as the witness could tell, she spoke out loud as did her "host." He answered her question about her little dog clearly and directly. This type of conversation qualifies as a CE-5 event. Of course there are a great many such conversations recorded in the abduction literature (Bullard 1987).

Reference: Author's file.

Case No. 141

Date: 88-11-29
Local Time: midnight to 0300
Duration: estimated 30 minutes
Location: Sherman Oaks, California

No. Witnesses: 2
DB: q   CB: a   PB-1: a   PB-2: x   UFOD: 0

Abstract: This alleged event involved Maria and Billy, her eight-year-old son (see case 239). The verbal narrative takes place at breakfast the morning after an alleged alien visit with both mother and son. Some time after midnight, Billy said he was taken outdoors, seemingly able to pass directly through the locked front door of their apartment. In a telephone conversation with Maria, I asked her about the claimed event. She claimed to have tingling, paralysis, and memory loss after the abduction. She asked her son the next morning about his rest, hesitant to directly ask about the encounter.[1] She described her conversation.

Maria:	Did you have a good night's sleep?
Billy:	Yes! And you know what?...My aliens came back!
Maria:	Is that right?
Billy:	Oh, yeah!
Maria:	(Billy proceeded to describe to me, using his fried eggs, the appearance of what the ground looked like where they took him during the night.)
Billy:	They landed in a big hole in the ground. (And he made a crater in the egg by popping this bubble.) We went way down, right into the ground, and parked it...in this big hole....They took me over to the playhouse.[2] But I do remember leaving here. And that's always the most fun.
Maria:	(I looked at the door, which still had the chain on it.) How did you leave? Look at the door. The door is still locked, the chain is on the door. (And he looked at me like I was crazy.)
Billy:	Mom, how do you think I went through it, like I always do.
Maria:	How did you go through the door?
Billy:	Well, you see, once you get half way through it it's not wood anymore.
Maria:	Oh, really?

402

| Billy: | Yeah, it's really neat! It changes into lots of little colored balls. And that's the most fun part and *the only time the aliens ever get mad at me* is because I like to stay in the middle and look at the balls for a really long time. But they tell me that that's dangerous to stay in the middle. I have to come out the other side. I can't stay too long and play there. |
| Maria: | (And so he…he was disappointed about that (laughs). But he was very emphatic and then he went on to ask what was for lunch later…and he dropped the whole subject. It was just as matter of fact as could be.) |

Comments: Billy's fascination with studying the balls of colored light inside the wood of the front door allegedly produced a response of anger in the aliens. This constitutes the CE-5 aspect of this case. Billy slept with his mom from time to time which was the case during this alleged event. Another interesting incident involving this woman is found in Haines (1997).

Notes: 1. Maria was referring to her own encounter the same night which she did not want to mention to her son. 2. Billy does not remember anything beyond this point because, as he said, "he slept through all of it."

Reference: Author's file.

Human motion without speech-evoked response(s)

Case No. 242

Date: 76-09-02
Local Time: afternoon
Duration: unknown
Location: Livermore, Colorado
No. Witnesses: 1
DB: kk   CB: a   PB-1: a   PB-2: x   UFOD: a

Abstract: This narrative occurred on July 21, 1984, in Livermore, Wyoming. It is the same event as described above in Case No. 240. Shirley is speaking while under hypnosis and is describing a circular room she has been taken into and its furniture.

Haines:	Is there any furniture in the room?
Shirley:	Um, Well I...I guess it's furniture. It's not like what you would see on Earth, but it's....there's....I guess that's what they use it for. There's like a big thing in the middle [of the room]...that comes up from the floor...It's just...built-in like, or it's molded into the rest of it.
Haines:	I see.
Shirley:	But there are people walking up to it and doing things. They're like...It has a kind of transparent top to it...It's like there's a grid thing on the top, printed or something, and then...these pictures come up from the bottom and they sort of stick on the bottom of that thing. I don't know what that does.
Haines:	Can you see any of the pictures?
Shirley:	They won't let me get that close.
Haines:	How far away are you?
Shirley:	"Um, about...6 or 8 feet, maybe.
Haines:	Would you like to go over and take a look?
Shirley:	Yeah, but they said...they won't let me...they just said I didn't need to look at that.

Comments: Here is an instance in which the human desired to do something that the aliens did not want her to do. It isn't clear how they prevented her from moving. Why they allowed even visual access of this object in the room is puzzling. Her description of the furniture is remarkably similar to a screen on which to project information from behind.

Reference: Author's file.

References

Bullard, T.E. 1987. UFO Abductions: The Measure of a Mystery. *Fund for UFO Research.* Mount Rainier, Md.

Bullard, T.E. 1994. *The influence of investigators on UFO abduction reports:* Results of a survey. In A. Pritchard, et al., 163-167, (op cit.).

Haines, R.F. 1989. *A Three Stage Technique (TST) to Help Reduce Biasing Effects During Hypnotic Regression.* Journal of UFO Studies 1, 163-167. Chicago: J. Allen Hynek Center for UFO Studies.

Haines, R.F. 1997. The Question. In *UFOs 1947-1997 From Arnold to the Abductions: Fifty Years of Flying Saucers.* Evans, ed. H. and D. Stacy. London: Fortean Tomes, 160-164.

Pritchard, A., D.E. Pritchard, J.E. Mack, P. Kasey, and C. Yapp, eds. 1994. *Alien Discussions: Proceedings of the Abduction Study Conference.* MIT, Cambridge, Mass, June 13-17, 1992. Cambridge, Mass.: North Cambridge Press.

PART III

Chapter 10

Analysis and Discussion of These Reports

Chapters three through six focused on the responses of the unidentified aerial objects to various kinds of overt human behavior. Most of these cases qualify as CE-5s. We noticed several things. First, UFOs appear to react to certain kinds of human behavior in a fairly consistent manner, as if they may be intelligently controlled or at least pre-programmed to respond in certain systematic ways. Second, human beings frighten easily and usually react with hostility of some kind. Third, buried within many of these detailed reports are interesting interrelationships which call for deeper analyses. This is the main objective of the rest of this chapter.

Chapters seven through nine dealt with the far more problematic subject of alleged humanoid responses to different kinds of deliberate human behavior. As before, it is essential to our understanding to classify the various kinds of humanoid responses observed, as well as what the humans did and their motives for behaving one way or another. Is there reasonable evidence to show that the human provoked an overt response from the humanoid?

Patterns in the data

Let us step back and take a broader look at the cases presented here. Are there any particular patterns in the data that help prove that "they" originate purely from within the realm of nature, from the mind and/or technology of the human, or from intelligent, non-terrestrial beings? We will consider temporal, response complexity, spatial, and miscellaneous patterns.

The reader will find a number of statistical tests presented in this chapter that will be familiar to those trained in this particular field of mathematics. Others may want to skip these particular statements and merely examine the tables provided.

Time of day results

As is shown in Table 3, the large majority of these CE-5 events took place during hours of darkness. The same finding has been reported by almost every other investigator of UFO evidence. This raises the question whether UFO are simply easier to see because of the dark visual environment or whether they are actually present in greater frequency at these times.

A two-way, mixed-model analysis of variance was conducted on the data of Table 3. While the local time main effect was not statistically significant ($p=0.073$; $df=47$), the main effect comprising three of the human-response classes (friendly, aggressive, thought) did differ ($F=5.81$; $df=2$; $p=0.004$). Their two-way interaction was not significant.

Table 3

Frequency of Reports by Local Time and Class of Human Behavior (Grand Total = 126)

Local Time	Part I (UFO)				Part II (Humanoid)		
	Friendly	Aggres.	Thought	Misc.	Friendly	Aggres.	Thought & Misc
Chapter	3	4	5	6	7	8	9
Midnight	3	0	0	3	0	0	0
11:30 P.M.	3	1	1	3	0	1	0
11:00 P.M.	4	2	2	3	0	1	0
10:30	4	0	0	1	0	0	0
10:00	1	2	2	0	0	0	0
9:30	8	1	1	1	0	1	0
9:00	4	2	0	2	0	2	1
8:30	5	0	1	2	0	3	0
8:00	3	2	1	1	0	0	0
7:30	2	1	0	2	0	2	0
7:00	0	1	0	0	0	1	0
6:30	1	1	0	0	0	0	0
6:00	0	4	0	0	0	1	0
5:30	0	0	0	1	0	1	0
5:00	1	0	1	0	0	0	0
4:30	0	0	0	1	0	0	0
4:00	0	0	1	0	0	1	0
3:30	0	0	0	0	0	0	0
3:00	0	0	0	0	0	0	0
2:30	0	1	0	1	0	0	0
2:00	0	1	0	0	0	0	0
1:30	1	0	0	0	0	0	0
1:00	0	0	0	0	0	0	0
0:30	0	0	0	0	0	0	0
Noon	0	0	0	1	0	0	0
11:30 A.M.	0	0	0	0	0	0	0
11:00	1	0	0	0	0	0	0
10:30	0	0	0	0	1	0	0

Table 3 (cont'd)

Local Time	Part I (UFO)				Part II (Humanoid)		
	Friendly	Aggres.	Thought	Misc.	Friendly	Aggres.	Thought & Misc
Chapter	3	4	5	6	7	8	9
10:00	0	2	0	0	0	0	0
9:30	0	0	0	0	0	0	0
9:00	0	0	0	0	0	0	0
8:30	0	0	0	0	0	0	0
8:00	0	0	0	0	0	0	0
7:30	0	0	0	0	0	0	0
7:00	0	2	0	0	0	0	0
6:30	0	0	0	0	0	0	0
6:00	0	0	0	0	0	0	0
5:30	0	0	0	0	0	0	0
5:00	0	0	0	1	0	0	0
4:30	1	0	0	0	0	0	0
4:00	2	1	0	2	0	1	0
3:30	0	0	0	1	0	2	0
3:00	4	1	0	0	0	0	0
2:30	0	1	0	1	0	2	0
2:00	3	1	1	2	1	1	0
1:30	0	3	0	0	0	1	2
1:00	6	0	0	1	0	0	0
0:30	1	0	0	0	0	0	0
"night"	20	6	1	5	1	1	0
"?"(unknown)	2	14	0	4	0	2	0
"other"	2	5	1	3	1	2	3
No. missing	0	6	0	0	0	0	0
Total	82	55	13	42	4	26	6

410

Frequency distribution by year

Table 4 presents a CE-5 report frequency distribution by year for all cases presented in Parts I and II. There are at least four periods of an increased number of events centered on the years 1957, 1967, 1977, and 1993.

Table 4

CE-5 Report Frequency Distribution by Year

Year	Part I	Part II	Year	Part I	Part II	Year	Part I	Part II
1996	2	0	1965	4	1	1934	0	0
1995	1	0	1964	1	1	1933	0	0
1994	3	0	1963	2	1			
1993	8	0	1962	2	0	1917	1	0
1992	7	0	1961	5	0			
1991	2	0	1960	2	0	1877	0	1
1990	3	0	1959	5	1			
1989	2	0	1958	3	0			
1988	1	2	1957	14	1			
1987	2	1	1956	8	0			
1986	0	0	1955	5	1			
1985	3	0	1954	5	5			
1984	1	0	1953	1	0			
1983	8	0	1952	2	0			
1982	2	1	1951	4	0			
1981	3	0	1950	1	0			
1980	4	0	1949	1	0			
1979	1	1	1948	3	0			
1978	5	3	1947	0	1			
1977	10	2	1946	0	0			
1976	7	3	1945	2	0			
1975	7	4	1944	1	0			
1974	7	1	1943	0	0			
1973	6	2	1942	1	0			
1972	9	0	1941	0	0			
1971	1	0	1940	0	0			
1970	3	0	1939	0	0			
1969	1	0	1938	0	0			
1968	2	0	1937	0	0			
1967	11	3	1936	0	0			
1966	3	0	1935	0	0			

411

Duration and number of witnesses results

The mean CE-5 event duration across all cases in Chapter 3 (friendly human behavior) is about six minutes, and the duration having the largest number of witnesses (fifty) lasted between ten and twelve minutes! Fifty-two of the sixty-eight cases (76.5 percent) where duration was available were multiple-witness events.

Table 5

Distribution of CE-5 Reports From Chapter Three by Number of Eyewitnesses and Event Duration

Event Duration	Number of Eyewitnesses									
	1	2	3	4	5	6-10	11-20	21-30	31-40	41-50
Seconds<1	0	0	0	0	0	0	0	0	0	0
1-5	0	1	0	0	0	0	0	0	0	0
6-15	1	1	0	0	0	0	0	0	0	0
16-59	0	0	0	0	0	0	0	0	0	0
Minutes<1	0	1	0	0	0	0	0	0	0	0
2	3	0	1	2	1	0	0	0	0	0
3	0	1	2	0	1	1	0	0	0	0
4	2	2	2	2	0	1	0	0	0	0
5	3	5	0	1	3	1	2	1	0	0
6-10	2	0	0	0	2	1	0	0	0	0
11-15	1	0	0	0	1	0	0	0	0	1
16-20	0	0	0	0	0	1	0	0	0	0
21-30	0	1	0	0	2	0	0	0	0	0
31-40	0	0	0	0	0	1	0	0	0	0
41-50	1	0	0	0	0	0	1	0	0	0
51-59	0	0	0	0	0	0	0	0	0	0
Hours<1	0	0	1	1	0	0	1	0	0	0
2	1	0	0	1	1	0	0	0	0	0
3	0	1	0	1	0	0	0	0	1	0
4	0	0	0	0	0	0	0	0	0	0
5	1	0	0	0	0	0	0	0	0	0
>5	1	0	0	1	0	0	0	0	0	0
Total	16	13	6	9	11	6	4	1	1	1

Grand Total = 68

Duration of behavior

As is described in Chapter 2, each case was scored with regard to its actual or estimated duration of behavior (DB). The duration of human behavior was placed into one of six categories and cross-referenced with the duration of the UFO's or humanoid's response also placed into the same six categories. Table 6 presents the number of cases falling within each cell of this matrix.

Table 6

Summary of the Number of Cases Possessing These Combinations of Behavioral Durations Across Part I and II (193 Cases Total)

		Friendly Human Behavior Duration					
		< 1s	1-5s	5-30s	.5-5m	5-30m	>30m
	>30m	0	0	0	0	0	11
Duration	5-30m	0	0	0	2	37	9
of UFO or	.5-5m	0	0	1	61	11	7
Humanoid	5-30s	0	0	6	16	17	4
Response	1-5s	0	2	5	3	0	0
	< 1s	0	1	0	0	0	0

A special statistical test known as "analysis of variance" was carried out on the data of Table 6. It showed that these six columns are very likely not the result a chance or random effect ($p = 0.0047$). The numbers in the six rows, however, were not found to be significantly different at the $p = 0.05$ level of confidence. The interaction of these rows and columns also was not statistically different.

An analysis was also performed on the number of cases occurring in each UFO duration category and each duration of human behavior and are summarized in Table 7.

Table 7

Summary of UFO Response and Human Behavior Durations Across all Chapters

For human behavior that lasts for	there is in the UFO/humanoid duration data
< one second	no definite trend of response duration
1 to 5 seconds	a slight trend toward longer UFO durations as human behavior lengthens
5 to 30 seconds	a slight trend toward longer UFO durations as human behavior lengthens
0.5 to 5 minutes	a prominent increase in number of cases having UFO/humanoid behavior lasting only from five to thirty seconds
5 to 30 minutes	a prominent increase in number of cases having UFO/humanoid behavior lasting only from one to five seconds
> 30 minutes	no definite trend (other than an increase in number of cases with UFO/humanoid behavior lasting longer than thirty minutes.

Tables 8 through 14 present the same data as above for each chapter separately.

Table 8

Number of Cases Possessing These Combinations of Behavioral Durations (81 Cases from Chapter 3)

		Friendly Human Behavior Duration					
		< 1s	1-5s	5-30s	.5-5m	5-30m	>30m
	>30m	0	0	0	0	0	3
Duration	5-30m	0	0	0	3	9	3
of UFO	.5-5m	0	0	0	26	7	1
Response	5-30s	0	0	5	7	5	1
	1-5s	1	2	5	3	0	0
	< 1s	0	1	0	0	0	0

(Note: Nine cases could not be scored)
Chi Sq. = 98.19; df = 25; p < 0.001

Table 9

Number of Cases Possessing These Combinations of Behavioral Durations (59 Cases from Chapter 4)

		Aggressive Human Behavior Duration					
		< 1s	1-5s	5-30s	.5-5m	5-30m	>30m
	>30m	0	0	0	0	0	2
Duration	5-30m	0	0	0	0	9	3
of UFO	.5-5m	0	0	0	14	1	2
Response	5-30s	0	0	1	5	1	1
	1-5s	0	0	1	0	0	0
	< 1s	0	0	0	0	0	0

(Note: Nineteen cases could not be scored)
Chi Sq. = 76.62; df = 25; p<0.001

Table 10

Number of Cases Possessing These Combinations of Behavioral Durations (13 Cases from Chapter 5)

		Human Thought Duration					
		< 1s	1-5s	5-30s	.5-5m	5-30m	>30m
	>30m	0	0	0	0	0	0
Duration	5-30m	0	0	0	0	0	0
of UFO	.5-5m	0	0	1	4	0	1
Response	5-30s	0	0	0	3	1	0
	1-5s	0	1	0	0	0	0
	< 1s	0	0	0	0	0	0

(Note: 2 cases could not be scored)
Chi Sq. test not possible

Table 11

Number of Cases Possessing These Combinations of Behavioral Durations (43 Cases from Chapter 6)

		Miscellaneous Human Behavior Duration					
		< 1s	1-5s	5-30s	.5-5m	5-30m	>30m
	>30m	0	0	0	0	0	4
Duration	5-30m	0	0	0	0	16	0
of UFO	.5-5m	0	0	0	11	2	2
Response	5-30s	0	0	0	1	4	2
	1-5s	0	0	1	0	0	0
	< 1s	0	0	0	0	0	0

(Note: All cases were scored)
Chi Sq. test conducted only on top 4 rows and 3 last cols.
Chi Sq. = 47.55; df = 6; p < 0.001

Table 12

Number of Cases Possessing These Combinations of Behavioral Durations (4 Cases from Chapter 7)

		Overtly Friendly Human Behavior Duration					
		< 1s	1-5s	5-30s	.5-5m	5-30m	>30m
	>30m	0	0	0	0	0	0
Duration	5-30m	0	0	0	0	0	1
of	.5-5m	0	0	0	2	0	0
Humanoid	5-30s	0	0	0	0	1	0
Response	1-5s	0	0	0	0	0	0
	< 1s	0	0	0	0	0	0

(Note: All cases were scored)
Chi Sq. test not possible

Table 13

Number of Cases Possessing These Combinations of Behavioral Durations (26 Cases from Chapter 8)

		Overtly Hostile Human Behavior Duration					
		< 1s	1-5s	5-30s	.5-5m	5-30m	>30m
	>30m	0	0	0	0	0	2
Duration	5-30m	0	0	0	0	3	1
of	.5-5m	0	0	0	7	1	0
Humanoid	5-30s	0	0	0	1	4	0
Response	1-5s	0	0	0	0	0	0
	< 1s	0	0	0	0	0	0

(Seven cases not scored)
Chi Sq. test conducted only on top 4 rows and 3 last cols.
Chi Sq. = 23.51; df = 6; $p < 0.001$

Table 14

Number of Cases Possessing These Combinations of Behavioral Durations (6 Cases from Chapter 9)

				Miscellaneous Human Behavior Duration			
		< 1s	1-5s	5-30s	.5-5m	5-30m	>30m
	>30m	0	0	0	0	0	0
Duration	5-30m	0	0	0	0	0	1
of	.5-5m	0	0	0	0	0	1
Humanoid	5-30s	0	0	1	0	1	0
Response	1-5s	0	0	0	0	0	0
	< 1s	0	0	0	0	0	0

(Two cases not scored)
Chi Sq. test not possible

418

Complexity of behavior

As with duration of behavior, each case was rated in terms of its complexity using a three-step scale. These ratings are highly subjective and the boundaries separating one step from the next also are not clear cut. Nevertheless, it was an attempt to quantify possible hidden relationships between the human behavior and the allegedly elicited UFO response. These results are presented in Tables 15 through 17.

Table 15

Number of Cases Possessing These Behavioral Complexity Levels Within Part I (N = 190 Cases)

Complexity of Human Behavior		Low				Medium				High			
Chapter		3	4	5	6	3	4	5	6	3	4	5	6
Complexity	High	0	0	0	1	2	1	0	1	0	0	1	0
of UFO	Medium	12	10	4	5	10	16	1	17	0	2	0	0
response	Low	53	16	4	13	4	10	1	5	1	0	0	0

Table 16

Number of Cases Possessing These Behavioral Complexity Levels Within Part II   (N = 36 Cases)

Complexity of Human Behavior		Low			Medium			High		
Chapter		7	8	9	7	8	9	7	8	9
Complexity	High	0	0	0	0	3	0	0	0	0
of humanoid	Medium	2	8	0	1	6	0	0	0	0
response	Low	1	7	6	0	2	0	0	0	0

Table 17

Percentage of Cases Possessing These Behavioral Complexities Across All Cases in Part I and II (N = 216)

Complexity of Human Behavior		Low	Medium	High
Complexity	High	1 (0.5%)	7 (3.2%)	1 (0.5%)
of UFO &	Medium	41 (19%)	41 (19%)	2 (0.9%)
humanoid	Low	100(46.2%)	22 (10.2%)	1 (0.5%)
response				

Chi Sq. = 37.98; df = 4; $p < 0.001$

UFO shapes reported

Shape names given in these cases correspond in every respect with those given by other witnesses over the years (Haines 1978, 1980). The perceived outline shape of an object varies in an almost infinite manner with viewing angle, lighting, and surface reflection effects, and other visual and optical factors (Haines 1980; Hall 1964). Such names as: airplane, ball, balloon, boomerang, cigar, circular, cylinder, dirigible, disc, domed craft, fireball, plate, point, saucer, sombrero, star, top, torpedo, and V-shaped were used here, among others.

Among the many kinds of analyses which might be carried out on the sketches included in CE-5 are width-to-thickness ratios, including any top protrusion ratios for side-view drawings. These measurements were carried out on the thirty-one sketches included here. The results are summarized in Table 18 along with selected comparison data. UFO outline ratios presented in this book are comparable with the earlier data. A separate Chi Square analysis was performed on the data across each row. It was found that these four means did not differ from each other at the p=0.05 level of statistical significance for any of these rows. That is, the present UFO shapes are considered to be relatively similar to the others analyzed.

Table 18

Selected UFO Drawing Outline Ratio Results

	CE-5	Haines, 1978[1]		
		1[2]	2[2]	3[2]
UFO "Body" Outline Measures				
Maximum width/height ratio	9.30	8.00	11.58	24.66
Mean width/height ratio (n = 31)	7.42	4.64	4.28	7.98[3]
Minimum width/height ratio	0.66	1.00	1.00	2.13
UFO "Dome" Outline Measures				
Maximum width/height ratio	5.77	17.40	4.42	n/a
Mean width/height ratio (n=7)	4.27	6.87	3.48	7.55
Minimum width/height ratio	2.36	2.42	1.14	n/a

[1] All values are from people who have seen a UFO as published in: Haines, R.F. 1978. *UFO Phenomena 2*, no. 1: 123-151.

[2] Group 1 consisted of fifteen people quite interested in UFO in the San Francisco Bay area. Group 2 consisted of twenty-eight attendees of the 1977 International UFO Conference, San Francisco, California. Group 3 consisted of seven professional architects in the San Francisco Bay area.

[3] None of the mean values in any of these six rows are statistically significantly different at the p=0.05 level by Chi Square analysis.

Signs of intelligence within the UFO response data

We have come to the crux of these analyses. One of the objectives of this review and analysis of the literature was to seek possible evidence for higher intelligence in the alleged responses of the UFO and/or humanoids. There are a number of findings which seem to point in this direction. They are listed in Table 19 and 20.

If these reports are taken seriously then we are confronted by non-trivial and important evidence supporting several interrelated postulates:

First, these UFO phenomena act as if they are intelligently controlled. Supporting evidence for this claim includes the fact that: (a) human behavior does appear to elicit an overt response from the UFOs, (b) the UFO's reported response exactly duplicates very complex human behavior in some instances, (c) a number of cases show that UFOs appear to go through similar response sequences when shot at and struck, and (d) the UFOs often depart or disappear from an airplane that is attempting to approach it and yet approaches and paces surface vehicles.

Second, these reports suggest that the UFO phenomenon exhibits a great degree of self-restraint in not responding in-kind to human aggressive behavior. In short, they seldom fire back when fired at. Does this simply indicate that it is a natural phenomenon incapable of doing so or is such restraint deliberate? If the latter, then this is another sign both of their intelligence and wisdom.

Third, in most of the human aggression cases, humans missed an opportunity to communicate in greater depth—not a sign of intelligence or wisdom.

Table 19

Selected Incidents from Part I that
Suggest UFO Response

(Note: Each UFO response occurred soon after the human's behavior.)

A. Friendly Human Behavior
- Craft rocked back and forth [21 (3); 44 (3)]
- Craft signaled with a light [29 (3); 41 (3); 52 (3); 64 (3); 70 (3); 197 (6)]
- Craft changed location or departed [14 (3); 36 (3); 37 (3); 46 (3); 47 (3); 62 (3); 63 (3); 67 (3); 72 (3); 75 (3); 79 (3); 80 (3)]
- Craft approached witness directly [40 (3); 59 (3); 68 (3); 73 (3); 78 (3); 84)(3); 85 (3); 88 (3); 89 (3); 199 (6)]
- Craft returned signal exactly
    - airplane landing lights [11 (3)]
    - flashlight 1x [15 (3)]
    - flashlight 2x [56 (3); 69 (3); (92-07-27)(3)]
    - flashlight 3x [34 (3); 69 (3); 90 (3)]
    - flashlight 4x [38 (3)]
    - car headlights [33 (3); 55 (3)]

Table 19 Cont'd
- Light source returned signal exactly [77 (3); 83 (3)]
- Light beam or ray shown back [64 (3); 70 (3); 87 (3)]

B. Hostile/Aggressive Human Behavior
- Flew away when struck by bullets [120 (4); 130 (4)]
- Immediate change in flight path of UFO [92 (4); 123 (4)]
- Rapid intensity increase when bullet hit it [94 (4); 98 (4); 115 (4)]
- Lights went out/vanished from sight [101 (4); 114 (4); 122 (4)]
- Light beam/ray emitted [39 (3); 58 0(3); 98 (4); 124 (4)]
- Outmaneuvered pursuing jet [110 (4)]

C. Passive and Miscellaneous Human Behavior
- Craft approached witness [45 (3); 88 (3); 155 (5); 156 (5); 178 (6); 180 (6); 185 (6); 192 (6); 197 (6)]
- Craft paced surface vehicle or airplane [86 (3); 171 (6); 174 (6); 175 (6); 179 (6); 184 (6); 187 (6); 188 (6); 191 (6); 192 (6); 194 (6); 196 (6); 204 (6)]
- Craft disappeared permanently [65 (3)]
- Craft disappeared, reappeared, disappeared [134 (4)]
- Craft changed in some way
    - Location [43 (3); 164 (6); 165 (6); 166 (6); 169 (6)]
    - Intensity dimmed and/or brightened [49 (3); 50 (3); 91 (3); 201 (6)]
    - Intensity brightened [22 (3)]
    - Color [26 (3); 54 (3)]

D. Miscellaneous Incidents
- Craft's edges became cloudy [63 (3)]

Table 20

Summary of Incidents from Part II that Suggest Humanoid Response Intelligence

Alleged Humanoid Responses to
A. Friendly Human Behavior
- Being waved [207 (7)]
- Being departed quickly [208 (7)]

Table 20 Cont'd

B. Hostile/Aggressive Human Behavior

- Beings walked or ran away [209 (7); 225 (8); 229 (8); 232 (8); 234 (8)]
- Light beam/ray emitted by being [211 (8); 214 (8); 223 (8); 230 (8)]
- Caused neuromuscular weakness in human [213 (8)]
- No apparent injury to being [218 (8); 228 (8); 235 (8)]
- Fog/smoke with anesthetic effect emitted by being [212 (8); 221 (8)]

C. Passive and Miscellaneous Human Behavior

- Being(s) try to abduct human by force [217 (8); 219 (8)]
- Beings continued working [227 (8)]

D. Miscellaneous behavior and incidents

- Sparks emitted from being [221 (8)]

Could all of these reported incidents have been pure coincidence, wishful thinking, hoaxes, imagination, or hallucination? The answer is not really up to us. Ultimately, it is up to the UFO phenomenon.

We see farther and much sooner by faith than by sight or intellect.

Epilogue

"I call our world Flatland, not because we call it so, but to make its nature clearer to you, my happy readers, who are privileged to live in Space." It was with these words that Edwin A. Abbott began his first chapter in *Flatland: A Romance of Many Dimensions,* a wonderfully thought-provoking and imaginative treatise that has become a source of some penetrating insights about how we see, and don't see, ourselves and our world. It is we humans who are truly privileged to live in Space—at least four dimensional space—and to have the mental capacity to realize it. Here, I include time as a dimension. God has given us an intellect sufficient to reflect on ourselves, a faculty which no other living creatures on earth seem to possess. So what has *Flatland* to do with *CE-5: Close Encounters of the Fifth Kind?*

While living within our Space, we are only vaguely aware of other possible dimensions. For all practical intents and purposes, they are invisible to us. Perhaps we are as limited as are the Flatlanders who knew only two dimensions. Does our inability to understand them translate into an intellectual blindness? Do we reason, since it can't be understood, it can't exist, so it can't be seen? Or, are other dimensions invisible only because we don't possess the sensory means to appreciate them? And what will happen when we do finally appreciate them one day? Will man's very powerful psychological denial mechanism, buried within us, keep these other dimensions hidden from view in the same way they have kept us from considering spiritual matters seriously? These kinds of questions are central to our study and understanding of CE-5 events.

This review and somewhat cursory examination of several hundred past CE-5 events have shown clearly that humans believe they have, on numerous occasions, initiated contact with UFO and/or with non-terrestrial beings. I cannot, in good conscience, keep these findings hidden. I am compelled to share them with you for whatever use you may find.

"It is part of the martyrdom which I endure for the cause of the Truth that there are seasons of mental weakness, when Cubes and Spheres flit away into the background of scarce-possible existence; when the Land of Three Dimensions seems almost as visionary as the Land of One or None; nay, when even this hard wall that bars me from my freedom, these very tables on which I am writing, and all the substantial realities of Flatland itself, appear no better than the offspring of a diseased imagination, or the baseless fabric of a dream" (103, Ibid.). With these tormented words, Abbott ended his book.

Having gently guided his reader through the imaginary society of a two-dimensional world, Abbott despaired of ever bursting through the "hard wall" into the third dimension.

Yet, we face the same sort of "hard wall." Sooner or later, each of us must confront the possibilities that lay hidden within the dimensions which probably permeate us at this very moment. Can we do it? Will we do it? Hopefully at least one of the cases presented in *CE-5* will have pushed or pulled you forward a bit toward the hard wall.

426 Reference

Abbott, E.A. 1952. *Flatland, A Romance of Many Dimensions.* New York: Dover Publications.

Appendix

The Center for the Study of Extraterrestrial Intelligence and its CE-5 Initiative

"The Center for the Study of Extraterrestrial Intelligence (CSETI) is an international non-profit scientific research and education organization dedicated to the understanding of Extraterrestrial Intelligence (ETI) and Extraterrestrial Civilizations. CSETI is especially interested in the ETI-Human relationship and in the peaceful furtherance of this relationship. A thorough review of existing data and documents concerning 'UFO's' indicates that the Earth has been visited by ETI and Extraterrestrial Spacecraft (ETS) for decades, if not centuries, and that this contact has intensified since 1947." This statement, comes from a CSETI document distributed in late 1996. It provides insight into the goals and objectives of its founder and director, Steven M. Greer.

A number of fundamental operating principles of the so-called Close Encounter of the Fifth Kind Initiative (CE-5 Initiative), developed by Greer, are presented below since they shed valuable light on an understanding of the CSETI program approach, and also provide useful guidance to other individuals and groups who may have the similar intentions. They are taken from an official publication of CSETI (Greer 1992, 22-23). Further information is available from: CSETI, P. O. Box 265, Crozet, VA 22932-0265.

Principles of the CE-5 Initiative:

1. ETI and Extraterrestrial Transports (UFOs) have been and currently are in contact with human society.

2. ETI has a net peaceful, benign, and probably protective motive for the relationship with humanity at this time.

3. The CE-5 Initiative is proactive, bilaterally communicative, and multi-disciplinary in nature and is not primarily motivated towards the current acquisition of ETI technology, except as mutually permitted by the ETI-Human relationship.

4. ETI's enigmatic and elusive behavior may be understood as human-protective when viewed from their perspective: a war-torn, aggressive, nuclear armed, and disunified earth civilization must not receive further potentially harmful technologies until a lasting world peace and unity is achieved, and international human goals become peaceful, cooperative, and unified in nature. Such a transformation will then indicate the readiness for a fuller contact and exchange between humans and ETI. We must respect and accept ETI's control and wisdom in this regard.

5. Notwithstanding the protective limits mentioned above, ETI is apparently desirous of an expanded contact with humans, and is open to voluntary, human initiated contact and exchange. There is strong evidence to suggest that ETI has been systematically introducing themselves to human civilization for the past forty-five years or longer, and that such contact has steadily deepened and intensified over this period.

8. All CE-5 Initiative contact will be for the benefit of all of humanity (as well as ETI) and will not redound to the benefit of only one nation or culture. This is essential.

9. All CE-5 Initiative contacts will be free of hostile intent, and will be free of the presence of any and all weaponry, defensive or offensive.

11. Both humans and ETI have physical and mental/spiritual aspects to their reality and contact and communication will proceed on all levels of our shared reality.

12. The CE-5 Initiative affirms that humans and ETI, as conscious beings, are essentially more alike than dissimilar. Regardless of how different we may externally or physically appear, the reality of humans as conscious and intelligent beings establishes the common basis for communication, deeper and bilateral contact, and mutual self-respect.

18. The activities of the CE-5 Initiative will maintain high standards of conduct and professionalism, while preserving an atmosphere of open-mindedness and creative "brain-storming." Anyone found to be spreading misinformation or falsified cases intentionally will be prohibited from participating in any of the activities of the CE-5 Initiative.

19. We hold that carefully planned Close Encounters of the Fifth Kind will unlock new frontiers in the relationship between ETI and humanity. All sincere researchers and theorists are welcome to join in this profound endeavor on behalf of mankind's growth and evolution. (Reprinted with permission.)

A privately published paper written by Greer (1992, 15-16) provides a clear position statement concerning the judged aggressiveness of E.T. He writes, "While some aspects of ETI behavior, research, and reconnaissance have been disturbing to some human observers, no data exists to indicate that ETI are motivated by net hostile intentions. It is our conclusion that, while ETI behavior does not always coincide with human values and expectations, the motives and ultimate intentions of the ETI currently visiting earth are non-hostile."

It is only prudent, however, to ask whether ETI will continue to be non-hostile. What if their past behavior toward humans is not representative of their future behavior? What if their present behavior is a part of a deliberate, long-term process of deception? What if humans provoke them to take aggressive action? These are serious questions which need to be asked by anyone who contemplates our future seriously.

Human motives—A call for change

The CSETI approach to contact with ETI is based upon the belief that they are "advanced conscious intelligent beings." This belief undergirds much of the methodology that is used in the out-of-doors to attempt to establish contact with them. Realizing that effective two-party communication depends upon the consciousness of each party, Greer is rightly concerned with man's motives and psychological makeup. Indeed, man cannot affect the motives or psychological makeup of the visitors. As Greer states (1992, 17-18), "For this reason, careful consideration must be given to not only our view of ETI's intentions, but more importantly, to our own intentions and attitudes. The human tendency—well-evidenced by both military and civilian reactions to ETS—for xenophobic, violent, and even paranoid reactions to the new or unknown must be addressed and rectified. The human predisposition to view anything which we do not understand or control as intrinsically hostile and threatening must be overcome." Many writers over many decades have repeated this same point of view.

Implications of an extraterrestrial disclosure

In an article titled, "Implications of an Extraterrestrial Disclosure," Greer wrote, in part, "The most profound pronounced immediate effect of such a disclosure (of the verified existence of E.T. life forms) will be that of altering the fundamental paradigm of how we view ourselves, the world of humanity and the universe. While the full effect of this will take some decades, if not centuries, to manifest fully, there will be an immediate realization that we are really all one people on the homeland of earth, among many worlds inhabited by other intelligent life forms.

"This realization will alter not only our view of ourselves and the earth, but I believe it will impact on the current state of geopolitical fractionation. In the post-cold war era, the world is becoming increasingly 'balkanized,' precisely at the time when one would hope for increasing world unity and coherence. The definitive disclosure that we are not alone in the universe, and that, more importantly, these advanced life forms are landing on terra firma, will provide significant impetus to the as-yet incomplete process of forming a truly global civilization....

"Even the process of coming to grips with the post-disclosure reality of an extraterrestrial presence will force the world to evolve the moral, spiritual, psychological and physical capacity to effectively meet the challenges of the situation....This, then, will provide both the practical and philosophical setting for the development of a world civilization" (1996, 2).

430 ──────────────────────────────

And, now that you have finished this book, hopefully you can make more intelligent decisions about how to respond should E.T. make their presence even more obvious than they already have. Eventually, you will have to confront your own personal beliefs about "their" existence as well as their motives toward mankind. But even more importantly, you may also want to assess your motives toward your fellow man. This may be turn out to be the most difficult, yet important, assessment of all!

References

Greer, S.M. 1992. *CSETI Center for the Study of Extraterrestrial Intelligence.* Asheville, N.C.: Privately published paper, 55.

Greer, S.M. 1996. *Implications of an Extraterrestrial Disclosure.* Asheville, N.C.: Privately published paper, 6.

Rating panel

I initially rated each case and then asked five impartial raters to verify my ratings on at least 50 percent of all cases on a random basis. Only a few of these volunteers needed to ask for clarification about possible scoring ambiguities. Rating differences were scored in favor of the panel member(s). These volunteers included:

Computer programmer with no prior knowledge of ufology (Bradley S.)
Computer graphic artist with moderate knowledge of ufology (Kurt S.)
Software expert very familiar with UFO literature (Larry H.)
High school teacher with no prior knowledge of ufology (Jeff T.)
Aerospace technician with very little prior knowledge of ufology (Jon. B. G.)

I am very much indebted to each of these persons for their dedicated efforts to cross check and verify my ratings.

Index

433